AF880125

ELEMENTS OF STOCHASTIC PROCESSES: A COMPUTATIONAL APPROACH

C. Douglas Howard

Baruch College

City University of New York

FE PRESS

New York

FE PRESS

New York
www.fepress.org

© C. Douglas Howard 2017

All rights reserved. No part of this publication may be reproduced, stored in a retrieval system, or transmitted, in any form or by any means, electronic, mechanical, photocopying, recording, or otherwise, without the prior written permission of the publisher.

First published 2017

Printed in the United States of America

ISBN 978-0-9797576-7-9

Typeset by $\mathcal{A}\mathcal{M}\mathcal{S}$-TEX

OTHER FE PRESS TITLES

A Primer for the Mathematics of Financial Engineering, Second Edition, by Dan Stefanica, FE Press 2011

Solutions Manual – A Primer for the Mathematics of Financial Engineering, Second Edition, by Dan Stefanica, FE Press 2011

150 Most Frequently Asked Questions on Quant Interviews, by Dan Stefanica, Rados Radoicic, and Tai-Ho Wang, FE Press 2013

Numerical Linear Algebra Primer for Financial Engineering, by Dan Stefanica, FE Press 2014

Solutions Manual – Numerical Linear Algebra Primer for Financial Engineering, by Dan Stefanica, FE Press 2014

TO MY FAMILY,

ANDREA AND ZOE

in Appendix II. The chapter includes a brief discussion of recurrence properties, in detail for dimension one and informally for higher dimensions.

In the final Chapter VII (A Glimpse of Stochastic Calculus), a simple version of Itô's formula is proved and the Ornstein-Uhlenbeck process is constructed from Brownian motion. Itô's formula is used to develop the Black-Scholes partial differential equation and derive its solution, yielding the much celebrated Black-Scholes formula for stock option prices. The Ornstein-Uhlenbeck process is used in an economics application to model the University of Michigan's Index of Consumer Sentiment.

Topic sections in this book are numbered sequentially throughout the book. For example, the first section of Chapter II is §8. Within a chapter, theorems, lemmas, corollaries, figures, examples, and tagged displays (e.g., equations) are *collectively* numbered sequentially. For example, in §6 of Chapter I, we find Figure 6.1 followed by Theorem 6.2, equation 6.3, and then Theorem 6.4. The intent of this unconventional scheme is to make things easier to find.

Acknowledgments. Over the years, many students have provided invaluable assistance in the writing of this book. They have found typos and mathematical errors and have given suggestions for improved readability, to name a few things. I thank them for that. It is sadly not possible to enumerate all contributors, however I wish to acknowledge the following students who contributed as the project neared completion: David Alter, Milen Bourilkov, Vincent Chen, Tao Cheng, Li Dong, Ori Golan, Michael Grasberg, Nafyal Ibrat, Kinsang Li, Weihao Li, Bogdan Nikitin, Prerak Paul, Taorui Peng, Tanay Rudra, John Scanlon, Kiranpal Singh, Sophie Song, Sergey Sorokin, Ovadia Sutton, Jonathan Toro, Zhongsuo Wu, Wei Yang. Any errors that remain are purely my own.

My thanks also to Elena Kosygina, whose comments and suggestions have improved the text. Additionally I would like to thank Dan Stefanića for all his help and Max Rumyantsev for creating the book's cover.

Finally I would like to acknowledge the support and encouragement of my beloved family, Andrea and Zoe. Without their support this book would not have pulled together.

C.D.H.

PREFACE

About this book. For the past five years I have had the pleasure and privilege of teaching *Introduction to Stochastic Processes* (MTH 4125) and *Computational Methods in Probability* (MTH 4135) at the City University of New York's Baruch College. During this time, these two courses have been offered annually during the fall and spring semesters, respectively. This book represents my class notes as developed for *Stochastic Processes* and is intended as a one-semester upper-level undergraduate course. A good one-semester course in probability is an essential prerequisite. Additionally, some comfort with 'ϵ-δ' and related arguments with a real analysis flavor is helpful. This is particularly true in Chapter II (Some Fundamental Limit Theorems) and Chapters VI and VII where Brownian motion and some related continuous-time stochastic processes are developed. The book contains an appendix (Appendix II) where requisite material from analysis is presented.

Stochastic Processes is the more "theoretical" of the two courses referenced above. In it, material concerning such stochastic processes as independent identically distributed (iid) random variables, Markov chains, random walks, Poisson processes, Brownian motion and other related processes is presented. The underlying probability space for all stochastic processes considered is the unit interval $\mathbf{\Omega} = (0, 1)$. The underlying "experiment" is to "randomly pick a number ω between 0 and 1." With such a number ω selected, a realization of any of these stochastic processes may be generated. The basic approach is to generate a sequence of numbers $U_1(\omega), U_2(\omega), \ldots$ with the property that the sequence of random variables $U_i : \mathbf{\Omega} \to \mathbb{R}$ are iid and uniformly distributed on the interval $(0, 1)$. With this sequence as a building block, each of the stochastic processes considered here may be constructed. This computational approach is a theoretical (i.e., mathematical) reflection of what is done in *Computational Methods*. There, a computer algorithmically generates the sequence $U_1(\omega), U_2(\omega), \ldots$ via random number generator. The stochastic processes (studied here from a theoretical standpoint) are brought to life on the computer in *Computational Methods.*

A guiding principle while writing this book was to be as rigorous as possible without the use of measure theory. In that spirit, issues of measurability are completely ignored. This is generally a technical rather than practical problem, however, as any construction that arises naturally yields a measurable object. The lack of measure theory does show up in a meaningful way in one place: independence cannot be properly understood without the notion of a σ-field and related classes of events. For example, if object A is constructed from a collection of objects $\mathcal{A}$ and object B is constructed from a collection of objects $\mathcal{B}$ that is independent of $\mathcal{A}$, then object B is independent of object A. As intuitive as this sounds, it takes measure theory even to make sense of the statement, let alone prove it. This also shows up in the study of continuous-time processes, where one often conditions improperly (technically) on events of probability zero.

Organization. Chapter I (Essentials of Probability) is a review of basic material from a first course in undergraduate probability (probability spaces, random variables, expectation, etc.). It should not be the student's first look at probability. Additionally, it contains some material that the student will probably not have seen. Some examples are: continuity of probability measure; countable sub-additivity of probability measure; and the inverse transform method for constructing random variables of a particular distribution. The importance of working on an uncountable (not discrete) probability space is emphasized.

Chapter II (Some Fundamental Limit Theorems) begins with the notions of convergence in probability and almost sure convergence. Versions of the weak and strong laws of large numbers are proved for iid sequences. A partial proof of the Central Limit Theorem for iid sequences is presented. The chapter ends with the monotone, dominated, and bounded convergence theorems for expectation. Only the monotone convergence theorem is used subsequently in the book, and then only in connection with computing the expectation of an infinite series of non-negative random variables.

Chapter III (Markov Chains with Finitely Many States) concerns itself with precisely that topic. First the two-state system is analyzed in detail with focus on the invariant distribution and its properties. With this as motivation, the general situation is discussed first in the ergodic case, then the irreducible periodic case, and finally in the case with multiple communication classes. An example from interacting particle systems graphically illustrates convergence to the invariant distribution in the ergodic case.

In Chapter IV (Random Walks), nearest-neighbor random walks on $\mathbb{Z}$ (the integers), $\mathbb{Z}^2$, and $\mathbb{Z}^3$ are discussed with a focus on recurrence versus transience. A version of Stirling's formula ($n! \approx c\cdot(n/e)^n\sqrt{n}$) is proved with exercises leading to the fact that the constant c in the formula is $\sqrt{2\pi}$. This sequence of exercises leads to LaPlace's version of the Central Limit Theorem.

Chapter V (Arrival Processes and Poisson Point Processes) studies the homogeneous and inhomogeneous Poisson process in one and two dimensions. In one dimension, this corresponds to an arrival process that is in common use in applications such as queueing theory. In higher dimensions, the process must be viewed as a random discrete subset of the plane or space. The chapter finishes with a discussion of the "waiting-time" process when inter-arrival times are iid but not necessarily exponential.

Chapter VI (Brownian Motion) presents some of the basic properties of Brownian motion. It is rigorously proved, for example, that Brownian paths are not rectifiable and in fact are nowhere differentiable. The Lévy construction of Brownian motion on the outcome space $\boldsymbol{\Omega} = (0, 1)$ is presented in detail. This construction requires a bit of analysis — the required background is presented

ELEMENTS OF STOCHASTIC PROCESSES:
A COMPUTATIONAL APPROACH

CONTENTS

CHAPTER I

ESSENTIALS OF PROBABILITY

§1. Probability Spaces

Probability theory is the mathematical modeling of real-world experiments that have unpredictable outcomes. For example, the experiment might be tossing a coin, rolling a pair of dice, dealing out a 5 card poker hand, or investing in IBM stock and holding it for a year. The mathematical model corresponding to any particular experiment consists of three objects: (i) the *outcome space*, typically called Ω; (ii) a collection of *events* $\mathcal{F}$; and (iii) a *probability measure* P. The triple $(\Omega, \mathcal{F}, P)$ collectively is called a *probability space*.

The Outcome Space. The outcome space Ω (also called the *sample space*) consists of all possible outcomes of the underlying experiment. For example, if the experiment consists of rolling a single die, we would have $\Omega = \{1, 2, 3, 4, 5, 6\}$. Individual elements of the outcome space are called *outcomes*. So 5 is an outcome in our current example. A generic outcome is typically denoted by the symbol ω.

Events. Events are subsets of the outcome space. Here, for example, $\{2, 4, 6\}$ corresponds to the event that an even number shows up on the die when it is rolled. If $A \subset \Omega$ is an event and the outcome of the experiment lies in A we say that "A occurs" or "A happens". In many situations, although for technical reasons not in every situation, we take the collection of events $\mathcal{F}$ to be *all* subsets of the outcome space Ω. In any case, $\mathcal{F}$ must contain Ω as an event and be closed under the basic set operations. For example, if A is an event (meaning $A \in \mathcal{F}$), then we require that A^c (the complement of A) is also an event. Furthermore, we require that if A and B are events, then so is their union $A \cup B$, which consists of all outcomes that are in either A or B or possibly both. For finite outcome spaces this is enough, but not in general. In general, we require the following: if $A_1, A_2, A_3, \ldots$ is a sequence of events, then

$$A = \bigcup_{n=1}^{\infty} A_n = A_1 \cup A_2 \cup A_3 \cup \cdots$$

is also an event. Summarizing, we have the following axioms for any probability space $(\Omega, \mathcal{F}, P)$:

$$\Omega \in \mathcal{F}; \tag{1.1}$$

$$A \in \mathcal{F} \quad \Longrightarrow \quad A^c \in \mathcal{F}; \tag{1.2}$$

$$A_1, A_2, A_3, \ldots \in \mathcal{F} \quad \Longrightarrow \quad \bigcup_{n=1}^{\infty} A_n \in \mathcal{F}. \tag{1.3}$$

This list of axioms is parsimonious, yet it insures that $\mathcal{F}$ is closed under all basic set operations. For example, (1.3) provides for the union of a sequence of events, but seems to ignore the simpler case of finite unions. However, (1.1) and (1.2) imply that $\emptyset = \Omega^c \in \mathcal{F}$, so if A and B are events, then so is

$$A \cup B \;=\; A \cup B \cup \emptyset \cup \emptyset \cup \emptyset \cup \cdots .$$

DeMorgan's laws insure that $\mathcal{F}$ is closed under intersection: if A and B are events, then so is $A \cap B \;=\; \left(A^c \cup B^c\right)^c$. Similarly, if each A_n is an event, then so is $\bigcap_{n=1}^{\infty} A_n$. Note that if *every* subset of Ω is an event, then axioms (1.1 – 1.3) are automatically satisfied.

The Probability Measure. The probability measure P assigns a number, denoted PA or $P[A]$, to each event A. Intuitively these probabilities are intended to measure the likelihood of the event happening. That is, they measure the probability that the outcome of the experiment, ω, lies in the subset of outcomes A. In our earlier example, with a fair die we would have $P\{2, 4, 6\} = 1/2$. A probability measure has its own set of axioms, which we list below:

$$P\Omega \;=\; 1; \tag{1.4}$$

$$\text{for any event } A,\ 0 \le PA \le 1; \tag{1.5}$$

$$\text{if } A_1, A_2, A_3, \ldots \text{ are disjoint events, then } P\Big[\bigcup_{n=1}^{\infty} A_n\Big] \;=\; \sum_{n=1}^{\infty} PA_n. \tag{1.6}$$

Axiom (1.4) corresponds to the real-world fact that the outcome of the experiment always lies in Ω. After all, Ω is the set of all possible outcomes of the experiment. Axiom (1.5) speaks for itself. Axiom (1.6) is called *countable additivity* for disjoint events.

Consequences of the Axioms. The following facts are easily derived from the axioms:

$$\begin{aligned} P[\emptyset] &= 0, \\ P[A \cup B] &= PA + PB - P[A \cap B], \\ A \cap B = \emptyset &\Longrightarrow P[A \cup B] = PA + PB, \\ P[A^c] &= 1 - PA, \\ B \subset A &\Longrightarrow P[A \setminus B] = PA - PB. \end{aligned} \tag{1.7}$$

In this last statement $A \setminus B$ denotes the event $A \cap B^c$ — start with A and remove all outcomes that are in B. The fact that $P[A \cup B] = PA + PB$ for two disjoint events generalizes to finite unions. If $A_1, \ldots, A_n$ are pairwise disjoint, then

$$P[A_1 \cup \cdots \cup A_n] = PA_1 + \cdots + PA_n.$$

This is called *finite additivity* for disjoint events.

The events $B_1, B_2, \ldots$ *partition* Ω if they are disjoint and their union is all of Ω. Then we have:

THEOREM 1.8 (PARTITION THEOREM). *Suppose $B_1, B_2, B_3, \ldots$ partition Ω. Then for any event A,*

$$P[A] = \sum_{n=1}^{\infty} P[A \cap B_n].$$

PROOF. This follows directly from the fact that

$$A = A \cap \Omega = A \cap \left(\bigcup_{n=1}^{\infty} B_n \right) = \bigcup_{n=1}^{\infty} (A \cap B_n),$$

where the events in the union on the right hand side are disjoint. The last equality is the distributive property. Disjointedness comes from that fact that the B_n's are disjoint: since B_m and B_n are disjoint (if $m \neq n$) so are $A \cap B_m$ (which is contained in B_m) and $A \cap B_n$ (which is contained in B_n). Axiom (1.6) then gives the theorem. □

Suppose we have events A and B with $P[B] > 0$. Then we define the *conditional probability of A given B* (or *conditioned on B*) as

$$P[A|B] = \frac{P[A \cap B]}{P[B]}.$$

Note that $A|B$ is not an event — A is the event whose probability (conditioned on B) is being measured. Noting that $P[A \cap B] = P[A|B]PB$, we have:

THEOREM 1.9 (PARTITION THEOREM WITH CONDITIONAL PROBABILITY). *If $B_1, B_2, B_3, \ldots$ is a partition of Ω with $P[B_n] > 0$ for each n, then*

$$P[A] = \sum_{n=1}^{\infty} P[A|B_n]\, P[B_n].$$

The sequence of events $A_1, A_2, \ldots$ *increases up to A* if $A_1 \subset A_2 \subset A_3 \subset \cdots$ (they are increasing) and $\bigcup_{n=1}^{\infty} A_n = A$. The sequence of events $B_1, B_2, \ldots$ *decreases down to B* if $B_1 \supset B_2 \supset B_3 \supset \cdots$ (they are decreasing) and $\bigcap_{n=1}^{\infty} B_n = B$. In this situation we write $A_n \uparrow A$ and $B_n \downarrow B$ as $n \to \infty$ and we have:

THEOREM 1.10 (CONTINUITY OF PROBABILITY MEASURES). *If $A_n \uparrow A$ as $n \to \infty$, then $P[A_n] \to P[A]$ as $n \to \infty$. If $B_n \downarrow B$ as $n \to \infty$, then $P[B_n] \to P[B]$ as $n \to \infty$.*

PROOF. We write $A = \bigcup_{n=1}^{\infty} A_n$ as a union of disjoint events as follows:

$$A = \bigcup_{n=1}^{\infty} (A_n \setminus A_{n-1}),$$

where we take A_0 to be the empty set. These sets are disjoint because if $m < n$ (so $m \leq n-1$) then $A_m \setminus A_{m-1} \subset A_m \subset A_{n-1}$ while $A_n \setminus A_{n-1} \subset A_{n-1}^c$. To see that their union is all of A, suppose $\omega \in A$. Then $\omega \in A_n$ for some $n \geq 1$. Let n^* be the smallest such n, so $\omega \in A_{n^*}$ but $\omega \notin A_{n^*-1}$. Then $\omega \in A_{n^*} \setminus A_{n^*-1}$ so $\omega \in \bigcup_{n=1}^{\infty} (A_n \setminus A_{n-1})$. Therefore

$$\begin{aligned} P[A] &= \sum_{n=1}^{\infty} P[A_n \setminus A_{n-1}] \\ &= \sum_{n=1}^{\infty} (P[A_n] - P[A_{n-1}]) \quad \text{by (1.7, last)} \\ &= \lim_{n\to\infty} [(P[A_1] - P[A_0]) + \cdots + (P[A_n] - P[A_{n-1}])] \\ &= \lim_{n\to\infty} (P[A_n] - P[A_0]) = \lim_{n\to\infty} P[A_n]. \end{aligned}$$

The fourth equality holds because all the other terms cancel; the last holds because $A_0 = \emptyset$. This proves the theorem's first statement.

Suppose $B_n \downarrow B$. Then $B_n^c \uparrow B^c$ so, by the first part, $P[B_n^c] \to P[B^c]$ as $n \to \infty$. But $P[B_n^c] = 1 - P[B_n]$ and $P[B^c] = 1 - P[B]$, so $P[B_n] \to P[B]$. □

Another useful fact is the following:

THEOREM 1.11 (SUB-ADDITIVITY OF PROBABILITY MEASURES). *If A_1, $A_2, \ldots$ is any sequence of events then*

$$P\Big[\bigcup_{n=1}^{\infty} A_n\Big] \leq \sum_{n=1}^{\infty} P[A_n].$$

PROOF. See exercises. □

Independence. Events A and B are *independent* if $P[A \cap B] = PA \cdot PB$. If $PB > 0$, this is equivalent to $P[A|B] = PA$. Informally, information about B's occurrence will not cause you to alter your assessment of A's likelihood of occurrence. Three events, A, B, and C, are independent if $P[A \cap B] = PA \cdot PB$, $P[A \cap C] = PA \cdot PC$, $P[B \cap C] = PB \cdot PC$ (these three conditions collectively are called *pairwise independence*), and $P[A \cap B \cap C] = PA \cdot PB \cdot PC$. Pairwise independence is a strictly weaker condition than independence. To see this, put $\Omega = \{1, 2, 3, 4\}$ with each outcome occurring with probability 1/4. Set $A = \{1, 2\}$, $B = \{1, 3\}$ and $C = \{1, 4\}$. Then $P[A \cap B] = P[A \cap C] = P[B \cap C] = P\{1\} = 1/4$ and $PA \cdot PB = PA \cdot PC = PB \cdot PC = 1/4$, so the events are pairwise independent. But $P[A \cap B \cap C] = P\{1\} = 1/4$ while $PA \cdot PB \cdot PC = 1/8$ and the events are not independent. If I tell you that both A and B have occurred, you know that the outcome is 1 so you know that C has also occurred. More generally, a sequence of events $A_1, A_2, A_3, \ldots$ are independent if, for any finite list of distinct indices $i_1, i_2, \ldots, i_n$, we have

$$P[A_{i_1} \cap A_{i_2} \cap \cdots \cap A_{i_n}] = PA_{i_1} \cdot PA_{i_2} \cdot \cdots \cdot PA_{i_n}.$$

Discrete Probability Spaces. A set A is *finite* if its elements can be presented in the form of a terminating list:

$$A = \{a_1, a_2, \ldots, a_n\}.$$

It is *countably infinite* if its elements can be presented in the form of a non-terminating list:

$$A = \{a_1, a_2, a_3, \ldots\}.$$

A set is *countable*, or *discrete*, if it is either finite or countably infinite. Probability spaces come in two flavors: *discrete* and *uncountable*. Probability measures on discrete and uncountable outcome spaces are fundamentally different. If an outcome space Ω is discrete, then any event A is also discrete. To list the outcomes in A simply cull the list of $\Omega = \{\omega_1, \omega_2, \ldots\}$, i.e., go through the list of Ω deleting any outcome that is not in A. Then either

$$A = \bigcup_{k=1}^{n} \{a_k\} \quad \text{or} \quad A = \bigcup_{k=1}^{\infty} \{a_k\},$$

and, by virtue of finite and countable additivity of probability measures for disjoint events, it follows that, depending on the situation, either

$$P[A] = \sum_{k=1}^{n} P\{a_k\} \quad \text{or} \quad P[A] = \sum_{k=1}^{\infty} P\{a_k\}.$$

We handle these cases simultaneously with the notation:

$$P[A] = \sum_{a \in A} P\{a\}. \tag{1.12}$$

In the discrete setting, the probability of *any* subset of Ω may be computed by summing the probabilities of the individual outcomes in the event. The key fact is that the set A is countable so axiom (1.6) applies.

REMARK. Here we have assumed that the *singletons* $\{\omega\}$, for all $\omega \in \Omega$, are events — that is, we assume that each $\{\omega\} \in \mathcal{F}$. In practice this is always the case. We make this assumption throughout the text.

Suppose, for example, that the underlying experiment is rolling a pair of dice. Suppose further that one die is red and the other green. We let

$$\Omega = \{(r,g) : r = 1,2,\ldots,6 \text{ and } g = 1,2,\ldots,6\},$$

where r is the number showing on the red die and g is the number showing on the green die. There are 36 outcomes and, if the dice are fair and independent, they are all equally likely: $P\{\omega\} = 1/36$ for each $\omega \in \Omega$. (We'll call this the *dice rolling space*.) So, for example,

$$\begin{aligned} P[\text{Roll a seven}] &= P\{(1,6),(2,5),(3,4),(4,3),(5,2),(6,1)\} \\ &= P\{(1,6)\} + P\{(2,5)\} + P\{(3,4)\} \\ &\qquad + P\{(4,3)\} + P\{(5,2)\} + P\{(6,1)\} \\ &= 6 \cdot \frac{1}{36} = \frac{1}{6}. \end{aligned}$$

Because of their simplicity, discrete probability spaces are the easiest to work with. Unfortunately, as we shall see, discrete probability spaces are not sufficiently commodious to support even the most basic stochastic processes.

EXERCISES

1. Prove the distributive laws: $A \cap (\bigcup_{n=1}^{\infty} B_n) = \bigcup_{n=1}^{\infty}(A \cap B_n)$, and $A \cup (\bigcap_{n=1}^{\infty} B_n) = \bigcap_{n=1}^{\infty}(A \cup B_n)$.

2. Prove DeMorgan's laws: $(\bigcup_{n=1}^{\infty} A_n)^c = \bigcap_{n=1}^{\infty} A_n^c$, and $(\bigcap_{n=1}^{\infty} A_n)^c = \bigcup_{n=1}^{\infty} A_n^c$.

3. Derive the facts in (1.7) from the axioms.

4. Show that $(A^c)^c = A$.

5. Show that $P[A \setminus B] = PA - P[A \cap B]$.

6. Show that $P[A \cap B] \geq PA + PB - 1$.

7. Suppose $PB > 0$ and, for any event A, define $Q[A] = P[A \mid B]$. Show that Q satisfies the axioms of a probability measure.

8. Suppose events A and B are disjoint and independent and $PA = 1/2$. What is PB?

9. Show that if A and B are independent, then so are: A^c and B; A and B^c; and A^c and B^c.

10. Show that $P\left[\bigcup_{n=1}^{\infty} A_n\right] \leq \sum_{n=1}^{\infty} PA_n$ (this is called *countable subadditivity*). Hint: let $B_n = A_n \setminus (A_1 \cup \cdots \cup A_{n-1})$. Show that the B_n are disjoint and that $\bigcup_n B_n = \bigcup_n A_n$.

11. Suppose A, B, and C are events with $P[A] = .42$, $P[B] = .45$, $P[C] = .35$, $P[A \cup B] = .75$, $P[A \cap B \cap C] = .07$. Calculate the indicated probabilities or indicate, where appropriate, that you have insufficient information to do the calculation: $P[A^c]$; $P[A \cap B]$; $P[A \cap B \cap C^c]$; $P[A \cup B \cup C]$.

12. Suppose A, B, and C are three events with $PA = .3$, $PB = .4$, $PC = .45$, $P[A \cup C] = .68$, $P[A \cup B] = .60$, $P[A \cap B \cap C^c] = .08$, and $P[A^c \cap B^c \cap C^c] = .2$. Are $A \cap B$ and $B \cap C$ independent?

13. Suppose the events B_1, B_2, B_3, and B_4 form a partition of a sample space Ω with $PB_1 = 0.15$, $PB_2 = 0.25$, and $PB_3 = 2PB_4$. What are PB_3 and PB_4?

14. Suppose an urn has 500 coins in it: 499 are fair coins with a head on one side and a tail on the other but one unfair coin has heads on both sides. Suppose I randomly select a coin and toss it nine times getting nine heads in a row. What is the probability that I selected the unfair coin?

15. Suppose 5% of the population has dread disease. Dr. Careful routinely tests all his patients for the disease. The test gives the correct diagnosis 99% of the time (whether you have the disease or not). What is the probability that any particular patient will test positive? If a patient does test positive for dread disease, what is the probability that he has it?

16. Anna has an infinite supply of marbles numbered 1, 2, 3, ... Anna places the marbles, two at a time and in numerical order, into an urn (i.e., first 1 and 2, then 3 and 4, etc.). Each time Anna puts two marbles into the urn, Louis reaches in and pulls a marble out (not necessarily one of the two just put in). This process goes on forever. We'll call each occurrence of putting two in and taking one out a "step", so after n steps there are n marbles in the urn. (a) If Louis always removes the marble with the lowest number, which marbles will remain in the urn forever? (b) If Louis always removes the marble with the highest number, which marbles will remain in the urn forever? Now suppose Louis always removes marbles randomly, with each marble in the urn equally likely to be removed. (c) With what probability will marble number 1 be in the urn after n steps? (d) With what probability will a particular marble of the two placed in the urn at step m still be in the urn after step n (here $n \geq m$)? (e) With what probability will any marble ever placed in the urn remain there forever?

§2. Random Variables and Expectation

A random variable is a real-valued function $X : \Omega \to \mathbb{R}$, i.e., X assigns a number, denoted $X(\omega)$, to each outcome ω. Random variables are usually denoted by capital letters. If X and Y are two random variables defined on the same outcome space Ω, we may form the sum $Z = X + Y$. Then Z is a new random variable defined, not surprisingly, by $Z(\omega) = X(\omega) + Y(\omega)$. Similarly, we may add a constant to a random variable, $Z = X + c$, where here $Z(\omega) = X(\omega) + c$. Indeed, we may perform all the usual arithmetic operations and make sense of random variables such as: XY, X/Y (provided $Y(\omega)$ is never 0), cX, X^2, X^n, and $h(X)$ where $h(x)$ is any function $h : \mathbb{R} \to \mathbb{R}$. In this latter example, if $Z = h(X)$, then $Z(\omega) = h(X(\omega))$.

For example, in games involving dice, one often sums the two showing faces. In the setting of the dice rolling space, if $\omega = (r, g)$, take $S(\omega) = r + g$. If we also take $R(\omega) = r$ and $G(\omega) = g$, then we have $S = R + G$. For example, if $\omega = (3, 1)$, then $S(\omega) = 4 = 3 + 1 = R(\omega) + G(\omega)$.

The range of a random variable X, denoted Range(X), is the set of real numbers that are taken as values of $X(\omega)$ as ω ranges over all of Ω. More formally,

$$\text{Range}\,(X) \;=\; \{x \in \mathbb{R} : x = X(\omega) \text{ for some } \omega \in \Omega\}.$$

For example, with S as above we have Range$(S) = \{2, 3, \ldots, 12\}$.

Discrete Random Variables. If Range(X) is countable, then X is called a *discrete* random variable. If $(\Omega, \mathcal{F}, P)$ is a discrete probability space (Ω is countable), then any random variable defined on Ω is automatically discrete. To see this, let $\{\omega_1, \omega_2, \ldots\}$ denote a listing of the outcomes in Ω. Then $\{X(\omega_1), X(\omega_2), \ldots\}$, with repetitions omitted, is a listing of Range(X).

For $x \in$ Range(X), let $X^{-1}(x)$ (read X *inverse* of x) denote the set of outcomes ω for which $X(\omega) = x$,

$$X^{-1}(x) \;=\; \{\omega \in \Omega : X(\omega) = x\}.$$

(This definition also makes sense if $x \notin$ Range(X), in which case $X^{-1}(x) = \emptyset$.) For example, in the dice rolling example, $S^{-1}(4) = \{(1,3), (2,2), (3,1)\}$. As x runs through Range(X), $X^{-1}(x)$ partitions Ω:

$$\Omega \;=\; \bigcup_{x \in \text{Range}\,(X)} X^{-1}(x),$$

where $X^{-1}(x)$ and $X^{-1}(y)$ are disjoint if $x \neq y$, since an outcome ω cannot satisfy both $X(\omega) = x$ and $X(\omega) = y$ unless $x = y$.

Discrete random variables have *probability mass functions* defined by

$$p_X(x) \;=\; P\{\omega : X(\omega) = x\} \;=\; P[X^{-1}(x)].$$

Since Range(X) is finite or countable, $p_X(x) > 0$ for only finitely or countably infinitely many $x \in \mathbb{R}$. Letting Range$(X) = \{x_1, x_2, \ldots\}$, we have

$$\sum_{x \in \text{Range}\,(X)} p_X(x) \;=\; \sum_i P[X^{-1}(x_i)] \;=\; P\Big[\bigcup_i X^{-1}(x_i)\Big] \;=\; P[\Omega] \;=\; 1.$$

For discrete X, we define the *expected value* of X, denoted EX or $E[X]$, by

$$EX = \sum_{x \in \text{Range}\,(X)} x\, p_X(x), \tag{2.1}$$

For example, with S from the dice rolling space we get that

$$\begin{aligned} ES &= 2\cdot\frac{1}{36} + 3\cdot\frac{2}{36} + 4\cdot\frac{3}{36} + 5\cdot\frac{4}{36} + 6\cdot\frac{5}{36} + 7\cdot\frac{6}{36} \\ &\quad + 8\cdot\frac{5}{36} + 9\cdot\frac{4}{36} + 10\cdot\frac{3}{36} + 11\cdot\frac{2}{36} + 12\cdot\frac{1}{36} = 7. \end{aligned}$$

A technical issue that arises is that some of the summands of (2.1) may be positive and others negative. This happens when Range(X) contains both positive and negative numbers. If the sum of the positive summands is $+\infty$ and the sum of the negative terms is $-\infty$, the conflict cannot be resolved and we must leave EX undefined. Except when otherwise indicated, we will restrict our attention to situations where both these sums are finite so the sum in (2.1) is absolutely convergent, i.e.,

$$\sum_{x \in \text{Range}\,(X)} |x|\, p_X(x) < \infty.$$

This condition is equivalent to $E|X| < \infty$ (see Theorem 2.5 below).

Continuous Random Variables. A *continuous random variable* X, by definition, possesses a density function $f(x) \geq 0$ where $\int_{-\infty}^{\infty} f(x)\,dx = 1$ and for $-\infty \leq c < d \leq \infty$

$$P[c < X < d] = \int_c^d f(x)\,dx. \tag{2.2}$$

The density function $f(x)$ is typically strictly positive and continuous on some interval $I = (a, b)$, and $f(x) = 0$ for $x \leq a$ or $x \geq b$. The interval I may either be finite, a half-line, or the entire line. Loosely speaking, $I = (a, b)$ is the smallest interval for which $P[a < X < b] = 1$. In the case where a and/or b are finite, we need not worry about the value of $f(a)$ or $f(b)$ since changing the value of $f(x)$ at one or two locations will not alter the value of the integral in (2.2). Note that continuous random variables take on particular values with probability 0 and we need not worry about whether inequalities are strict or not:

$$P[c < X < d] = P[c < X \leq d] = P[c \leq X < d] = P[c \leq X \leq d].$$

This, of course, is not true for discrete random variables, which *do* take on particular values with positive probability. For a continuous random variable X with density function $f(x)$, the expectation of X is given by

$$EX = \int_{-\infty}^{\infty} x f(x)\,dx. \tag{2.3}$$

The same technical issue that we had in the discrete case arises here. In particular, if

$$\int_0^\infty xf(x)\,dx = \infty \quad \text{and} \quad \int_{-\infty}^0 xf(x)\,dx = -\infty,$$

then the conflict cannot be resolved and EX is undefined. Except as noted, we will restrict our attention to the situation where both these integrals are finite and

$$E|X| = \int_{-\infty}^\infty |x|f(x)\,dx < \infty.$$

In either the discrete or the continuous setting, if $E|X| < \infty$ we will say that X has *finite first moment.* Note that all the distributions of Appendix I satisfy this condition.

Here we state without proof some of the basic properties of expectation. These properties hold in both the discrete and continuous setting.

THEOREM 2.4 (PROPERTIES OF EXPECTATION). *If X and Y have finite first moment then the following statements hold: for any constants a and b, $aX + bY$ has finite first moment and $E[aX+bY] = aEX + bEY$; $|EX| \leq E|X|$; if $X \leq Y$ in the sense that $X(\omega) \leq Y(\omega)$ for all $\omega \in \Omega$, then $EX \leq EY$, in particular, if $0 \leq X$, then $0 \leq EX$.*

Particular cases arise when Y is a constant c: $E[X + c] = EX + c$; and $EX \leq c$, provided $X(\omega) \leq c$ for all outcomes ω. That $E[aX+bY] = aEX+bEY$ is called *linearity of expectation.* The following is also often useful for computing expectations:

THEOREM 2.5. *Suppose $g : \mathbb{R} \to \mathbb{R}$ is any function. If X is a discrete random variable with mass function $p_X(x)$ and $g(X)$ has finite first moment then*

$$E[g(X)] = \sum_{x \in \text{Range}\,(X)} g(x)p_X(x).$$

If X is a continuous random variable with density function $f(x)$ and $g(X)$ has finite first moment then

$$E[g(X)] = \int_{-\infty}^\infty g(x)f(x)\,dx.$$

We will freely, and without reference, use Theorems 2.4 and 2.5.

Indicator Random Variables. If A is an event, there is an associated random variable called the *indicator* random variable, denoted by I_A. It is defined as

$$I_A(\omega) = \begin{cases} 1 & \text{if } \omega \in A \\ 0 & \text{otherwise.} \end{cases}$$

The value of I_A, 1 or 0, indicates whether or not (respectively) the event A occurs. Its expected value is:

$$E[I_A] \;=\; 0 \cdot P[I_A = 0] + 1 \cdot P[I_A = 1] \;=\; P[I_A = 1] \;=\; PA.$$

EXERCISES

1. Suppose X is uniformly distributed on the integers $\{1, 2, 3\}$ and let $Y = 1/X$. What are EX and EY? Does EY equal $1/EX$?

2. The shelf life in hours of a perishable product is a random variable with density function

$$f(x) \;=\; \begin{cases} 20{,}000/(x+100)^3 & \text{for } x > 0 \\ 0 & \text{otherwise.} \end{cases}$$

Find the probabilities that the product will have a shelf life: (a) of at most 100 hours; (b) somewhere between 100 and 200 hours.

3. The life-time in years of a light bulb, L, is a random variable with density function

$$f(x) \;=\; \begin{cases} 4/x^5 & \text{for } x > 1 \\ 0 & \text{otherwise.} \end{cases}$$

Find the expected life-time of the bulb.

4. The lifetime of a sound system is exponentially distributed with mean 5 years. The product guarantee will pay a refund of \$1000 if the system fails during the first year, \$500 if it fails during the second year, \$300 if it fails during the third year, and nothing if it lasts 3 years or longer. What is the expected payout under this guarantee?

5. A continuous random variable X has density function given by

$$f(x) \;=\; \begin{cases} cxe^{-x} & \text{if } x > 1 \\ 0 & \text{otherwise.} \end{cases}$$

Determine c and compute EX and $E[X^2]$.

6. We play the following "fair" game. I wager \$$k$ and we toss a fair coin. If it comes up heads you pay me \$$k$, if it comes up tails I pay you \$$k$. Suppose I pursue the following strategy. My first wager is \$1. Every time I lose the game we play again and I double my wager. I stop playing when I first win. With this strategy of play what are my expected winnings? Why couldn't I implement this strategy in real life?

7. With the set-up of problem 6, suppose I only have \$50 to play with. I pursue the same strategy with the following real-world modification. If I lose five games in a row, I can only wager \$(50 − 1 − 2 − 4 − 8 − 16) = \$19. If I also lose that game, I've lost all \$50 and I have to quit. What are my expected winnings?

8. Re-do problems 6 and 7 where my strategy is to *triple* my wager with each loss.

9. A 52 card poker deck has 12 "face" cards: 4 kings, 4 queens, and 4 jacks. Suppose I deal out 10 cards from a well shuffled deck. What is the probability that the 5th card dealt is a face card? (This requires virtually no computation.) Let N denote the number of face cards appearing among the 10 cards dealt. What is EN? Hint: expectation is linear.

10. Suppose X is a non-negative integer valued random variable, in other words $\text{Range}\,(X) \subset \mathbb{N}$. Show that $EX = \sum_{n=1}^{\infty} P[X \geq n]$. Hints: $n = \sum_{k=1}^{n} 1$; exchange order of summation.

11. Suppose X is a non-negative continuous random variable with density function $f(x)$, so $f(x) = 0$ for $x \leq 0$. Show that $EX = \int_0^\infty P[X \geq x]\,dx$.

§3. Moments, Variance, Covariance, and Correlation

If $E|X|^n < \infty$ for some positive integer n we say that X *has finite* n^{th} *moment*, and EX^n is called the nth *moment of* X. If X has finite n^{th} moment and $1 \leq m < n$ then X also has finite m^{th} moment. To see this, observe that $|X|^m \leq 1$ whenever $|X| \leq 1$ and $|X|^m \leq |X|^n$ whenever $|X| > 1$. In either case, then, we have that $|X|^m \leq 1 + |X|^n$, so $E|X|^m \leq 1 + E|X|^n < \infty$.

For any two non-negative numbers a and b, $a + b \leq 2\max(a, b)$ and

$$(a+b)^n \;\leq\; 2^n(\max(a,b))^n \;=\; 2^n \max(a^n, b^n) \;\leq\; 2^n(a^n + b^n),$$

so

$$|X+Y|^n \;\leq\; (|X| + |Y|)^n \;\leq\; 2^n(|X|^n + |Y|^n),$$

and $X + Y$ has finite n^{th} moment if both X and Y have finite n^{th} moment. A special case of this is when Y is any constant c: if X has finite n^{th} moment then so does $X + c$.

Variance and Covariance. If X has finite second moment, then it also has finite first moment and $\mu = EX$ is defined and finite. Also, $X - \mu$ has finite second moment and the *variance* of X, denoted $\text{Var}\,(X)$ or σ_X^2, is defined as

$$\text{Var}\,(X) \;=\; \sigma_X^2 \;=\; E\big[(X-\mu)^2\big].$$

The number σ_X, the square root of the variance, is called the *standard deviation* of X and is also denoted $\text{Std}\,(X)$. Now $(X-\mu)^2 = X^2 - 2\mu X + \mu^2$, so

$$\text{Var}\,(X) \;=\; E[X^2] - 2\mu EX + \mu^2 \;=\; E[X^2] - \mu^2 \;=\; E[X^2] - (EX)^2.$$

If X has mean 0, then $\mathrm{Var}\,(X)$ is just the second moment of X.

If X and Y have finite second moments, then XY has finite first moment (we will show this shortly) and the *covariance* of X and Y, denoted $\mathrm{Cov}\,(X,Y)$ or σ_{XY}, is defined by

$$\mathrm{Cov}\,(X,Y) \;=\; \sigma_{XY} \;=\; E[(X-EX)(Y-EY)].$$

This is a measure of how the two random variables move together. If X tends to be bigger (smaller) than its mean when Y is bigger (smaller) than its mean, then the product $(X-EX)(Y-EY)$ will tend to be positive (positive times positive or negative times negative) and we will have $\mathrm{Cov}\,(X,Y)>0$. Conversely, if X tends to be smaller (bigger) than its mean when Y is bigger (smaller) than its mean, then $(X-EX)(Y-EY)$ will tend to be negative (negative times positive or positive times negative) and we will have $\mathrm{Cov}\,(X,Y)<0$. Multiplying out the right hand side and using the linearity of expectation gives that

$$\mathrm{Cov}\,(X,Y) \;=\; E[XY]-EX\,EY, \tag{3.1}$$

a formula that is often computationally more convenient. This brings us back to the matter of showing that $E|XY|<\infty$ provided X and Y have finite second moments. But if X and Y do have finite second moments, then $|X|+|Y|$ has finite second moment and

$$E|XY| \;\le\; E[X^2]+2E|XY|+E[Y^2] \;=\; E(|X|+|Y|)^2 \;<\; \infty.$$

Covariance is *symmetric* and *bilinear* in the sense that

$$\begin{aligned}
\mathrm{Cov}\,(X,Y) &= \mathrm{Cov}\,(Y,X),\\
\mathrm{Cov}\,(aX+bY,Z) &= a\mathrm{Cov}\,(X,Z)+b\mathrm{Cov}\,(Y,Z), \text{ and}\\
\mathrm{Cov}\,(Z,aX+bY) &= a\mathrm{Cov}\,(Z,X)+b\mathrm{Cov}\,(Z,Y).
\end{aligned}$$

Here a and b are constants and X, Y, and Z are presumed to have finite second moments. Note that $\mathrm{Var}\,(X)=\mathrm{Cov}\,(X,X)$, so

$$\begin{aligned}
\mathrm{Var}\,(aX+bY) &= \mathrm{Cov}\,(aX+bY,aX+bY)\\
&= \mathrm{Cov}\,(aX,aX)+\mathrm{Cov}\,(aX,bY)\\
&\qquad +\mathrm{Cov}\,(bY,aX)+\mathrm{Cov}\,(bY,bY)\\
&= a^2\mathrm{Var}\,X+2ab\mathrm{Cov}\,(X,Y)+b^2\mathrm{Var}\,Y.
\end{aligned}$$

Similarly, repeated application of bilinearity shows that

$$\begin{aligned}
\mathrm{Var}\left[\sum_{i=1}^n a_iX_i\right] &= \sum_{i=1}^n\sum_{j=1}^n a_ia_j\mathrm{Cov}\,(X_i,X_j)\\
&= \sum_{i=1}^n a_i^2\mathrm{Var}\,(X_i) \;+\; 2\sum_{1\le i<j\le n} a_ia_j\mathrm{Cov}\,(X_i,X_j).
\end{aligned}$$

The random variables X and Y are said to be *uncorrelated* if $\text{Cov}(X, Y) = 0$ (this terminology will be explained below). If X and Y are uncorrelated then $\text{Var}(X + Y) = \text{Var}(X) + \text{Var}(Y)$. Furthermore, if X_i, $i = 1, 2, \ldots, n$, are pairwise uncorrelated, then we have

$$\text{Var}\left[\sum_{i=1}^{n} X_i\right] = \sum_{i=1}^{n} \text{Var}(X_i).$$

Correlation. The *correlation* of X and Y, denoted $\rho(X, Y)$ or ρ_{XY}, is covariance scaled to be a number between -1 and $+1$:

$$\rho(X, Y) = \frac{\sigma_{XY}}{\sigma_X \sigma_Y}. \tag{3.2}$$

Here we assume that neither σ_X nor σ_Y are zero. To see that $|\rho(X, Y)| \le 1$, let t be any number and consider the random variable $Z = tX - Y$. Now

$$\begin{aligned} 0 \le \text{Var}(Z) &= t^2\text{Var}(X) - 2t\text{Cov}(X, Y) + \text{Var}(Y) \\ &= \sigma_X^2 t^2 - 2\sigma_{XY} t + \sigma_Y^2. \end{aligned}$$

This latter expression is a quadratic in t that is always non-negative and therefore has at most one zero. The quadratic's discriminant (a.k.a. $b^2 - 4ac$) is therefore less than or equal to 0:

$$(-2\sigma_{XY})^2 - 4\sigma_X^2\sigma_Y^2 \le 0,$$

which is equivalent to

$$|\rho(X, Y)| = \left|\frac{\sigma_{XY}}{\sigma_X \sigma_Y}\right| \le 1.$$

Working backward, if $|\rho(X, Y)| = 1$ then the discriminant is 0 and there is exactly one zero for the quadratic: for some number m, $\text{Var}(mX - Y) = 0$. The random variable $mX - Y$ is therefore a constant and $Y = mX + b$, for some numbers m and b. If this is true, then

$$\sigma_{XY} = \text{Cov}(X, Y) = \text{Cov}(X, mX + b) = m\text{Cov}(X, X) = m\sigma_X^2,$$

and

$$\text{Var}(Y) = \text{Var}(mX + b) = m^2\text{Var}(X),$$

so $\sigma_Y = |m|\sigma_X$ (note: $\sqrt{m^2} = |m|$, not m). Setting these facts into (3.2) gives that

$$\rho(X, Y) = \frac{m\sigma_X^2}{\sigma_X |m| \sigma_X} = \frac{m}{|m|}.$$

Hence $m > 0$ if $\rho(X, Y) = +1$ and $m < 0$ if $\rho(X, Y) = -1$. (Note that we cannot have $m = 0$ for in that situation $\text{Var}(Y) = 0$.)

To summarize, we have:

THEOREM 3.3. *The correlation $\rho(X,Y)$ is a number between -1 and $+1$. If $|\rho(X,Y)| = 1$, then Y is a linear function of X. If $\rho(X,Y) = +1$, the linear function has positive slope; if $\rho(X,Y) = -1$, the linear function has negative slope.*

EXERCISES

1. Suppose $\operatorname{Var} X = 3$, $\operatorname{Var} Y = 4$, $\operatorname{Var} Z = 5$, and that $\operatorname{Cov}(X,Y) = 1$, $\operatorname{Cov}(X,Z) = 2$, and $\operatorname{Cov}(Y,Z) = .5$. Calculate: (a) $\operatorname{Cov}(3X, Y + 2Z)$, (b) $\operatorname{Var}(Y + 2Z)$.

2. Suppose two stocks each cost \$1 per share today. Let the random variables U and V denote their values per share one year from now. Suppose I know that $\operatorname{Var} U = .15$, $\operatorname{Var} V = .2$, and U and V are uncorrelated. I have \$1000 to invest in these two stocks; if I invest x in stock U and $1000 - x$ in stock V then one year from now my portfolio will be worth $P = xU + (1000 - x)V$. How much should I invest in stocks U and V to minimize the variance of P?

3. How would your answer change if U and V had correlation 0.5?

4. How large must $\rho(U,V)$ be before your entire investment would be in U?

5. Suppose $\operatorname{Var}(X) = 2$, $\operatorname{Var}(Y) = 5$, and $\operatorname{Cov}(X,Y) = 1$. Compute $\rho(2X + Y, 3X - Y)$.

6. If $X_1, \ldots, X_n$ are random variables defined on the same probability space Ω, their *covariance matrix* is an $n \times n$ matrix whose i-j^{th} entry is given by $\sigma_{ij} = \operatorname{Cov}(X_i, X_j)$. Suppose N_1, N_2, N_3 are uncorrelated standard normals and $X_1 = 2N_1 + N_2 + N_3$, $X_2 = N_1 + 2N_2 + 3N_3$, and $X_3 = N_1 + 4N_2$. Compute the covariance matrix for X_1, X_2, X_3.

7. Suppose X and Y are exponential random variables with means x and y, respectively. Show that $\rho(X,Y) = -1$ is not possible.

8. Suppose I randomly and independently place n marbles into m urns, where $m \geq 2$. Let N_i be the number of marbles placed in urn number i, so N_i is a binomial random variable with parameters n and $p = \frac{1}{m}$. Calculate $\rho(N_i, N_j)$ for $i \neq j$. Explain your answer in the case $m = 2$.

9. Suppose X has finite second moment, so that its variance σ^2 and mean μ are defined and finite. Show that $P[X = \mu] = 1$ if $\sigma^2 = 0$. Hint: show that $P[|X - \mu| > \frac{1}{n}] = 0$ for $n = 1, 2, \ldots$ and use that $\{X \neq \mu\} \subset \bigcup_n \{|X - \mu| > \frac{1}{n}\}$. Then use problem 10 from §1.

§4. Jointly Distributed Random Variables

Independent Random Variables. Two random variables X and Y defined on a common probability space Ω are independent if, for any numbers x and y, we have

$$P[X \le x, Y \le y] = P[X \le x] \cdot P[Y \le y].$$

This generalizes to more than two random variables: $X_1, X_2, \ldots, X_n$ are independent if for all numbers $x_1, \ldots, x_n$,

$$P\big[X_1 \le x_1, X_2 \le x_2, \ldots, X_n \le x_n\big] = P[X_1 \le x_1] \cdot P[X_2 \le x_2] \cdot \cdots \cdot P[X_n \le x_n].$$

The probability on the left side is the *joint distribution function* for $(X_1, \ldots, X_n)$ which we denote here by $F(x_1, x_2, \ldots, x_n)$. The individual probabilities on the right side are the *marginal* distribution functions $F_{X_i}(x) = P[X_i \le x]$. Independence means that

$$F(x_1, x_2, \ldots, x_n) = F_{X_1}(x_1) \cdot F_{X_2}(x_2) \cdot \cdots \cdot F_{X_n}(x_n),$$

so random variables $X_1, \ldots, X_n$ are independent if their joint distribution function factors into their marginal distribution functions. An infinite sequence of random variables $X_1, X_2, \ldots$ are independent if $X_1, \ldots, X_n$ are independent for all $n \ge 2$.

If the X_i's are discrete random variables, let

$$p(x_1, x_2, \ldots, x_n) = P\big[X_1 = x_1, X_2 = x_2, \ldots, X_n = x_n\big]$$

denote their *joint mass function.* To establish independence it is enough to verify that

$$p(x_1, x_2, \ldots, x_n) = p_{X_1}(x_1) \cdot p_{X_2}(x_2) \cdot \cdots \cdot p_{X_n}(x_n) \tag{4.1}$$

holds for all numbers $x_1, \ldots, x_n$. That is, it is enough that the joint mass function factors into the marginal mass functions.

Suppose the X_i's are jointly continuous random variables. If they are independent, taking the derivative of the joint distribution function $F(x_1, \ldots, x_n)$ with respect to x_1 (using that $F_{X_2}(x_2) \cdot \cdots \cdot F_{X_n}(x_n)$ is constant with respect to x_1) gives

$$\frac{\partial}{\partial x_1} F(x_1, \ldots, x_n) = f_{X_1}(x_1) \cdot \big[F_{X_2}(x_2) \cdot \cdots \cdot F_{X_n}(x_n)\big].$$

Repeated additional differentiation with respect to first x_2, then x_3, etc., gives

$$\frac{\partial^n}{\partial x_n \cdots \partial x_1} F(x_1, \ldots, x_n) = f_{X_1}(x_1) \cdot f_{X_2}(x_2) \cdot \cdots \cdot f_{X_n}(x_n).$$

The left side of the above equality is the *joint density function* of $(X_1, \ldots, X_n)$ and is denoted here by $f(x_1, \ldots, x_n)$. If the X_i's are jointly continuous and independent,

$$f(x_1, x_2, \ldots, x_n) = f_{X_1}(x_1) \cdot f_{X_2}(x_2) \cdot \cdots \cdot f_{X_n}(x_n), \tag{4.2}$$

which is the obvious continuous analog of (4.1). The argument reverses in the sense that, if (4.2) holds for jointly continuous random variables X_i, then $X_1, \ldots, X_n$ are independent.

The next theorem relates independence to correlation and is proved below for the discrete case only.

THEOREM 4.3. *If X and Y are independent and have finite second moments, they are uncorrelated.*

PROOF (DISCRETE CASE). In view of (3.1) it suffices to show that $E[XY] = EX \cdot EY$. With $Z = XY$, we have

$$\begin{aligned}
EX \cdot EY &= \Big(\sum_x xP[X=x]\Big) \cdot \Big(\sum_y yP[Y=y]\Big) \\
&= \sum_{(x,y)} \Big(xy \cdot P[X=x] \cdot P[Y=y]\Big) \\
&= \sum_{(x,y)} xyP[X=x, Y=y] \quad \text{(by independence)} \\
&= \sum_z \sum_{(x,y):xy=z} xyP[X=x, Y=y] \quad \text{(re-group the summands)} \\
&= \sum_z \sum_{(x,y):xy=z} zP[X=x, Y=y] \\
&= \sum_z z \sum_{(x,y):xy=z} P[X=x, Y=y] \;=\; \sum_z zP[Z=z] \;=\; EZ. \quad \square
\end{aligned}$$

While independent random variables are uncorrelated, uncorrelated random variables are not necessarily independent — see exercises.

Expectation By Conditioning. There is an analog of the partition theorem for expectation. For discrete jointly distributed X and Y, we have that

$$EX = \sum_{y \in \text{Range}(Y)} E[X \mid Y = y]\, p_Y(y).$$

Here the conditional expectation of X given $\{Y = y\}$ is computed in the usual way, except the conditional distribution of X is used:

$$E[X \mid Y = y] = \sum_{x \in \text{Range}(X)} x p_{(X \mid Y=y)}(x),$$

where the conditional mass function of X is given by

$$p_{(X \mid Y=y)}(x) = \frac{p_{(X,Y)}(x,y)}{p_Y(y)}.$$

This is verified as follows:

$$\begin{aligned}\sum_y E[X\,|\,Y=y]\,p_Y(y) &= \sum_y\sum_x x p_{(X\,|\,Y=y)}(x)\,p_Y(y)\\ &= \sum_y\sum_x x p_{(X,Y)}(x,y)\\ &= \sum_x x\sum_y p_{(X,Y)}(x,y)\\ &= \sum_x x p_X(x) \;=\; EX. \;\checkmark\end{aligned}$$

For jointly continuous X and Y, we have:

$$EX = \int E[X\,|\,Y=y] f_Y(y)\,dy,$$

where in this setting

$$E[X\,|\,Y=y] = \int x f_{(X\,|\,Y=y)}(x)\,dx,$$

where the conditional density function of X is given by

$$f_{(X\,|\,Y=y)}(x) = \frac{f_{(X,Y)}(x,y)}{f_Y(y)}.$$

EXERCISES

• For problems 1 – 8, compute the marginal mass functions (or density functions, as appropriate) for the random variables X and Y.

1. The joint mass function for (X,Y) is: $p(1,0) = p(0,1) = 1/2$, otherwise $p(x,y) = 0$.

2. The joint mass function for (X,Y) is: $p(0,1) = p(1,0) = 1/2$, otherwise $p(x,y) = 0$.

3. A coin biased to produce heads with probability $\frac{3}{4}$ is tossed three times. Let X denote the number of heads obtained (in all three tosses) and let Y denote the number of heads obtained the first two tosses.

4. The joint mass function for (X,Y) is given by:

	y = 0	1	2
x = 2	.10	.15	.05
4	.20	.10	.05
6	.05	.15	.15

Also calculate $P[X+Y \text{ is an even number}]$ and $P[Y \geq 1 | X = 2]$.

5. Continuous X and Y have joint density function

$$f(x,y) = \begin{cases} x+y & \text{for } 0 < x < 1 \text{ and } 0 < y < 1 \\ 0 & \text{otherwise.} \end{cases}$$

Also calculate $P[X+Y \leq 1/2]$.

6. Continuous X and Y have joint density function

$$f(x,y) = \begin{cases} 20x^{-3}y^{-4} & \text{for } x > 1,\ y > 1 \\ 0 & \text{otherwise.} \end{cases}$$

Also compute $E[XY^2]$.

7. Continuous X and Y have joint density function

$$f(x,y) = \begin{cases} 24xy & \text{for } 0 < x,\ 0 < y,\ x+y < 1 \\ 0 & \text{otherwise.} \end{cases}$$

8. The joint density function of X and Y is

$$f(x,y) = \begin{cases} 1/y & \text{for } 0 < x < y \text{ and } 0 < y < 1 \\ 0 & \text{elsewhere.} \end{cases}$$

Also calculate $P[Y > 2X]$.

9. Calculate $\operatorname{Cov}(X,Y)$ and $\rho(X,Y)$ for the joint distributions of problems 1 – 8.

10. Suppose X and Y are discrete random variables with joint mass function $p = p_{(X,Y)}$ given by $p(1,-1) = p(0,0) = p(1,1) = \frac{1}{3}$. What is $\operatorname{Cov}(X,Y)$? Are X and Y independent?

§5. IID Sequences of Random Variables

A *stochastic process* is an indexed collection of random variables typically of the form $(X_n : n = 0,1,2,3,\ldots)$ or $(X_t : t \geq 0)$. Usually they describe the random evolution of some system over time. In the first instance, time is measured in discrete increments, or periods. In the second, time is thought of as a continuous variable. For example, X_n may measure the amount of rainfall in your neighborhood on day n of some study. The value of X_t might represent the value of one share of IBM stock t seconds into a particular trading day.

The most elementary of stochastic processes are *independent, identically distributed* sequences. (They are so elementary that they are sometimes not even thought of as stochastic processes.) A sequence of random variables, $X_1, X_2, \ldots$, defined on a common probability space, are independent and identically distributed (abbreviated "iid") if they are independent and all have the same distribution function: $F_{X_i}(x) = P[X_i \le x] = F(x)$ for some common distribution function $F(x)$. In this case, we will write that "the X_i are iid with each $X_i \sim F$". (The "$\sim$" symbol is read "is distributed like" and indicates that X_i has distribution function F. Sometimes we will also write, for example, $X_i \sim X$, or, $X_i \sim \text{Binomial}\,(3, \frac{1}{2})$, which mean, respectively, that X_i and X have the same distribution function and that the distribution of X_i is binomial with parameters $n = 3$ and $p = \frac{1}{2}$.) Clearly the random variables in an iid sequence all have the same range, mean, and variance, as these features of a random variable depend only on its distribution. Iid sequences will be the building blocks for the stochastic processes studied in subsequent chapters.

As we have seen, probability spaces come in two flavors: discrete and uncountable. Unfortunately, discrete probability spaces are not sufficiently commodious to support iid sequences, except in the most uninteresting way. Indeed, the following theorem says that the only iid sequences on a discrete probability space are constant sequences!

THEOREM 5.1. *Suppose Ω is a discrete probability space and $X_1, X_2, \ldots$ is an iid sequence of random variables defined on Ω. Then, for some number r, $P[X_i = r] = 1$.*

PROOF. Since Ω is discrete, the random variables X_i must be discrete as well. Also, the X_i's all have the same range and probability mass function as they are identically distributed. Let $\mathcal{R} = \{r_1, r_2, \ldots\}$ denote the range of the X_i's and put $p_k = P[X_i = r_k]$, so each $p_k > 0$. Since $\sum_k p_k = 1$, only finitely many of the p_k's can equal or exceed p_1. Among these finitely many probabilities, there is a largest probability — call it p. Then each $p_k \le p$ and for some number $r \in \mathcal{R}$ we have $P[X_i = r] = p$. [For example, if the X_i's are Binomial$\,(2, \frac{1}{2})$ then $r_1 = 0$, $r_2 = 1$, $r_3 = 2$, with $p_1 = \frac{1}{4}$, $p_2 = \frac{1}{2}$, $p_3 = \frac{1}{4}$, so $r = r_2 = 1$ and $p = p_2 = \frac{1}{2}$.]

Clearly we always have $0 < p \le 1$; presently we show that if the X_i's are iid we must have $p = 1$. This will yield $P[X_i = r] = 1$, as required. Suppose, to the contrary, that $p < 1$. For any sequence of numbers $x_1, x_2, \ldots$, we have $P[X_i = x_i] \le p < 1$, so

$$\begin{aligned} P[X_1 = x_1, X_2 = x_2, \ldots] &\le P[X_1 = x_1, X_2 = x_2, \ldots, X_n = x_n] \\ &= P[X_1 = x_1] \cdot P[X_2 = x_2] \cdot \ldots \cdot P[X_n = x_n] \\ &\le p^n. \end{aligned}$$

Since $p < 1$, the right side above can be made arbitrarily small by choosing n sufficiently large. It follows that $P[X_1 = x_1, X_2 = x_2, \ldots] = 0$. Now pick *any* outcome $\omega^* \in \Omega$ and, for each i, let $x_i^* = X_i(\omega^*)$. Then, by construction, we must have

$$\omega^* \in \{X_1 = x_1^*, X_2 = x_2^*, \ldots\}.$$

Since the event on the right side above has 0 probability, it follows that $P\{\omega^*\} = 0$. Since this holds for all outcomes, an application of (1.12) then gives that

$$1 = P[\Omega] = \sum_{\omega \in \Omega} P\{\omega\} = \sum_{\omega \in \Omega} 0 = 0,$$

a contradiction. Hence we must have $p = 1$ and therefore $P[X_i = r] = 1$. □

EXERCISES

• Suppose the distribution function $F(x)$ has no discontinuities (a normal distribution, e.g.). Here we modify the proof of Theorem 5.1 to show that it is impossible for a random variable X defined on a discrete probability space Ω to have $F(x)$ as its distribution.

1. Fix any number x. Use that $\{X = x\} \subset \{x - \epsilon < X \leq x\}$ for all $\epsilon > 0$ to show that $P[X = x] = 0$.

2. Fix any ω^* and use that $\omega^* \in \{\omega : X(\omega) = X(\omega^*)\}$ to prove the assertion.

§6. The Unit Interval $\mathbf{\Omega} = (0, 1)$ as Outcome Space

It should be clear from Theorem 5.1 that the most interesting and useful problems in probability theory cannot be modeled on a discrete probability space. Rather, an outcome space with *uncountably* many outcomes is necessary. Fortunately, there is one particular outcome space that is intuitive and easy to visualize upon which all probabilistic models of practical importance can be constructed. The underlying experiment is to "randomly pick a real number between 0 and 1." (We can only do that in our imagination — but that is where mathematics takes place!). Here the outcome space is the interval $\mathbf{\Omega} = (0, 1)$ — all real numbers between 0 and 1 (we exclude 0 and 1 for convenience only). We may associate each number ω on this interval with its decimal expansion, e.g. $\omega = 0.83553\cdots$. One technical issue that arises is that some numbers have two different expansions, for example $0.6230000\cdots = 0.6229999\cdots$. We choose to represent such an ω with the first version, the one terminating in a sequence of 0's. With this convention in place, every number ω on the interval $(0, 1)$ has a unique decimal expansion. This outcome space cannot be listed. To see this, suppose $\{\omega_1, \omega_2, \omega_3, \ldots\}$ is *any* list of outcomes from $\mathbf{\Omega} = (0, 1)$. For example, this list might begin as shown in Figure 6.1. We produce a new number in $(0, 1)$, call it $\tilde{\omega}$, as follows. Go along the diagonal, shown with underscores, to produce the sequence of digits $.4281\cdots$. Now change every digit to something other than 0 or 9. For example, change d to $d + 1$ unless $d = 8$ or 9 and change 8 and 9 to 1. This rule produces the number $\tilde{\omega} = .5312\cdots$. (Note that the resulting sequence of digits cannot terminate with an infinite sequence of 9's because of our rule for 8; nor can we get $\tilde{\omega} = .000\cdots$.) Then $\tilde{\omega} \neq \omega_1$, because they differ at the first digit. (The digits obviously differ elsewhere as well, but by construction they *must* differ at least at the first digit.) Also $\tilde{\omega} \neq \omega_2$, because they differ at the second digit. Also $\tilde{\omega} \neq \omega_3$, because they differ at the third digit. In general, $\tilde{\omega} \neq \omega_k$ because they differ at the kth digit. Clearly $\tilde{\omega}$, which *is* an element of $\mathbf{\Omega}$, is not present in the list. Since this argument works for *any* listing of elements from $\mathbf{\Omega}$ we are forced to conclude that the elements of $\mathbf{\Omega}$ cannot be presented exhaustively in a list. This beautiful argument is due to Georg Cantor (1845–1918), and is called Cantor's diagonalization argument.

Figure 6.1. A hypothetical listing of outcomes from $\mathbf{\Omega} = (0, 1)$.

$$\begin{aligned} \omega_1 &= .\,\underline{4}\,6\,1\,0\,0\,8\,2\,5\,6\,\cdots \\ \omega_2 &= .\,0\,\underline{2}\,7\,5\,0\,8\,7\,9\,3\,\cdots \\ \omega_3 &= .\,6\,7\,\underline{8}\,7\,0\,2\,6\,1\,4\,\cdots \\ \omega_4 &= .\,5\,3\,8\,\underline{1}\,1\,0\,0\,4\,7\,\cdots \\ &\vdots \qquad\quad \vdots \end{aligned}$$

With uncountable outcome spaces, the situation is considerably more complicated than (1.12). In our current construction, intuitively any particular individual number ω between 0 and 1 will be picked with probability 0. If we were to attempt to compute the probability of the entire outcome space by summing the probabilities of individual outcomes we would obtain:

$$P[\mathbf{\Omega}] \;=\; \sum_{\omega \in \mathbf{\Omega}} P\{\omega\} \;=\; \sum_{\omega \in \mathbf{\Omega}} 0 \;=\; 0,$$

which is obviously incorrect. The second and third equalities above are correct, it is the first that is not right. The axioms of probability (specifically, axiom (1.6)) require that this first inequality hold only for countable sums; the set $\mathbf{\Omega}$ above is uncountable.

The question then arises: for any subset $A \subset \mathbf{\Omega}$, how do we compute $P[A]$, the probability that a 'randomly selected' number lies in the set A? Of equal importance is the question: for what subsets A does it even make sense to ask what $P[A]$ is? In other words, what may we take as the events, $\mathcal{F}$? Certainly we wish for the events to satisfy axioms (1.1, 1.2, and 1.3) and the probability measure P to satisfy axioms (1.4, 1.5, and 1.6). We also want the outcome ω to be *uniformly* distributed on $(0, 1)$ in the sense of the following examples. We want both $P[0 < \omega < \frac{1}{2}]$ and $P[\frac{1}{2} < \omega < 1]$ to be $\frac{1}{2}$. (Let's not worry about writing '$<$' or '$\leq$', because the probability of picking any specific number is 0.) Similarly, we would like $P[0 < \omega < \frac{1}{3}]$, $P[\frac{1}{3} < \omega < \frac{2}{3}]$, and $P[\frac{2}{3} < \omega < 1]$ to all equal $\frac{1}{3}$. These are special cases of the following rule: if $0 \leq a < b \leq 1$, then we would like $P[a < \omega < b] = b - a$. The intuition here is that $b - a$ is the fraction of the length of $(0, 1)$ that is occupied by the interval (a, b), so the probability that a number randomly selected on $(0, 1)$ lies within (a, b) should be $b - a$. With all this in mind, we state without proof the following:

THEOREM 6.2. *There is a collection,* $\mathcal{M}$, *of subsets of* $(0,1)$ *that satisfy axioms* (1.1, 1.2, and 1.3) *and a probability measure* P *defined on all events* $A \in \mathcal{M}$ *satisfying axioms* (1.4, 1.5, and 1.6) *such that:* (i) *all subintervals of* $(0,1)$ *are in* $\mathcal{M}$; *and* (ii), *if* $A = (a,b) \subset (0,1)$, *then* $P[A] = b - a$.

Throughout this text, when we refer to the sample space $\mathbf{\Omega}$ (i.e., a 'bold face' Ω), it will be the interval $(0,1)$ equipped with the events $\mathcal{M}$ and the uniform probability measure P of Theorem 6.2. That is, $\mathbf{\Omega}$ will be our shorthand notation for the probability space $((0,1), \mathcal{M}, P)$. Construction of $\mathcal{M}$ and the probability measure P is difficult. It involves the fundamentals of measure theory and represents the boundary between undergraduate and graduate probability.

Properties of ω's Digits. A number's decimal expansion begins with 0.623 if and only if that number is in the interval $[0.623, 0.624)$, so

$$\begin{aligned} P\{\omega \in \mathbf{\Omega} : \text{the first 3 digits of } \omega \text{ are } 6,2,3\} &= P[0.623 \le \omega < 0.624] \\ &= .001 = 10^{-3}. \end{aligned}$$

In general, for $\omega \in \mathbf{\Omega}$ the first n digits of ω's decimal expansion are $d_1, \ldots, d_n$, i.e,

$$\omega = 0.d_1 d_2 \cdots d_n \cdots$$

if and only if

$$0.d_1 d_2 \cdots d_n \le \omega < 0.d_1 d_2 \cdots d_n + 10^{-n},$$

that is, if and only if ω lies in the interval $[a, a + 10^{-n})$ where $a = 0.d_1 d_2 \cdots d_n$. Since this latter interval is of length 10^{-n}, it follows that

$$P\{\omega \in \mathbf{\Omega} : \text{the first } n \text{ digits of } \omega \text{ are } d_1, \ldots, d_n\} = 10^{-n}. \tag{6.3}$$

Equation (6.3) is just what we need to prove:

THEOREM 6.4. *Let* $D_n(\omega)$ *denote the* n^{th} *digit in the decimal expansion of the number* $\omega \in \mathbf{\Omega}$. *Then the sequence* $D_1, D_2, D_3, \ldots$ *is iid with each* $D_n \sim$ Uniform $\{0, 1, \ldots, 9\}$.

PROOF. Let $d_1, d_2, \ldots$ be any sequence of numbers from $\{0, 1, \ldots, 9\}$ and put

$$\begin{aligned} A &= \{\omega \in \mathbf{\Omega} : \text{the } n^{\text{th}} \text{ digit of } \omega \text{ is } d_n\}, \text{ and} \\ B &= \{\omega \in \mathbf{\Omega} : \text{the first } n-1 \text{ digits of } \omega \text{ are } d_1, \ldots, d_{n-1}\}. \end{aligned}$$

We will show that $PA = \frac{1}{10}$ and, furthermore, $P[A|B] = \frac{1}{10}$. This will suffice, as the first statement says that the digits in ω's decimal expansion are each uniformly distributed on $\{0, 1, \ldots, 9\}$ while the second statement says that each digit is independent of previous digits.

We begin with the second statement. In view of (6.3), we see that $PB = 10^{-(n-1)}$. Also,

$$A \cap B = \{\omega \in \boldsymbol{\Omega} : \text{the first } n \text{ digits of } \omega \text{ are } d_1, \dots, d_n\},$$

so $P[A \cap B] = 10^{-n}$. Then

$$P[A|B] = \frac{P[A \cap B]}{PB} = \frac{10^{-n}}{10^{-(n-1)}} = \frac{1}{10}. \tag{6.5}$$

Turning to the first statement, there are 10^{n-1} possibilities for the first $n-1$ digits of a number $\omega \in (0,1)$. Let $B_1, \dots B_{10^{n-1}}$ denote the 10^{n-1} disjoint subsets (sub*intervals*, in fact) of $(0,1)$ that correspond to these different possibilities. They form a partition of $(0,1)$ and, from the above reasoning, are all equally likely: $P[B_k] = 10^{-(n-1)}$. Using (6.5) and the partition theorem we see that

$$PA = \sum_{k=1}^{10^{n-1}} P[A|B_k]PB_k = 10^{n-1} \cdot \frac{1}{10} \cdot 10^{-(n-1)} = \frac{1}{10}. \quad \square$$

This theorem has a converse, which we state as:

THEOREM 6.6. *Suppose $D_1, D_2, D_3, \dots$ are iid with each D_i uniformly distributed on the numbers $\{0, 1, \dots, 9\}$. Let $U = 0.D_1D_2D_3\cdots$, so the digits in U's decimal expansion are $D_1, D_2, D_3, \dots$. Then $U \sim$ Uniform$(0,1)$, that is, $P[U \le x] = x$ for all x with $0 \le x \le 1$.*

A proof of this is left to the exercises.

EXERCISES

• Here we prove Theorem 6.6.

1. Show that $P[U < \frac{k}{10}] = \frac{k}{10}$ for $k = 0, 1, \dots, 10$. Hint: $\{U < \frac{k}{10}\} = \{10U < k\} = \{D_1 < k\}$.

2. By considering the first two digits, show that $P[U < \frac{k}{100}] = \frac{k}{100}$ for $k = 0, 1, \dots, 100$.

3. By considering the first n digits, show that $P[U < \frac{k}{10^n}] = \frac{k}{10^n}$ for $k = 0, 1, \dots, 10^n$.

4. Fix $x \in (0,1)$. Let x_n be the smallest fraction of the form $\frac{k}{10^n}$ that exceeds x. (For example, if $x = \frac{1}{3}$, $x_1 = .4$, $x_2 = .34$, $x_3 = .334$, $x_4 = .3334$, etc.) Let $y_n = x_n - \frac{1}{10^n}$ and show that $y_n \le x < x_n$. Use problem 3 and the fact that

$$\{U < y_n\} \subset \{U \le x\} \subset \{U < x_n\}$$

to prove the theorem.

§7. Constructing Stochastic Processes on **Ω**

Independence is a fairly tenacious property, as we illustrate here. For example, if X and Y are independent, so are X^2 and Y^3. More generally, $g(X)$ and $h(Y)$ are independent for any two functions $g(x)$ and $h(y)$. This idea generalizes substantially. In a slightly more complicated construction, suppose X_1, X_2, X_3, and X_4 are independent. Under these circumstances, if $g(x, y)$ and $h(x, y)$ are any two functions, then $U = g(X_1, X_2)$ and $V = h(X_3, X_4)$ are independent. For example, $U = X_1 + X_2$ and $V = X_3X_4$ are independent. In general, random variables constructed from independent collections of other random variables are themselves independent. Stating and proving this in full generality requires measure theory and is beyond the scope of this text. Yet this clearly should be true for any acceptable definition of independence.

We shall use this fact to generate an entire sequence of independent random variables $U_1, U_2, U_3, \ldots$ defined on $\mathbf{\Omega}$ with each $U_i \sim$ Uniform$\,(0, 1)$. To accomplish this, suppose $\omega = 0.D_1D_2D_3D_4\cdots$ is the decimal expansion of the outcome $\omega \in \mathbf{\Omega}$. That is, each $D_n(\omega)$ is the nth digit in ω's decimal expansion. Our approach here is to present these digits in an infinite *array*, rather than in a one dimensional list. This is shown in Figure 7.1 below, where the arrows illustrate the pattern in which the digits are laid out.

Figure 7.1. The digits of ω presented in an array.

$$
\begin{array}{llllllll}
D_1 \rightarrow & D_2 & D_4 & D_7 & D_{11} & D_{16} & D_{22} & D_{29} \\
D_3 & D_5 & D_8 & D_{12} & D_{17} & D_{23} & \cdot^{\cdot^{\cdot}} & \\
D_6 & D_9 & D_{13} & D_{18} & D_{24} & & & \\
D_{10} & D_{14} & D_{19} & D_{25} & & & & \\
D_{15} & D_{20} & D_{26} & & \ddots & & & \\
D_{21} & D_{27} & & & & & & \\
D_{28} & & & & & & &
\end{array}
$$

We then use the digits in the first row to form $U_1(\omega)$, the digits in the second row to form $U_2(\omega)$, etc.:

$$
\begin{aligned}
U_1(\omega) &= 0.D_1D_2D_4D_7D_{11}\cdots, \\
U_2(\omega) &= 0.D_3D_5D_8D_{12}D_{17}\cdots, \\
U_3(\omega) &= 0.D_6D_9D_{13}D_{18}D_{24}\cdots.
\end{aligned}
$$

For example, suppose $\omega = 0.638031684634792760376503\cdots$. The reader will verify that this produces:

$$\begin{aligned} U_1(\omega) &= 0.63063\cdots, \\ U_2(\omega) &= 0.83846\cdots, \\ U_3(\omega) &= 0.14703\cdots, \end{aligned}$$

and so forth.

From Theorem 6.4 the D_n's are iid and uniformly distributed on $\{0, 1, \ldots, 9\}$. It follows from Theorem 6.6 that the U_k's are all distributed like Uniform $(0, 1)$. Furthermore, the U_k form an independent sequence of random variables. This is because each successive U_k is constructed from a set of digits that are independent of the digits used earlier in the construction, i.e., for the construction of the U_i's with $i < k$. The motto for the Seal of the United States is "E pluribus unum" ("from many, one" — look at the back of any dollar bill above the eagle). Here we have a case of E unum pluribus!

Other Distributions. The sequence $U_1, U_2, \ldots$ constructed above is the basic building block from which we will construct other stochastic processes. Presently we illustrate how, given any distribution function $F(x)$, we may use $U_1, U_2, \ldots$ to construct an iid sequence $X_1, X_2, \ldots$ on $\mathbf{\Omega}$ with each $X_i \sim F$. We first illustrate the technique, known as *inverse transform*, for discrete distributions.

Inverse Transform for Discrete Distributions. Suppose we seek to construct a discrete random variable with range $\{x_1, x_2, \ldots\}$. This list may terminate or continue indefinitely, according to whether the range is finite or countably infinite. Suppose further that the desired probability mass function is $p(x)$, so we seek X with $P[X = x_i] = p(x_i)$. Accumulate these numbers as follows:

$$\begin{aligned} P_0 &= 0, \quad \text{and} \\ P_i &= p(x_1) + p(x_2) + \cdots + p(x_{i-1}) + p(x_i) \\ &= P_{i-1} + p(x_i), \quad i \geq 1. \end{aligned}$$

Since $\sum_i p(x_i) = 1$, we see that $0 = P_0 < P_1 < P_2 < \cdots \to 1$. Now, for $0 < y < 1$ put

$$H(y) = \begin{cases} x_1 & \text{if } P_0 < y \leq P_1 \\ x_2 & \text{if } P_1 < y \leq P_2 \\ x_3 & \text{if } P_2 < y \leq P_3 \\ \ldots & \ldots \end{cases} \tag{7.2}$$

and let $X = H(U)$, where $U \sim$ Uniform $(0, 1)$. Then $X = x_i \iff P_{i-1} < U \leq P_i$. Hence $P[X = x_i] = P_i - P_{i-1} = p(x_i)$, as desired. If we put $X_i = H(U_i)$, then the sequence $X_1, X_2, \ldots$ is iid with the probability mass function $p(x)$.

EXAMPLE 7.3. Suppose we wish to generate X with range $\{1, 2, 3, 4\}$ and

$$P[X = x] \;=\; \begin{cases} 0.125 & \text{if } x = 1 \\ 0.125 & \text{if } x = 2 \\ 0.5 & \text{if } x = 3 \\ 0.25 & \text{if } x = 4. \end{cases}$$

The reader will verify that

$$H(y) \;=\; \begin{cases} 1 & \text{if } 0 < y \le 0.125 \\ 2 & \text{if } 0.125 < y \le 0.25 \\ 3 & \text{if } 0.25 < y \le 0.75 \\ 4 & \text{if } 0.75 < y < 1, \end{cases}$$

which gives

$$X \;=\; H(U) \;=\; \begin{cases} 1 & \text{if } 0 < U \le 0.125 \\ 2 & \text{if } 0.125 < U \le 0.25 \\ 3 & \text{if } 0.25 < U \le 0.75 \\ 4 & \text{if } 0.75 < U < 1. \end{cases}$$

If $U \sim \text{Uniform}(0,1)$ and $X = H(U)$, we will have $P[X = 1] = 0.125$, $P[X = 2] = 0.25 - 0.125 = 0.125$, $P[X = 3] = 0.75 - 0.25 = 0.5$, and $P[X = 4] = 1.0 - 0.75 = 0.25$, as desired.

Inverse Transform for Continuous Distributions. Recall that a continuous random variable possesses a density function $f(x)$ that is related to the distribution function $F(x)$ by

$$F(x) \;=\; \int_{-\infty}^{x} f(t)\,dt.$$

The density function typically is strictly positive on some interval (a, b), with $f(x) = 0$ for $x \le a$ or $x \ge b$. The interval (a, b) is the range of the random variable and we may have $a = -\infty$ and/or $b = +\infty$.

EXAMPLE 7.4. Uniform$(0, 1)$ random variables assume values between 0 and 1, so $a = 0$ and $b = 1$. A Normal$(0, 1)$ random variable can assume any value, so $a = -\infty$ and $b = +\infty$. Exponential(λ) random variables are non-negative, so and $a = 0$ and $b = +\infty$.

For the distribution function we have

$$F(x) \;=\; \begin{cases} 0 & \text{if } x \le a \\ \int_a^x f(t)\,dt & \text{if } a < x < b \\ 1 & \text{if } b \le x. \end{cases}$$

Suppose we seek to construct $X \sim F$. If we restrict the domain of F to the interval (a,b), the function $F:(a,b)\to(0,1)$ is onto, continuous, and strictly increasing and hence has an inverse $F^{-1}:(0,1)\to(a,b)$. $F(x)$ is strictly increasing on (a,b) because $F'(x)=f(x)>0$ for $x\in(a,b)$. Therefore if x_1 and x_2 lie on the interval (a,b), we have $x_1<x_2 \iff F(x_1)<F(x_2)$. Now put $X=F^{-1}(U)$, where $U\sim$ Uniform $(0,1)$. Then for $a<x<b$,

$$\begin{aligned} P[X\le x] &= P[F^{-1}(U)\le x] \\ &= P[F(F^{-1}(U))\le F(x)] \\ &= P[U\le F(x)] \;=\; F(x). \end{aligned}$$

The last equality holds because $0<F(x)<1$ and $P[U\le y]=y$ for any $y\in[0,1]$. Hence, the random variable $X=F^{-1}(U)$ has distribution function $F(x)$ and density function $f(x)$, as desired. If we put $X_i=F^{-1}(U_i)$, then the sequence $X_1,X_2,\ldots$ is iid with distribution function F.

EXAMPLE 7.5. Suppose we seek to construct $X\sim$ Exponential (1). Then $(a,b)=(0,\infty)$ and

$$f(x) = \begin{cases} e^{-x} & \text{if } x>0 \\ 0 & \text{otherwise,} \end{cases}$$

and

$$F(x) = \begin{cases} 1-e^{-x} & \text{if } x>0 \\ 0 & \text{if } x\le 0. \end{cases}$$

Here $F^{-1}(y)=-\ln(1-y)$ for $0<y<1$. If $U\sim$ Uniform $(0,1)$, then $X=-\ln(1-U)\sim$ Exponential (1). Since $1-U\sim$ Uniform $(0,1)$, we alternatively may take $X=-\ln(U)$. Refer to Figure 7.6 below.

Figure 7.6. The distribution function for $X\sim$ Exponential (1).

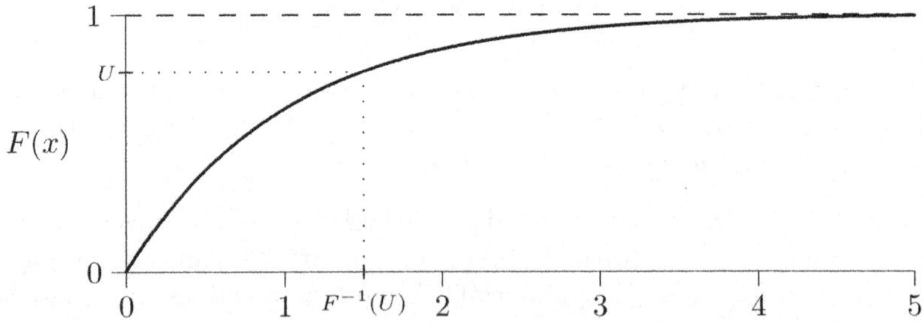

Inverse Transform for General Distributions. We use the following properties that hold for all distribution functions (see exercises). First, they are monotonic in the sense that $F(x_1)\le F(x_2)$ whenever $x_1\le x_2$. Second, $\lim_{x\to\infty}F(x)=1$ and $\lim_{x\to-\infty}F(x)=0$. Third, distribution functions are right continuous in the sense that

$$\lim_{x\to a^+} F(x) = F(a),$$

but they may have jump discontinuities from the left.

Suppose, then, we have an arbitrary distribution function $F(x)$ and we seek to construct a random variable X with $X \sim F$. For $0 < y < 1$ we define

$$S_y \;=\; \{x : y \le F(x)\}. \tag{7.7}$$

Note that S_y is non-empty, as $y < 1$ and $F(x) \to 1$ as $x \to \infty$. Also, S_y is bounded from below as $0 < y$ and $F(x)$ monotonically goes to 0 as $x \to -\infty$. Now put

$$H(y) \;=\; \min(S_y). \tag{7.8}$$

We illustrate this below in Figure 7.9 for the case where $F(x)$ is continuous at $H(y)$ and where $F(x)$ has a jump discontinuity at $H(y)$. In either case we clearly have $y \;\le\; F(H(y))$.

Figure 7.9. S_y and $H(y)$ — two cases illustrated.

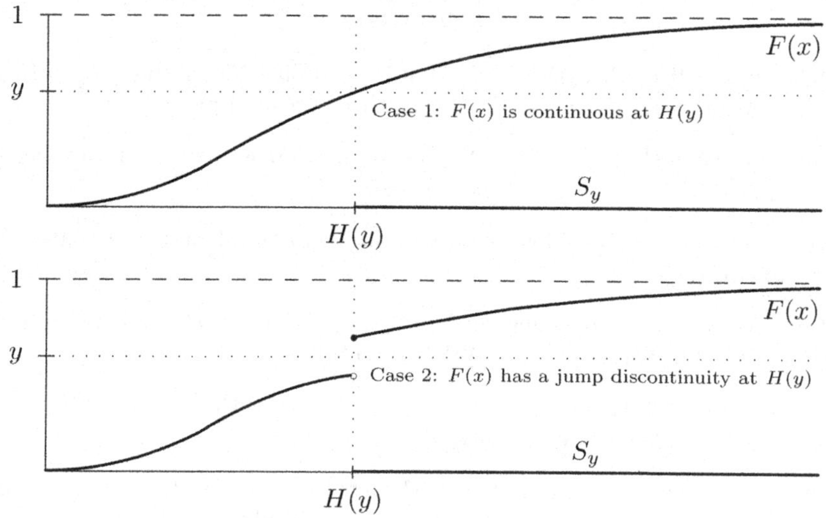

We claim that, for all $0 < y < 1$ and $x \in \mathbb{R}$,

$$H(y) \le x \iff y \le F(x). \tag{7.10}$$

Indeed, if $H(y) \le x$, then $F(H(y)) \le F(x)$ by the monotonicity of F. Since $y \le F(H(y))$, we also have $y \le F(x)$. Hence $H(y) \le x \implies y \le F(x)$. On the other hand, if $y \le F(x)$ then $x \in S_y$ so $H(y) \le x$ as $H(y)$ is a lower bound for S_y. That is, $y \le F(x) \implies H(y) \le x$. These two implications establish (7.10).

Letting $X = H(U)$, where $U \sim$ Uniform$(0, 1)$, we have:

$$\begin{aligned} P[X \le x] &= P[H(U) \le x] \\ &= P[U \le F(x)] \quad \text{by (7.10)} \\ &= F(x), \end{aligned}$$

where the last equality holds because $U \sim$ Uniform$(0,1)$ and $0 \leq F(x) \leq 1$. Hence, $X \sim F$, as desired. We summarize all this with:

THEOREM 7.11. *There exists a sequence $U_1, U_2, \ldots$ of iid random variables on $\mathbf{\Omega}$ with $U_i \sim$ Uniform$(0,1)$. For any distribution function $F(x)$ there corresponds a function $H(y)$ as defined in (7.7) and (7.8). With $X_i = H(U_i)$, the sequence of random variables $X_1, X_2, \ldots$ defined on $\mathbf{\Omega}$ are iid with each $X_i \sim F$.*

Once an iid sequence of Uniform$(0,1)$ random variables $U_1, U_2, \ldots$ is constructed, all manner of random objects follow. We have seen that we may generate iid sequences with any specified distribution. Other, more complicated, stochastic processes may be also constructed. Indeed, in subsequent chapters we will use the sequence $U_1, U_2, \ldots$ to construct Markov chains, arrival processes, random walks, Brownian motion, and other stochastic processes. All this springs from one number randomly selected from the interval $(0,1)$!

EXERCISES

1. Suppose $\omega = 0.71430692537187935086 1\cdots$. Using the methodology of Figure 7.1, what are $U_1(\omega)$, $U_2(\omega)$ and $U_3(\omega)$ to three decimal places?

2. Suppose we wish to generate $X_1, X_2, X_3, \ldots$ iid with $X_i \sim$ Binomial$(2, \frac{1}{2})$. What is $H(y)$?

3. With the setup of problem 2 and the ω from problem 1, what are $X_1(\omega)$, $X_2(\omega)$ and $X_3(\omega)$?

4. Suppose we wish to generate $X_1, X_2, X_3, \ldots$ iid with each X_i uniformly distributed on the interval (a, b). What is $H(y)$?

5. Suppose we wish to generate $X_1, X_2, X_3, \ldots$ iid with each $X_i \sim$ Poisson(3). If $U_4(\omega) = 0.75324\cdots$, what is $X_4(\omega)$?

6. Suppose we wish to generate X with density function $f(x) = \frac{3}{2}x^{1/2}$ on the interval $(0,1)$ and $f(x) = 0$ elsewhere. What is $H(y)$?

7. In the discrete case, show that (7.2) and (7.7, 7.8) are equivalent (they produce the same H) if, in (7.2), the range is presented in increasing order: $x_1 < x_2 < \cdots$.

8. Let $F(x) = P[X \leq x]$ denote the distribution function of the random variable X. (a) Show that $F(x_1) \leq F(x_2)$ whenever $x_1 < x_2$. (b) Show that $\lim_{x\to\infty} F(x) = 1$ and $\lim_{x\to-\infty} F(x) = 0$. (c) Show that $F(x)$ is right-continuous: $\lim_{x\to a+} F(x) = F(a)$. Hint for (b) and (c): Continuity of probability measures.

9. Show that $F(x)$ has a jump discontinuity from the left at $x = a$ if and only if $P[X = a] > 0$. Show that $P[X = a] = F(a) - \lim_{x\to a^-} F(x)$.

CHAPTER II

SOME FUNDAMENTAL LIMIT THEOREMS

§8. Two Types of Convergence

In probability theory, there are many senses in which a sequence of random variables $X_1, X_2, \ldots$ can converge to a number c or to another random variable X. In this section we describe two such modes of convergence.

Almost Sure Convergence. Suppose $X_1, X_2, X_3, \ldots$ is a sequence of random variables defined on a common probability space Ω. For each outcome $\omega \in \Omega$, $X_1(\omega), X_2(\omega), X_3(\omega), \ldots$ is a sequence of *numbers.* This sequence of numbers may or may not have a limit. Suppose that for some number c we have

$$P\{\omega : X_n(\omega) \to c \text{ as } n \to \infty\} \;=\; 1. \tag{8.1}$$

Then, with probability one, the sequence of numbers $X_1(\omega), X_2(\omega), \ldots$ converges to the number c and we say that the sequence X_n converges *almost surely*, or *pointwise*, or *strongly* to c and we write $X_n \to c$ as $n \to \infty$. We note that there may be *exceptional outcomes* ω for which $X_n(\omega)$ does not converge to c, but collectively they have probability 0. That is, we have $PE = 0$ where $E = \{\omega : X_n(\omega) \not\to c\}$. By deleting these exceptional outcomes from the outcome space — i.e., by working with $\tilde{\Omega} = \Omega \setminus E$ — we may assume that convergence occurs for all ω. The following is convenient for establishing almost sure convergence:

THEOREM 8.2. *If, for all* $\epsilon > 0$,

$$P\{\omega : |X_n(\omega) - c| < \epsilon \textit{ for all sufficiently large } n\} \;=\; 1,$$

then $X_n \to c$ *as* $n \to \infty$.

This may seem obvious, but there really is something to prove here. If we let $A_\epsilon = \{\omega : |X_n(\omega) - c| < \epsilon$ for all sufficiently large $n\}$ then $PA_\epsilon = 1$ and

$$X_n(\omega) \to c \iff \omega \in A_\epsilon \text{ for all } \epsilon > 0.$$

We must show that the theorem's hypotheses imply that $P[\bigcap_{\epsilon>0} A_\epsilon] = 1$. The difficulty is that this is an *uncountable* intersection, so the conclusion is not automatic. See the exercises for more on this.

PROOF OF THEOREM. Instead, for $m = 1, 2, 3, \ldots$, let

$$A_m = \left\{\omega : |X_n(\omega) - c| < \frac{1}{m} \text{ for all sufficiently large } n\right\},$$

so, by hypotheses, each $P[A_m] = 1$. With $A = \bigcap_{m=1}^{\infty} A_m$, we also have $P[A] = 1$ as this is a *countable* intersection (see exercises). But, for any $\omega \in A$, $X_n(\omega) \to c$ as $n \to \infty$, establishing the theorem. We verify this latter statement carefully. Suppose $\omega \in A$. We wish to see that $X_n(\omega) \to c$ as $n \to \infty$. Fix any $\epsilon > 0$ and choose m so that $\frac{1}{m} < \epsilon$. Now $A \subset A_m$ so $\omega \in A_m$ and $|X_n(\omega) - c| < \frac{1}{m}$ for all sufficiently large n. But then also $|X_n(\omega) - c| < \epsilon$ for all sufficiently large n. Since $\epsilon > 0$ was arbitrarily small, $X_n(\omega) \to c$ as $n \to \infty$. □

Convergence in Probability. We say that the sequence X_n converges to the number c *in probability* and write $X_n \xrightarrow{\text{P}} c$ as $n \to \infty$, if for any $\epsilon > 0$,

$$P\{\omega : |X_n(\omega) - c| \geq \epsilon\} \to 0 \quad \text{as} \quad n \to \infty. \tag{8.3}$$

Equivalently, we require that, for any $\epsilon > 0$,

$$P\{\omega : |X_n(\omega) - c| < \epsilon\} \to 1 \quad \text{as} \quad n \to \infty. \tag{8.4}$$

At first it might seem that almost sure convergence and convergence in probability are the same. They are not. Rewriting them slightly, almost sure converge means that

$$P\big[\lim_{n\to\infty} X_n = c\big] = 1,$$

while convergence in probability means that, for any $\epsilon > 0$,

$$\lim_{n\to\infty} P\big[|X_n - c| < \epsilon\big] = 1.$$

For convergence in probability, the limit is outside the probability; for almost sure convergence, the limit is inside the probability. Roughly speaking, almost sure convergence means that, for large n, X_n *is* close to c, while convergence in probability means that, for large n, X_n *is probably* close to c.

EXAMPLE 8.5. We illustrate that it is, in fact, possible to have a sequence of random variables $X_1, X_2, \ldots$ with limit-in-probability 0 but, surprisingly, where $P[X_n \to 0] = 0$. The construction is odd, but it makes the point. Partition the natural numbers into blocks $B_1 = \{1\}$, $B_2 = \{2, 3\}$, $B_3 = \{4, 5, 6\}$, $B_4 = \{7, 8, 9, 10\}$, etc., always putting the next k consecutive integers in block number k. Now randomly choose a "distinguished" number from each block. For example, the distinguished numbers might begin with 1, 3, 5, 9. If $k(n)$ denotes the block number that the number n lies in (e.g., $k(9) = 4$), then the probability that n is distinguished is $1/k(n)$ since that block has $k(n)$ numbers in it. Let $X_n = 1$ if n is a distinguished number and $X_n = 0$ otherwise. For distinguished numbers $1, 3, 5, 9 \ldots$, we would have the corresponding sequence of X_n's $1, 0, 1, 0, 1, 0, 0, 0, 1, 0, \ldots$. Presently we show that 0 is the limit in probability, i.e., that for all $\epsilon > 0$, $P[|X_n| \geq \epsilon] \to 0$. Now $|X_n| \geq \epsilon \implies X_n \neq 0 \implies X_n = 1$, so $P[|X_n| \geq \epsilon] \leq P[X_n = 1] = 1/k(n)$. Since the blocks are all of finite length, $k(n) \to \infty$ as $n \to \infty$ and $P[|X_n| \geq \epsilon] \to 0$. On the other hand, the sequence (X_n) has infinitely many 0's and infinitely many 1's (even though the 1's become sparse) and therefore cannot have a limit in the usual (almost sure) sense. So $P[X_n \to 0] = 0$.

Almost Sure Convergence Implies Convergence in Probability. It is the case that almost sure convergence implies convergence in probability, and hence is a stronger convergence. This is a little harder to see and uses the continuity of probability measures. We will show the contrapositive: if you don't have convergence in probability you can't have almost sure convergence (so if you have almost sure convergence you must have convergence in probability). Suppose, then, that it is NOT true that X_n converges in probability to c, i.e., it is NOT true that for all $\epsilon > 0$, $P[|X_n - c| \geq \epsilon] \to 0$ as $n \to \infty$. Then for some $\epsilon > 0$, there is a number $p > 0$ and an increasing sequence $n_1 < n_2 < n_3 < \cdots$ with each $P[|X_{n_i} - c| \geq \epsilon] \geq p$. For positive integers m, define the event

$$C_m = \{\omega : |X_n(\omega) - c| \geq \epsilon \text{ for some } n \geq m\}.$$

Then for each m we may choose an $n_i \geq m$ and, for that n_i,

$$\{|X_{n_i} - c| \geq \epsilon\} \subset C_m,$$

giving that $P[C_m] \geq p$.

Now the C_m are a decreasing sequence of events: $C_1 \supset C_2 \supset C_3 \supset \cdots$. (If $|X_n - c| \geq \epsilon$ for some $n \geq m+1$ then certainly $|X_n - c| \geq \epsilon$ for some $n \geq m$ — the same n will do! — and $C_{m+1} \subset C_m$.) Letting C be the intersection of all of them, $C = \bigcap_{m=1}^{\infty} C_m$, we have, by the continuity of probability measures, that $P[C_m] \to P[C]$ as $m \to \infty$. Since each $P[C_m] \geq p$, it follows that $P[C] \geq p$ as well. For any $\omega \in C$, it is impossible for $X_n(\omega) \to c$. To see this, suppose $\omega \in C$ and choose any positive integer m. Regardless of how large m is, $\omega \in C_m$. So for some $n \geq m$ we have $|X_n(\omega) - c| \geq \epsilon$. Therefore $X_n(\omega) \not\to c$ and $C \subset \{\omega : X_n(\omega) \not\to c\}$ giving

$$P\{\omega : X_n(\omega) \text{ does not converge to } c\} \geq P[C] \geq p > 0.$$

The sequence X_n therefore does not converge almost surely to c.

When the Limit is Random. These two modes of convergence generalize easily to the case where the limiting value is random. We say that the sequence $X_1, X_2, \ldots$ converges almost surely to the random variable X and we write $X_n \to X$ if

$$P\{\omega : X_n(\omega) \to X(\omega) \text{ as } n \to \infty\} = 1.$$

This is equivalent to the statement that the sequence $X_n - X \to 0$ as $n \to \infty$. Similarly, the sequence $X_1, X_2, \ldots$ converges in probability to the random variable X and we write $X_n \xrightarrow{\text{P}} X$ if, for all $\epsilon > 0$,

$$P\{\omega : |X_n(\omega) - X(\omega)| \geq \epsilon\} \to 0 \quad \text{as} \quad n \to \infty.$$

This is equivalent to the statement that the sequence $X_n - X \xrightarrow{\text{P}} 0$ as $n \to \infty$.

REMARK. The statement that $X_n \xrightarrow{\text{P}} X$ is equivalent to the statement that for all $\epsilon > 0$ there is a number n_0 such that $P[|X_n - X| \geq \epsilon] \leq \epsilon$ for $n \geq n_0$. (See exercises.)

We conclude this section with the following facts, which we will use in subsequent chapters. Here and below we assume that all random variables are real-valued — that is the values $+\infty$ and $-\infty$ are precluded.

THEOREM 8.6. *If $X_n \xrightarrow{\text{P}} X$ and $X_n \xrightarrow{\text{P}} Y$, then $P[X = Y] = 1$ (limits-in-probability are unique). If $X_n \xrightarrow{\text{P}} X$ and $Y_n \xrightarrow{\text{P}} Y$, then $X_n + Y_n \xrightarrow{\text{P}} X + Y$ and $X_n Y_n \xrightarrow{\text{P}} XY$.*

PROOF. First we show that limits-in-probability are unique. With $X_n \xrightarrow{\text{P}} X$ and $X_n \xrightarrow{\text{P}} Y$, let $A_m = \{\omega : |X(\omega) - Y(\omega)| \geq \frac{1}{m}\}$. Now $|X - Y| \leq |X - X_n| + |X_n - Y|$, so

$$A_m \subset \left\{\omega : |X_n(\omega) - X(\omega)| \geq \frac{1}{2m}\right\} \cup \left\{\omega : |X_n(\omega) - Y(\omega)| \geq \frac{1}{2m}\right\}.$$

By hypothesis, the probability of both events in the union above go to 0 as $n \to \infty$ and hence $P[A_m] = 0$. But $A_m \uparrow \{\omega : X(\omega) \neq Y(\omega)\}$ (see exercises) and hence, by continuity of probability measures, $P[X \neq Y] = 0$.

Now suppose $X_n \xrightarrow{\text{P}} X$ and $Y_n \xrightarrow{\text{P}} Y$. We will show that $X_n + Y_n \xrightarrow{\text{P}} X + Y$. Now

$$|(X_n + Y_n) - (X + Y)| = |(X_n - X) + (Y_n - Y)| \leq |X_n - X| + |Y_n - Y|,$$

so for any $\epsilon > 0$,

$$\{|(X_n + Y_n) - (X + Y)| \geq \epsilon\} \subset \left\{|X_n - X| \geq \frac{\epsilon}{2}\right\} \cup \left\{|Y_n - Y| \geq \frac{\epsilon}{2}\right\}.$$

By hypothesis, the probability of both events in the union above go to 0 as $n \to \infty$ and hence $P[|(X_n + Y_n) - (X + Y)| \geq \epsilon] \to 0$ as $n \to \infty$. This gives the result.

Finally, if $Y_n \xrightarrow{\text{P}} Y$ and X is any real-valued random variable, we show that $XY_n \xrightarrow{\text{P}} XY$ as $n \to \infty$ (this version of the theorem is all we use in subsequent chapters, we leave the more general result as an exercise). Fix any $\epsilon > 0$. Since

X is real-valued, $\{\omega : |X(\omega)| \geq m\} \downarrow \emptyset$ as $m \to \infty$. By the continuity of probability measures, $P[|X| \geq m] \to 0$ as $m \to \infty$ and we may choose m with $P[|X| \geq m] \leq \frac{\epsilon}{2}$. Now $|XY_n - XY| = |X| \cdot |Y_n - Y|$, so

$$\{|XY_n - XY| \geq \epsilon\} \subset \{|X| \geq m\} \cup \left\{|Y_n - Y| \geq \frac{\epsilon}{m}\right\},$$

and

$$P[|XY_n - XY| \geq \epsilon] \leq \frac{\epsilon}{2} + P\left[|Y_n - Y| \geq \frac{\epsilon}{m}\right].$$

Since $Y_n \xrightarrow{P} Y$, we may choose n_0 so that $n \geq n_0 \implies P\left[|Y_n - Y \geq \frac{\epsilon}{m}\right] \leq \frac{\epsilon}{2}$. Then $n \geq n_0 \implies P[|XY_n - XY| \geq \epsilon] \leq \epsilon$. We now appeal to the remark just prior to the theorem. $\square$

Remark. Suppose the sequence of numbers a_n converges to a as $n \to \infty$. If $X_n \xrightarrow{P} X$, then $a_n X_n \xrightarrow{P} aX$ as $n \to \infty$. To see this, take $Y_n = a_n$ and $Y = a$ in Theorem 8.6.

EXERCISES

1. Prove the assertion make in the remark just prior to Theorem 8.6.

2. Show that if $X_n \xrightarrow{P} X$ and $Y_n \xrightarrow{P} Y$, then $X_n Y_n \xrightarrow{P} XY$ as $n \to \infty$.

3. Show that $X_n \to c$ almost surely if and only if $X_n - c \to 0$ almost surely. Do the same for convergence in probability.

4. Show that $\{\omega : |X(\omega) - Y(\omega)| \geq \frac{1}{m}\} \uparrow \{\omega : X(\omega) \neq Y(\omega)\}$ as $m \to \infty$.

5. Prove that on a discrete probability space Ω, if $X_n \xrightarrow{P} c$ as $n \to \infty$ then also $X_n \to c$ almost surely, i.e. $P[X_n \to c] = 1$. Since almost sure convergence always implies convergence in probability, on a discrete space, convergence in probability and almost sure convergence are the same thing.

6. Suppose $U_1, U_2, \ldots$ are iid Uniform$(0,1)$ random variables. Let $M_n = \max(U_1, \ldots, U_n)$. Show that $M_n \xrightarrow{P} 1$.

7. Show that if $A_1, A_2, \ldots$ have $PA_n = 1$, then $P[\bigcap_{n=1}^{\infty} A_n] = 1$. Hint: De Morgan's law and subadditivity of probability measures.

8. For $x \in (0,1)$, let $A_x = (0,x) \cup (x,1)$. As an event in $\boldsymbol{\Omega}$, show that $PA_x = 1$. Show that $P[\bigcap_{x \in (0,1)} A_x] = 0$. Why does this not contradict problem 7?

§9. The Weak Law of Large Numbers

Suppose $X_1, X_2, X_3, \ldots$ is a sequence of random variables all with the same mean μ. Let

$$\overline{X}_n = \frac{X_1 + X_2 + \cdots + X_n}{n}$$

denote the arithmetic average of the first n of them, a quantity known as the *sample mean*. (The cleaner notation $\overline{X}$ is standard for this, but it suppresses the fact that the sample mean depends on the sample size.) Various versions of the Law of Large Numbers assert conditions under which, as n gets large, the sample means $\overline{X}_n$ converge to the true mean μ. There are several variations of the Law of Large Numbers involving different hypotheses about the underlying X_i's and different conclusions about the nature of the convergence of $\overline{X}_n$ to μ.

We begin with some basic tools for estimating the probability mass in the "tail" of a probability distribution. The estimate from which all else springs is:

THEOREM 9.1 (MARKOV'S INEQUALITY). *If $X \geq 0$ and $EX < \infty$ then, for any $a > 0$, $P[X \geq a] \leq \frac{EX}{a}$.*

PROOF. Let

$$Y = \begin{cases} 0 & \text{if } X < a \\ a & \text{if } X \geq a, \end{cases}$$

so either way $Y \leq X$ and therefore $EY \leq EX$. But Y is either 0 or a, so

$$EX \geq EY = 0 \cdot P[Y = 0] + a \cdot P[Y = a] = aP[X \geq a].$$

Rearranging gives $P[X \geq a] \leq \frac{EX}{a}$. □

A "two-sided" version of this is:

COROLLARY 9.2. *If $E|X| < \infty$, then, for $a > 0$, $P[|X| \geq a] \leq \frac{E|X|}{a}$.*

This corollary follows immediately as $|X| \geq 0$. The "two-sided" description refers to the fact that $\{|X| \geq a\} = \{X \leq -a\} \cup \{X \geq +a\}$. If more is known about X, then more can be said about tail size. In particular, with finite higher moments we have:

COROLLARY 9.3. *If $X \geq 0$ and $EX^n < \infty$ for some $n \geq 1$, then, for any $a > 0$, $P[X \geq a] \leq \frac{EX^n}{a^n}$. If X is any random variable with $E|X|^n < \infty$, then $P[|X| \geq a] \leq \frac{E|X|^n}{a^n}$.*

PROOF. If $X \geq 0$ and $a > 0$ then $X \geq a \iff X^n \geq a^n$, so applying Markov's inequality to X^n (with a^n instead of a) gives

$$P[X \geq a] = P[X^n \geq a^n] \leq \frac{EX^n}{a^n}.$$

The second statement follows, as in Corollary 9.2, because $|X| \geq 0$. □

COROLLARY 9.4 (CHEBYSHEV'S INEQUALITY). *Suppose X has finite second moment, so $\operatorname{Var} X$ and $\mu = EX$ are finite. Then*

$$P[|X - \mu| \geq a] \leq \frac{\operatorname{Var} X}{a^2}.$$

PROOF. Put $Y = (X - \mu)^2$ so $Y \geq 0$ and $EY = \operatorname{Var} X < \infty$ so, by Markov's inequality, for any $a > 0$

$$P[Y \geq a^2] \leq \frac{EY}{a^2} = \frac{\operatorname{Var} X}{a^2}.$$

But $Y \geq a^2 \iff |X - \mu| \geq a$. □

We are now in a position to prove our first Law of Large Numbers.

THEOREM 9.5 (WEAK LAW OF LARGE NUMBERS). *Suppose $X_1, X_2, X_3, \ldots$ are uncorrelated with mean μ and variance bounded by some number b, i.e., $\operatorname{Var} X_i \leq b$ for all i. Then for any $\epsilon > 0$,*

$$P\left[|\overline{X}_n - \mu| \geq \epsilon\right] \to 0 \quad \text{as } n \to \infty.$$

In other words, $\overline{X}_n \xrightarrow{\text{P}} \mu$ as $n \to \infty$.

REMARK. Here we do *not* require that the X_i are iid, merely uncorrelated. However, in applications the X_i are often iid with a (common) finite variance. Such a sequence satisfies the theorem's hypotheses.

PROOF OF THEOREM 9.5. This is an easy application of Chebyshev's inequality. Since the X_i are uncorrelated,

$$\operatorname{Var}[X_1 + \cdots + X_n] = \operatorname{Var} X_1 + \cdots + \operatorname{Var} X_n \leq nb,$$

and

$$\operatorname{Var} \overline{X}_n = \operatorname{Var}\left[\frac{X_1 + \cdots + X_n}{n}\right] = \frac{1}{n^2}\operatorname{Var}[X_1 + \cdots + X_n] \leq \frac{b}{n}.$$

Additionally,

$$E\overline{X}_n = E\left[\frac{X_1 + \cdots + X_n}{n}\right] = \frac{1}{n}(EX_1 + \cdots + EX_n) = \frac{1}{n}n\mu = \mu.$$

Applying Chebyshev's inequality, we obtain

$$P\left[|\overline{X}_n - \mu| \geq \epsilon\right] \leq \frac{b/n}{\epsilon^2} = \frac{b}{\epsilon^2} \cdot \frac{1}{n} \to 0 \quad \text{as } n \to \infty. \qquad \square$$

EXAMPLE 9.6. Suppose we repeatedly toss a fair coin with independent tosses. Let H_n be the fraction of the first n tosses that come up heads. Then $H_n \xrightarrow{P} \frac{1}{2}$. If we let

$$X_i = \begin{cases} 1 & \text{if the } i^{\text{th}} \text{ toss is heads} \\ 0 & \text{if the } i^{\text{th}} \text{ toss is tails,} \end{cases}$$

then $H_n = \frac{X_1 + \cdots + X_n}{n} = \overline{X}_n$. The X_i are uncorrelated since they are independent, and $EX_i = \frac{1}{2}$ and $\operatorname{Var} X_i = \frac{1}{4}$, so with $b = \frac{1}{4}$ we get

$$P\left[\left|H_n - \frac{1}{2}\right| \geq \epsilon\right] \leq \frac{1}{4\epsilon^2} \cdot \frac{1}{n} \to 0 \quad \text{as } n \to \infty.$$

EXERCISES

• In problems 1 – 5, use the Markov or Chebyshev inequality to bound the indicated probabilities.

1. Suppose $X \geq 0$ with $EX = 10$. What can be said about $P[X \geq 5]$? What about $P[X \geq 20]$?

2. Boxes of a certain kind of cookie have an average weight of 12 ounces. With what probability can we assert that the weight of a randomly chosen box will be less than 15 ounces?

3. Suppose $EX = 70$ and $\operatorname{Std} X = 5$. What can be said about $P[60 \leq X \leq 80]$?

4. With reference to problem 2, suppose we also know that the standard deviation of box weight is 1.5 ounces. With what probability can we assert that the weight of a randomly chosen box will be between 10 and 14 ounces?

5. A fair coin is tossed 50 times obtaining H heads. With what probability can we assert that H will fall between 20 and 30?

6. Suppose, in problem 3, we know that X is normally distributed. What can be said about $P[60 \leq X \leq 80]$? Why is this estimate better (i.e., larger)?

7. Suppose $X_1, X_2, \ldots$ is any sequence of random variables with bounded absolute first moment: for some number b, $E|X_n| \leq b$ for all n. Show that the sequence $\frac{X_1}{1}, \frac{X_2}{2}, \frac{X_3}{3}, \ldots$ converges to 0 in probability.

§10. The Strong Law of Large Numbers

The Borel-Cantelli Lemma. In probability theory, we are sometimes interested in establishing that something cannot happen infinitely many times. For example, suppose that we wish to apply Theorem 8.2 to establish that $X_n \to c$ almost surely. We note that

$$\begin{aligned}\{|X_n - c| < \epsilon \text{ for all sufficiently large } n\}^c \\ &= \{|X_n - c| \geq \epsilon \text{ for arbitrarily large values of } n\} \\ &= \{|X_n - c| \geq \epsilon \text{ for infinitely many } n\}.\end{aligned}$$

To see the last equality above, note that a subset of the natural numbers contains arbitrarily large numbers if and only if that subset is infinite. To establish almost sure convergence, it therefore suffices to show that

$$P\{|X_n - c| \geq \epsilon \text{ for infinitely many } n\} = 0$$

holds for all positive ϵ.

In a more concrete setting, suppose we toss a fair coin repeatedly. We claim that we will, with probability one, eventually toss a head. To see this, let

$$\begin{aligned}A_n &= \{\text{first } n \text{ tosses are tails}\}, \quad \text{and} \\ A &= \{\text{never toss a head}\},\end{aligned}$$

so $A \subset A_n$. Then $0 \leq PA \leq PA_n = \frac{1}{2^n}$. Now $\frac{1}{2^n} \to 0$ as $n \to \infty$, hence $PA = 0$. In fact, if we produce an infinite sequence of coin tosses, we will obtain a head infinitely often. Each time we obtain a head, as we continue tossing the coin we will, with probability one, eventually obtain another head. This reasoning is predicated on a fair coin — but what if the coin is not fair?

Example 10.1. Suppose we toss a coin independently infinitely many times. But each time we toss it, it becomes more biased toward tails. Specifically, suppose the nth toss is a head with probability $\frac{1}{n^2}$. So the first toss is heads with probability 1, the second toss is heads with probability $\frac{1}{4}$, the third with probability $\frac{1}{9}$, etc. Under these circumstances, the incidence of heads becomes sparser with each additional toss of the coin. The question is: will we toss infinitely many heads? This turns out to be a subtle question.

Suppose $A_1, A_2, \ldots$ is any sequence of events (they don't have to be independent) and put

$$\begin{aligned}A &= \{A_n \text{ happens for infinitely many } n\} \\ &= \{\omega : \omega \in A_n \text{ for infinitely many } n\}.\end{aligned}$$

In our current example, we would put $A_n = \{\text{the } n^{\text{th}} \text{ toss is heads}\}$, in which case A is the event that we toss infinitely many heads. Now, for any $m \geq 1$, let

$$B_m = \bigcup_{n=m+1}^{\infty} A_n.$$

In our example, B_m is the event that we toss at least one head sometime after the mth toss. Clearly, if we toss infinitely many heads, we must toss at least one head sometime after the mth toss. In other words,

$$A \subset B_m, \quad \text{for each } m \geq 1. \tag{10.2}$$

It follows from this that

$$\omega \in A \implies \omega \in \bigcap_{m=1}^{\infty} B_m. \tag{10.3}$$

On the other hand, suppose we toss only finitely many heads, i.e., $\omega \notin A$. If the last head is obtained on the $m^{*\text{th}}$ toss, we have that $\omega \notin B_{m^*}$ and therefore

$$\omega \notin A \implies \omega \notin \bigcap_{m=1}^{\infty} B_m. \tag{10.4}$$

The statements (10.3) and (10.4) together mean that

$$\begin{aligned} A &= \{A_n \text{ happens for infinitely many } n\} \\ &= \bigcap_{m=1}^{\infty} B_m \\ &= \bigcap_{m=1}^{\infty} \Big[\bigcup_{n=m+1}^{\infty} A_n \Big]. \end{aligned}$$

Here we have expressed the event A in terms of the more familiar set operations of union and intersection, although the expression is messy. Now we state:

THEOREM 10.5 (THE BOREL-CANTELLI LEMMA). *With notation as above, if* $\sum_{n=1}^{\infty} P[A_n] < \infty$, *then* $P[A] = 0$.

PROOF. The sum of the tail of a convergent series goes to zero, so under the theorem's hypotheses:

$$\sum_{n=m+1}^{\infty} P[A_n] \to 0 \quad \text{as } m \to \infty.$$

Using (10.2) and the sub-additivity of probability measures,

$$0 \leq P[A] \leq P[B_m] = P\left[\bigcup_{n=m+1}^{\infty} A_n \right] \leq \sum_{n=m+1}^{\infty} P[A_n].$$

This holds for *any* m, and the right hand side can be made arbitrarily close to 0 by choosing m sufficiently large. We therefore must have $P[A] = 0$. $\square$

EXAMPLE 10.6. In Example 10.1 we have $\sum_{n=1}^{\infty} \frac{1}{n^2} < \infty$, and the probability of tossing infinitely many heads is zero. Eventually, you will toss your last head. Where in the sequence of coin tosses this happens is random, but it will eventually happen. Note that if the coin were biased to produce heads with probability $\frac{1}{n}$ (as opposed to $\frac{1}{n^2}$), then the lemma would not apply because $\sum_{n=1}^{\infty} \frac{1}{n}$ diverges. In fact, in this situation (with independent coin tosses) you *would* observe infinitely many heads with probability one, even though the probability of getting a head approaches 0 with each successive toss.

The Strong Law of Large Numbers. The Strong Law of Large Numbers asserts that under appropriate hypotheses the sample means converge almost surely to the true mean. The Borel-Cantelli Lemma is the main ingredient.

THEOREM 10.7 (STRONG LAW OF LARGE NUMBERS). *Suppose $X_1, X_2, \ldots$ are iid with finite fourth moment. Let $\mu = EX_i$. Then $\overline{X}_n \to \mu$ almost surely as $n \to \infty$.*

REMARK. One may prove this theorem under the much relaxed hypothesis that the *first* moment is finite. However, the proof is substantially more difficult.

PROOF. Assume, first, that $\mu = 0$, and let $\sigma^2 = \operatorname{Var} X_i = EX_i^2$ and $a = EX_i^4$. Then put

$$S_n = X_1 + \cdots + X_n,$$

so, $ES_n = 0$. If we take $S_0 = 0$, then for $n \geq 1$ we have $S_n = S_{n-1} + X_n$, so

$$\begin{aligned} ES_n^4 &= E\left[(S_{n-1} + X_n)^4\right] \\ &= E\left[S_{n-1}^4 + 4S_{n-1}^3 X_n + 6S_{n-1}^2 X_n^2 + 4S_{n-1} X_n^3 + X_n^4\right] \\ &= ES_{n-1}^4 + 0 + 6\sigma^2 ES_{n-1}^2 + 0 + a. \end{aligned} \tag{10.8}$$

The last equality holds because $ES_{n-1} = EX_n = 0$ and X_n and S_{n-1} are independent, so: $E[S_{n-1}^3 X_n] = E[S_{n-1}^3] \cdot E[X_n] = 0$; $E[S_{n-1}^2 X_n^2] = E[S_{n-1}^2] \cdot E[X_n^2] = \sigma^2 ES_{n-1}^2$; and $E[S_{n-1} X_n^3] = E[S_{n-1}] \cdot E[X_n^3] = 0$. Since $E[S_{n-1}] = 0$,

$$ES_{n-1}^2 = \operatorname{Var} S_{n-1} = \operatorname{Var}[X_1 + \cdots + X_{n-1}] = (n-1)\sigma^2 \leq n\sigma^2.$$

Setting this into (10.8) and using that $n \geq 1$ and $a \geq 0$ gives

$$ES_n^4 \leq ES_{n-1}^4 + 6n\sigma^4 + a \leq ES_{n-1}^4 + n(6\sigma^4 + a) = ES_{n-1}^4 + nb,$$

where $b = 6\sigma^4 + a$. Recall that $S_0 = 0$. With $n = 1$, then 2, and then 3, we see that

$$\begin{aligned} ES_1^4 &\leq ES_0^4 + b = b, \\ ES_2^4 &\leq ES_1^4 + 2b \leq b + 2b, \text{ and} \\ ES_3^4 &\leq ES_2^4 + 3b \leq b + 2b + 3b. \end{aligned}$$

Repeated application of this yields that

$$ES_n^4 \leq b + 2b + 3b + \cdots + nb \leq n^2 b.$$

Now $\overline{X}_n = \frac{S_n}{n}$, so for any $\epsilon > 0$, $|\overline{X}_n| \geq \epsilon \iff |S_n| \geq n\epsilon \iff S_n^4 \geq n^4\epsilon^4$, and

$$P[|\overline{X}_n| \geq \epsilon] = P\left[S_n^4 \geq n^4\epsilon^4\right] \leq \frac{ES_n^4}{n^4\epsilon^4} \leq \frac{n^2 b}{n^4\epsilon^4} = \frac{b}{\epsilon^4} \cdot \frac{1}{n^2}.$$

Since $\sum_{n=1}^{\infty} \frac{1}{n^2} < \infty$, the Borel-Cantelli Lemma gives that

$$P[|\overline{X}_n| \geq \epsilon \text{ for infinitely many } n] = 0,$$

or

$$P[|\overline{X}_n| < \epsilon \text{ for all sufficiently large } n] = 1.$$

By Theorem 8.2, we see that $\overline{X}_n \to 0$ almost surely, establishing the theorem for the case $\mu = 0$.

If $\mu \neq 0$, put $Y_i = X_i - \mu$. Then $Y_1, Y_2, \ldots$ satisfy the hypotheses of the theorem and $EY_i = 0$. With $\overline{Y}_n = \frac{Y_1 + \cdots + Y_n}{n}$, we have from the above that $P[\overline{Y}_n \to 0] = 1$. But $\overline{Y}_n = \overline{X}_n - \mu$, so $\overline{Y}_n \to 0 \iff \overline{X}_n \to \mu$ and therefore $P[\overline{X}_n \to \mu] = 1$, i.e., $\overline{X}_n \to \mu$ almost surely. □

EXERCISES

1. Show that if $X_n \xrightarrow{\text{P}} c$, then for some subsequence $X_{n_1}, X_{n_2}, \ldots$ we have $X_{n_i} \to c$ almost surely as $i \to \infty$. (Here $n_1 < n_2 < \cdots$.) Hint: Borel-Cantelli.

2. Suppose $X_1 \leq X_2 \leq X_3 \leq \cdots$ and $X_n \xrightarrow{\text{P}} c$. Show that $X_n \to c$ almost surely. Hint: Use problem 1. Apply this to problem 6 from §8.

3. Suppose $X_1, X_2, \ldots$ is a sequence of random variables with

$$P[X_n = x] = \begin{cases} \frac{1}{n^2} & \text{if } x = n^3 \\ 1 - \frac{1}{n^2} & \text{if } x = 0 \\ 0 & \text{otherwise.} \end{cases}$$

What is $\lim_{n\to\infty} EX_n$? Identify the limiting random variable $X = \lim_{n\to\infty} X_n$. What is EX?

4. Show that $\bigcup_{m=1}^{\infty} \left[\bigcap_{n=m+1}^{\infty} A_n\right] = \{A_n \text{ happens for all but finitely many } n\}$.

5. Suppose $X_1, X_2, \ldots$ is any sequence of random variables with bounded second moment: for some number b, $EX_n^2 \leq b$ for all n. Show that the sequence $\frac{X_1}{1}, \frac{X_2}{2}, \frac{X_3}{3}, \ldots$ converges almost surely to 0.

§11. The Central Limit Theorem

Moment Generating Functions. Since $e^x > 0$ for all real x, $e^{\theta X}$ is a positive-valued random variable for any real θ and random variable X. Its expectation, $Ee^{\theta X}$, therefore always exists but is possibly infinite. For values of θ for which this expectation is finite, it defines the *moment generating function* of X:

$$M_X(\theta) \;=\; Ee^{\theta X}, \quad \text{provided this expectation is finite.}$$

The domain of the moment generating function, by definition, is $\{\theta \in \mathbb{R} : Ee^{\theta X} < \infty\}$. Note that we always have $M_X(0) = 1$, so 0 is always in the domain. It is possible that 0 is the *only* value of θ for which $Ee^{\theta X}$ is finite. This is not the case, however, for most of the usual distributions. In any event, the domain of M_X is either just $\{0\}$ or is an interval containing 0.

EXAMPLE 11.1. Suppose $X \sim \text{Exponential}\,(\lambda)$, where $\lambda > 0$. Then

$$M_X(\theta) \;=\; \int_0^\infty e^{\theta x}\frac{1}{\lambda}e^{-x/\lambda}\,dx \;=\; \frac{1}{\lambda}\int_0^\infty e^{(\theta-\frac{1}{\lambda})x}\,dx.$$

If $\theta > 1/\lambda$, the integrand grows exponentially fast in x and the integral is infinite. If $\theta = 1/\lambda$, the integrand is always 1 and the integral is also infinite. If $\theta < 1/\lambda$, the integrand decreases exponentially fast and simple calculus yields

$$M_X(\theta) \;=\; \frac{1}{1-\lambda\theta}.$$

Here the domain of M_X is $(-\infty, \frac{1}{\lambda})$.

EXAMPLE 11.2. If $N \sim \text{Normal}\,(0,1)$. Then

$$\begin{aligned} M_N(\theta) \;&=\; \int_{-\infty}^\infty e^{\theta x}\frac{1}{\sqrt{2\pi}}e^{-x^2/2}\,dx \\ &=\; \int_{-\infty}^\infty \frac{1}{\sqrt{2\pi}}e^{-(x-\theta)^2/2}e^{\theta^2/2}\,dx \\ &=\; e^{\theta^2/2}\int_{-\infty}^\infty \frac{1}{\sqrt{2\pi}}e^{-(x-\theta)^2/2}\,dx \;=\; e^{\theta^2/2}, \end{aligned}$$

where the last equality holds because $\frac{1}{\sqrt{2\pi}}e^{-(x-\theta)^2/2}$ is the density function of a Normal $(\theta,1)$ random variable and hence integrates to 1. This holds for any $\theta \in (-\infty,\infty)$.

It is not hard to see that for any $c > 0$,

$$Ee^{\theta X} < \infty \text{ for all } \theta \in [-c,c] \iff Ee^{c|X|} < \infty. \tag{11.3}$$

For suppose the left side of (11.3) holds. Taking $\theta = \pm c$ shows that both Ee^{-cX} and Ee^{cX} are finite. Now

$$e^{c|X|} = \begin{cases} e^{cX} & \text{if } X \geq 0 \\ e^{-cX} & \text{if } X < 0, \end{cases}$$

and in either case $e^{c|X|} \leq e^{-cX} + e^{cX}$ (since always $e^x > 0$) so $Ee^{c|X|} \leq Ee^{-cX} + Ee^{cX} < \infty$. Suppose, on the other hand, that $Ee^{c|X|}$ is finite for some $c > 0$. If $\theta \in [-c, c]$, then $\theta X \leq |\theta X| \leq c|X|$ so $e^{\theta X} \leq e^{c|X|}$ and $Ee^{\theta X} \leq Ee^{c|X|} < \infty$.

If this obtains, i.e., if $Ee^{c|X|} < \infty$ for some $c > 0$, we say that X has *finite exponential moment*. If X has finite exponential moment, then it has finite moments of all orders:

THEOREM 11.4. *Suppose X has finite exponential moment. Then $E|X|^n < \infty$ for all positive integers n.*

PROOF. By comparison to the Taylor series expansion, we see that $|x|^n \leq n!e^{|x|}$. Hence, for any number $c > 0$,

$$E|X|^n = \frac{1}{c^n} E|cX|^n \leq \frac{1}{c^n} n! Ee^{|cX|}.$$

By hypotheses, the right hand side above is finite for an appropriate $c > 0$. □

In view of this theorem, it should not be surprising that the moment generating function yields good information about the tails of a random variable's probability distribution. Suppose, for example, that $M_X(\theta) < \infty$ for some $\theta > 0$. Then, for $a > 0$, an application of Markov's inequality gives

$$\begin{aligned} P[X \geq a] &= P[\theta X \geq \theta a] = P[e^{\theta X} \geq e^{\theta a}] \\ &\leq \frac{E[e^{\theta X}]}{e^{\theta a}} = M_X(\theta) e^{-\theta a}, \end{aligned} \tag{11.5}$$

and the probability mass in the (positive) tail of X's distribution falls off exponentially fast in a. Sometimes we can do even better by choosing θ to minimize the right hand side of (11.5). For example, suppose $N \sim \text{Normal}\,(0, 1)$. Here $M_N(\theta) = e^{\theta^2/2}$ so, for $a > 0$ and $\theta > 0$,

$$P[N \geq a] \leq e^{\theta^2/2} e^{-\theta a} = e^{-\theta a + \frac{\theta^2}{2}}.$$

Now this holds for *any* $\theta > 0$; to make the bound on $P[N \geq a]$ as tight as possible, we choose θ to minimize $f(\theta) = -\theta a + \frac{\theta^2}{2}$. The vertex of this upward-opening parabola is at $\theta = a$. Since $f(a) = -\frac{a^2}{2}$ we get that $P[N \geq a] \leq e^{-a^2/2}$. The standard normal distribution is symmetric about 0, so we also obtain

$$P[|\text{Normal}\,(0, 1)| \geq a] \leq 2e^{-a^2/2}. \tag{11.6}$$

Finally, we establish the following useful estimate for $M_X(\theta)$, which is good for small values of θ.

THEOREM 11.7. *Suppose $Ee^{c|X|} < \infty$ for some $c > 0$. Let $\mu = EX$, $\sigma^2 = \operatorname{Var} X$ and put $b = Ee^{c|X|}/c^3$. Then, for $|\theta| \le c$ we have*

$$M_X(\theta) \;=\; 1 + \mu\theta + \left(\frac{\sigma^2 + \mu^2}{2}\right)\theta^2 + \epsilon(\theta),$$

where the error term satisfies $|\epsilon(\theta)| \le b|\theta|^3$.

PROOF. Using the Taylor series expansion for e^x, we see that for $\theta \in [-c, c]$

$$M_X(\theta) \;=\; Ee^{\theta X} \;=\; E\left[1 + \theta X + \frac{\theta^2 X^2}{2} + T_\theta\right] \tag{11.8}$$

where

$$|T_\theta| \;=\; \left|\sum_{n\ge3} \frac{\theta^n X^n}{n!}\right| \;\le\; \sum_{n\ge3} \frac{|\theta|^n |X|^n}{n!} \;\le\; \frac{|\theta|^3}{c^3} \sum_{n\ge3} \frac{|cX|^n}{n!} \;\le\; \frac{|\theta|^3}{c^3} e^{c|X|}.$$

The second inequality above holds because $|\theta| \le c$ and $n - 3 \ge 0$ in the range of summation, so

$$|\theta|^n \;=\; |\theta|^3 |\theta|^{n-3} \;\le\; |\theta|^3 c^{n-3} \;=\; \frac{|\theta|^3}{c^3} c^n.$$

Hence

$$|ET_\theta| \;\le\; E|T_\theta| \;\le\; b|\theta|^3,$$

where $b = \frac{Ee^{c|X|}}{c^3} < \infty$ by hypothesis. Setting this into (11.8) gives that

$$\begin{aligned} M_X(\theta) &= 1 + \theta EX + \frac{\theta^2 EX^2}{2} + ET_\theta \\ &= 1 + \mu\theta + \left(\frac{\sigma^2 + \mu^2}{2}\right)\theta^2 + \epsilon(\theta), \end{aligned}$$

where $|\epsilon(\theta)| \le b|\theta|^3$, as required. □

COROLLARY 11.9. *If X has finite exponential moment then $EX = M'_X(0)$.*

PROOF. It is easy to verify that $\epsilon(\theta)$ is differentiable at $\theta = 0$ and $\epsilon'(0) = 0$. Hence $M'_X(0) = 0 + \mu + (\sigma^2 + \mu^2) \cdot 0 + 0 = \mu$. □

The Central Limit Theorem. Here we state a version of the Central Limit Theorem and provide a partial proof.

THEOREM 11.10 (CLT). *Suppose that $X_1, X_2, \cdots$ are iid with finite exponential moment and that $Z \sim$ Normal $(0,1)$. Let $\mu = EX$ and $\sigma^2 = \operatorname{Var} X$. Then, for every number z,*

$$P\left[\frac{X_1 + \cdots + X_n - n\mu}{\sigma\sqrt{n}} \leq z\right] \rightarrow P[Z \leq z] \text{ as } n \rightarrow \infty.$$

REMARK. If we let

$$Z_n = \frac{X_1 + \cdots + X_n - n\mu}{\sigma\sqrt{n}},$$

the CLT says that the distribution function of Z_n converges to the distribution function of a standard normal. With $S_n = X_1 + \cdots + X_n$, we note that $ES_n = n\mu$ and $\operatorname{Var}[S_n] = n\sigma^2$, so $\operatorname{Std}[S_n] = \sigma\sqrt{n}$. The random variable Z_n is just S_n with its mean subtracted off (so the mean of Z_n is 0) and divided by its standard deviation (so the standard deviation and variance of Z_n are 1). This is called *normalizing* the random variable S_n. The resulting random variable Z_n has mean 0 and variance 1. It is not normally distributed, as the term "normalizing" suggests. But in this instance the CLT says that it is approximately normal, at least for large n. The CLT holds with the substantially weaker hypotheses that the X_i have finite second moment.

A PARTIAL PROOF OF THE CLT. Assume that $\mu = 0$ and $\sigma^2 = 1$ (the general case easily reduces to this — see exercises). By hypotheses, for some $c > 0$ we have $Ee^{c|X|} < \infty$; let $b = Ee^{c|X|}/c^3$. Then, by Theorem 11.7, for $|\theta| \leq c$ we have $M_{X_i}(\theta) = 1 + \frac{1}{2}\theta^2 + \epsilon(\theta)$ where $|\epsilon(\theta)| \leq b|\theta|^3$. Put $Z_n = (X_1 + \cdots + X_n)/\sqrt{n}$ and let θ be a fixed number. If n is large enough so that $\left|\frac{\theta}{\sqrt{n}}\right| \leq c$, we have

$$\begin{aligned} M_{Z_n}(\theta) = Ee^{\theta Z_n} &= Ee^{(\theta X_1 + \cdots + \theta X_n)/\sqrt{n}} \\ &= E\left[\prod_{i=1}^{n} e^{(\theta/\sqrt{n})X_i}\right] \\ &= \prod_{i=1}^{n} Ee^{(\theta/\sqrt{n})X_i} \quad \text{(the } X_i \text{ are independent)} \\ &= \left[M_{X_1}(\theta/\sqrt{n})\right]^n \quad \text{(each } X_i \sim X_1) \\ &= \left[1 + \frac{1}{2}\frac{\theta^2}{n} + \epsilon\left(\frac{\theta}{\sqrt{n}}\right)\right]^n = \left[1 + \frac{\frac{1}{2}\theta^2 + \epsilon_n}{n}\right]^n, \end{aligned}$$

where $\epsilon_n = n\epsilon\left(\frac{\theta}{\sqrt{n}}\right)$ satisfies

$$|\epsilon_n| \leq nb\left(\frac{|\theta|}{\sqrt{n}}\right)^3 = b|\theta|^3 \cdot \frac{1}{\sqrt{n}}.$$

Hence $|\epsilon_n| \to 0$ as $n \to \infty$. If $a_1, a_2, \ldots$ is a sequence of numbers with $a_n \to a$ as $n \to \infty$, then also $\left(1 + \frac{a_n}{n}\right)^n \to e^a$ (see Theorem 45.4, Appendix II). In our setting, $\frac{1}{2}\theta^2 + \epsilon_n \to \theta^2/2$ as $n \to \infty$, so

$$M_{Z_n}(\theta) \;\to\; e^{\theta^2/2} \;=\; M_Z(\theta), \text{ as } n \to \infty,$$

where, recall, $Z \sim$ Normal$(0,1)$. That $P[Z_n \le z] \to P[Z \le z]$ then follows from:

LEMMA 11.11. *Suppose X is a continuous random variable with finite exponential moment and $X_1, X_2, \ldots$ are random variables such that $M_{X_n}(\theta) \to M_X(\theta)$. Then, for every number x, $P[X_n \le x] \to P[X \le x]$ as $n \to \infty$.*

The proof of this lemma unfortunately involves analysis beyond the scope of this text, which is why the proof is only "partial". However, exercise 8 below illustrates its applicability in an entirely different setting.

EXERCISES

1. Show that if $X \sim$ Uniform$(0,1)$, then, for $\theta \neq 0$, $M_X(\theta) = \frac{e^\theta - 1}{\theta}$. Use this to compute EX.

2. If a and b are constants and $Y = aX + b$, show that $M_Y(\theta) = e^{b\theta} M_X(a\theta)$.

3. If X and Y are independent, show that $M_{X+Y}(\theta) = M_X(\theta) M_Y(\theta)$.

4. Show that if $X \sim$ Poisson(λ), then $M_X(\theta) = e^{\lambda(e^\theta - 1)}$. Use this to compute EX.

5. Generalize Theorem 11.4 as follows. Suppose that, for some $\epsilon > 0$, the random variable $|X|^\epsilon$ has finite exponential moment. Show that X has finite moments of all orders ($E|X|^n < \infty$ for all positive integers n).

6. Suppose $X \sim$ Exponential(1) and let $Y = X^2$. What is the domain of $M_Y(\theta)$? Show that Y has finite moments of all orders.

7. Use the case $\mu = 0$, $\sigma^2 = 1$ to establish the Central Limit Theorem for general μ and $\sigma^2 > 0$. Hint: normalize the X_i.

8. Verify by direct computation that Lemma 11.11 holds in the case where $X_n \sim$ Uniform$\left\{\frac{1}{n}, \frac{2}{n}, \ldots, \frac{n}{n}\right\}$ (so $P[X_n = \frac{k}{n}] = \frac{1}{n}$ for $1 \le k \le n$) and $X \sim$ Uniform$(0,1)$. That is, verify that the hypotheses of the theorem hold and that the conclusion does as well. Hint: you will need to use that $e^{\theta/n} \to 1$ and $n(e^{\theta/n} - 1) \to \theta$ as $n \to \infty$.

§12. The Monotone Convergence Theorem

In general, it is not the case that, whenever $X_n \to X$ as $n \to \infty$ (almost sure convergence), one necessarily has that $EX_n \to EX$ as $n \to \infty$. To see this, define the random variables X_n on $\mathbf{\Omega}$ by

$$X_n(\omega) = \begin{cases} n & \text{if } 0 < \omega < \frac{1}{n} \\ 0 & \text{if } \frac{1}{n} \leq \omega < 1. \end{cases}$$

Then, for any fixed outcome ω, $X_n(\omega) = 0$ provided $n > \frac{1}{\omega}$. Therefore, $X_n \to X$ where $X(\omega) = 0$ for *all* outcomes ω. The limiting random variable has expectation $EX = 0$. On the other hand, $EX_n = 0 \cdot (1 - \frac{1}{n}) + n \cdot \frac{1}{n} = 1$ so $EX_n \to 1$ (not 0) as $n \to \infty$. Problem 3 of §10 provides another example.

There are three main theorems in probability theory that establish settings in which $EX_n \to EX$ when $X_n \to X$ (almost surely). We will need a restricted version of one of them. Our theorem concerns the case where the random variables are non-negative ($X_n(\omega) \geq 0$) and increasing ($X_{n+1}(\omega) \geq X_n(\omega)$). Any increasing sequence of numbers has a limiting value. If the sequence is bounded, the limiting value is a real number. If the sequence is not bounded, the limiting value is $+\infty$. Hence if X_n is an increasing sequence of random variables, it has a pointwise or almost sure limit X. The value $X(\omega)$ is simply the limiting value of the numbers $X_n(\omega)$, as $n \to \infty$. We use the notation $X_n \uparrow X$ in this situation. Even though each X_n is real valued, as noted above we must allow for the situation that $X(\omega) = +\infty$. Let us agree that if X is a non-negative random variable with $P[X = \infty] > 0$, then $EX = \infty$ (what else?). If $P[X = \infty] = 0$, then EX is computed as usual. Before we state our convergence theorem for expectation, we establish the following lemma, which is useful in many contexts.

LEMMA 12.1. *Suppose X assumes values in $\mathbb{N} \cup \{+\infty\}$. Then $EX = \sum_{k=1}^{\infty} P[X \geq k]$.*

PROOF. *Case 1,* $P[X = \infty] > 0$. Here $EX = \infty$. But $X = \infty \implies X \geq k$, so $0 < P[X = \infty] \leq P[X \geq k]$ and $\sum_{k=1}^{\infty} P[X \geq k] = \infty$.

Case 2, $P[X = \infty] = 0$. Then

$$\begin{aligned} EX &= \sum_{j=0}^{\infty} jP[X = j] = \sum_{j=1}^{\infty} jP[X = j] \\ &= \sum_{j=1}^{\infty} \sum_{k=1}^{j} P[X = j] \\ &= \sum_{k=1}^{\infty} \sum_{j=k}^{\infty} P[X = j] = \sum_{k=1}^{\infty} P[X \geq k]. \quad \square \end{aligned}$$

EXAMPLE 12.2. Suppose $X \sim$ Geometric (p), which we view as the number of trials of an experiment required for a "success" if each trial is independently successful with probability p. Then $P[X \geq k] = P[k-1$ consecutive failures$] = q^{k-1}$ (where $q = 1-p$), giving

$$EX = \sum_{k=1}^{\infty} q^{k-1} = \frac{1}{1-q} = \frac{1}{p}.$$

Lemma 12.1 is just what we need to prove:

THEOREM 12.3 (MONOTONE CONVERGENCE THEOREM). *If $X_n \geq 0$ and $X_n \uparrow X$, then $EX_n \to EX$ as $n \to \infty$.*

PROOF. We prove this in detail when the X_n, and hence X, are integer-valued. This is the only case needed in this text. For the general case, see the exercises below.

Step 1. Here we prove that $\{X_n \geq k\} \uparrow \{X \geq k\}$ as $n \to \infty$. Here k is a fixed non-negative integer. Clearly the events $\{X_n \geq k\}$ are increasing in n as $X_{n+1} \geq X_n$ so $X_n \geq k \implies X_{n+1} \geq k$ and therefore $\{X_n \geq k\} \subset \{X_{n+1} \geq k\}$. Let $A = \bigcup_{n=1}^{\infty} \{X_n \geq k\}$. We must show that $A = \{X \geq k\}$.

Suppose $\omega \in \{X \geq k\}$, so $X(\omega) \geq k$. Since $X_n(\omega) \to X(\omega)$, we have that $X_n(\omega) > k-1$ for $n \geq n_0$, for some number n_0. Then also $X_n(\omega) \geq k$ for $n \geq n_0$, as X_n is integer valued. Hence $\omega \in \{X_{n_0} \geq k\}$ and therefore $\omega \in A$.

Now suppose $\omega \in A$. Then $\omega \in \{X_{n_0} \geq k\}$ for some n_0. Hence $X_{n_0}(\omega) \geq k$. Since each $X_n \leq X$, we see that $X(\omega) \geq k$ and therefore $\omega \in \{X \geq k\}$.

We have shown that $\{X \geq k\} \subset A$ and that $A \subset \{X \geq k\}$, establishing step 1.

Step 2. Here we complete the proof. Since the X_n are an increasing sequence of random variables, EX_n is an increasing sequence of numbers. The sequence EX_n therefore converges to a limit L, which may be infinite. Note that each $X_n \leq X$, so $EX_n \leq EX$ and therefore $L \leq EX$. We must show that $L = EX$.

Pick any number $a < EX$ and choose another number b with $a < b < EX$. By Lemma 12.1, $\sum_{k=1}^{\infty} P[X \geq k] = EX > b$, so we may choose a number k^* such that $\sum_{k=1}^{k^*} P[X \geq k] > b$. By Step 1 and the continuity of probability measures, for each fixed k, $P[X_n \geq k] \to P[X \geq k]$ as $n \to \infty$. Since there are only *finitely many* k with $1 \leq k \leq k^*$, we may choose n_0 such that $n \geq n_0$ implies

$$P[X_n \geq k] \geq P[X \geq k] - \frac{b-a}{k^*} \quad \text{for each } k \text{ with } 1 \leq k \leq k^*.$$

Then for $n \geq n_0$ we have

$$\begin{aligned} EX_n \geq \sum_{k=1}^{k^*} P[X_n \geq k] &\geq \left(\sum_{k=1}^{k^*} P[X \geq k] \right) - (b-a) \\ &> b - (b-a) = a. \end{aligned}$$

Hence we have $a < L \leq EX$. Since a is an arbitrary number less than EX we must have $L = EX$. □

We will actually use the following important corollary:

COROLLARY 12.4. *If $X_1, X_2, \ldots$ are non-negative random variables then*

$$E\left[\sum_{k=1}^{\infty} X_k\right] = \sum_{k=1}^{\infty} EX_k.$$

REMARK. This establishes *countable additivity* for expectation in this setting. Finite additivity always holds for expectation; countable additivity does not.

PROOF OF COROLLARY. Let $S_n = X_1 + \cdots + X_n$, so $0 \leq S_1 \leq S_2 \leq \cdots$ and $S_n \uparrow S = \sum_{k=1}^{\infty} X_k$. Then by the Monotone Convergence Theorem $ES_n \to ES$ as $n \to \infty$. That is

$$\begin{aligned} E\left[\sum_{k=1}^{\infty} X_k\right] &= ES = \lim_{n\to\infty} ES_n \\ &= \lim_{n\to\infty} (EX_1 + \cdots + EX_n) \quad \text{(by finite additivity)} \\ &= \sum_{k=1}^{\infty} EX_k. \quad \square \end{aligned}$$

EXERCISES

• Here we complete the proof of the (general) Monotone Convergence Theorem.

1. Suppose that the X_n are non-negative, increasing, and assume values in $\{0, h, 2h, 3h, \ldots\}$ for some number $h > 0$ ($h = 1$ in the proof of the MCT above). Establish the theorem in this case. There are two ways to do this. Method 1: Repeat the argument presented in the Theorem's proof for general $h > 0$ — there is nothing special about $h = 1$. Method 2: Let $Y_n = \frac{1}{h} X_n$ and show that the Y_n satisfy the conditions of the MCT as proved above. Use the conclusion to establish the case $h \neq 1$.

2. Now suppose only that the X_n are non-negative and increasing and let X be the limiting value. Round each X_n down to the nearest integer multiple of h: put $\widehat{X}_n = mh$ where m is the non-negative integer with $mh \leq X_n < (m+1)h$. Show that the $\widehat{X}_n$'s satisfy the conditions of problem 1. Show that either (i) $E\widehat{X}_n = EX_n = \infty$ or (ii) $|E\widehat{X}_n - EX_n| \leq h$.

3. Let $\widehat{X}$ be the limiting value of the $\widehat{X}_n$'s. Show that either (i) $E\widehat{X} = EX = \infty$ or (ii) $E\widehat{X}$ and EX are both finite and $|E\widehat{X} - EX| \leq h$.

4. Use that $h > 0$ can be taken arbitrarily small to prove the MCT with the hypotheses as stated in the theorem.

§13*. Two More Limit Theorems for Expectation

In this section we prove two additional limit theorems for expectation — the Dominated Convergence Theorem and the Bounded Convergence Theorem. With the exception of the next section (which is an application of the Dominated Convergence Theorem) we will not use these theorems in this text. The next two sections can be safely skipped.

LEMMA 13.1. *If $X_n \geq 0$, $EX_1 < \infty$, and $X_n \downarrow 0$, then $EX_n \to 0$ as $n \to \infty$.*

PROOF. Put $D_n = X_1 - X_n$ for $n = 1, 2, \ldots$. Then each $D_n \geq 0$ and $D_n \uparrow X_1$. From the Monotone Convergence Theorem we get that $ED_n \to EX_1$, that is, $EX_1 - EX_n \to EX_1$, as $n \to \infty$. Since $EX_1 < \infty$, we must have $EX_n \to 0$. □

LEMMA 13.2. *If $X \geq 0$, $EX < \infty$, and A_n is a sequence of events with $P[A_n] \to 0$, then $E[XI_{A_n}] \to 0$.*

PROOF. Let $B_n = \{\omega : X(\omega) > n\}$ so $B_n \downarrow \emptyset$. First we treat the case where $P[B_{n^*}] = 0$ for some n^*. Then $P[X > n^*] = 0$ and $E[XI_{A_n}] \leq E[n^* I_{A_n}] = n^* P[A_n] \to 0$ as $n \to \infty$.

Assume, then, that $P[B_n] > 0$ for all n. With $X_n = XI_{B_n}$ we have $X_n \geq 0$, $EX_1 \leq EX < \infty$, and $X_n \downarrow 0$. By Lemma 13.1, $EX_n \to 0$. Choose any $\epsilon > 0$ and fix k with $EX_k < \epsilon$. Now pick n_0 so that $P[A_n] < P[B_k]$ provided $n \geq n_0$. (We may do this because $P[B_k] > 0$ and $P[A_n] \to 0$ as $n \to \infty$.) Then, for $n \geq n_0$,

$$\begin{aligned}
E[XI_{A_n}] &= E[XI_{A_n \setminus B_k}] + E[XI_{A_n \cap B_k}] \\
&\leq k \cdot P[A_n \setminus B_k] + E[XI_{A_n \cap B_k}] \\
&\leq k \cdot P[B_k \setminus A_n] + E[XI_{A_n \cap B_k}] \\
&\leq E[XI_{B_k \setminus A_n}] + E[XI_{A_n \cap B_k}] = E[XI_{B_k}] = EX_k < \epsilon,
\end{aligned}$$

establishing the Lemma. The first equality holds because A_n is the disjoint union of $A_n \setminus B_k$ and $A_n \cap B_k$. The first inequality holds because $X \leq k$ on B_k^c while the third inequality holds because $k < X$ on B_k. The second inequality holds because, in general $PA \leq PB \implies P[A \setminus B] \leq P[B \setminus A]$ (draw a Venn diagram). □

THEOREM 13.3 (DOMINATED CONVERGENCE THEOREM). *If $X_n \to X$ and $|X_n| \leq Y$ where $EY < \infty$, then $EX_n \to EX$.*

PROOF. Fix any $\epsilon > 0$ and let $A_n = \{\omega : |X_n(\omega) - X(\omega)| \geq \epsilon\}$. Since almost sure convergence implies convergence in probability, $P[A_n] \to 0$ as $n \to \infty$. Now

$$\begin{aligned}
|EX_n - EX| = |E(X_n - X)| &= |E(X_n - X)I_{A_n^c} + E(X_n - X)I_{A_n}| \\
&\leq E\big[|X_n - X|I_{A_n^c}\big] + E\big[|X_n - X|I_{A_n}\big] \\
&\leq \epsilon + 2E\big[YI_{A_n}\big],
\end{aligned}$$

where the second inequality holds because $|X_n - X| < \epsilon$ on A_n^c and $|X_n - X| \le |X_n| + |X| \le 2Y$. (We have by hypotheses that $|X_n| \le Y$ but also $X_n \to X$ so $|X| \le Y$.) By Lemma 13.2, $E[YI_{A_n}] \to 0$ so we may choose n_0 such that $E[YI_{A_n}] < \epsilon$ whenever $n \ge n_0$. Then, for $n \ge n_0$ we have $|EX_n - EX| < 3\epsilon$. Since ϵ can be chosen arbitrarily small, $|EX_n - EX| \to 0$ and $EX_n \to EX$ as $n \to \infty$. □

THEOREM 13.4 (BOUNDED CONVERGENCE THEOREM). *If $X_n \to X$ and each $|X_n| \le B$ for some number B, then $EX_n \to EX$.*

PROOF. Simply take $Y(\omega) = B$ for all $\omega \in \Omega$, and apply the Dominated Convergence Theorem. □

§14*. Recovering the Moments of X from $M_X(\theta)$

Here we apply the Dominated Convergence Theorem to rigorously prove that a random variable's moments may be recovered from its moment generating function. Suppose a random variable X has a finite exponential moment, so for some $c > 0$, $Ee^{c|X|} < \infty$ and $M_X(\theta)$ is defined on the interval $[-c, c]$. By formally differentiating through the expectation operator we get

$$M'_X(\theta) = \frac{d}{d\theta}Ee^{\theta X} = E\frac{d}{d\theta}e^{\theta X} = E[Xe^{\theta X}], \tag{14.1}$$

and repeated differentiation gives $M_X^{(n)}(\theta) = E[X^n e^{\theta X}]$. Evaluating this at $\theta = 0$ gives $E[X^n] = M_X^{(n)}(0)$. This is, in fact, correct. But exchanging the expectation and differentiation operators (i.e., the second equality in (14.1)) must be justified. This is because both differentiation and the computation of expectation involve limiting processes. First an example.

EXAMPLE 14.2. A standard normal N has moment generating function $M(\theta) = e^{\theta^2/2}$, so $M'(\theta) = \theta e^{\theta^2/2}$ and $M''(\theta) = (1 + \theta^2)e^{\theta^2/2}$. Hence $EN = M'(0) = 0$ and $EN^2 = M''(0) = 1$, as expected.

JUSTIFICATION OF (14.1). The key here is to show that

$$\frac{d}{d\theta}E[X^n e^{\theta X}] = E[X^{n+1}e^{\theta X}] \tag{14.3}$$

holds for $|\theta| < c$. Repeated application of this, beginning with $n = 0$, gives $M_X^{(n)}(\theta) = E[X^n e^{\theta X}]$.

For the remainder of the proof, fix θ with $|\theta| < c$ and take $\alpha = \frac{1}{2}(c - |\theta|) > 0$, so $|\theta| + 2\alpha = c$. First we establish that $X^n e^{\theta X}$ has finite first moment. By comparison to the Taylor series for $e^{\alpha|X|}$ we see that

$$\frac{\alpha^n |X|^n}{n!} \le e^{\alpha|X|},$$

so $|X|^n \le \alpha^{-n} n! e^{\alpha|X|}$, and

$$|X^n e^{\theta X}| \le |X|^n e^{|\theta|\cdot|X|} \le \alpha^{-n} n! e^{\alpha|X|} e^{|\theta|\cdot|X|} = \alpha^{-n} n! e^{(\alpha+|\theta|)|X|}. \quad (14.4)$$

Since $\alpha + |\theta| < c$,

$$E|X^n e^{\theta X}| \le \alpha^{-n} n! E[e^{c|X|}] < \infty,$$

establishing that $E\big[X^n e^{\theta X}\big]$ is defined.

Now

$$\begin{aligned} \frac{d}{d\theta} E\big[X^n e^{\theta X}\big] &= \lim_{h\to 0} \frac{E\big[X^n e^{(\theta+h)X}\big] - E\big[X^n e^{\theta X}\big]}{h} \\ &= \lim_{h\to 0} E\left[X^n e^{\theta X} \frac{e^{hX} - 1}{h}\right], \end{aligned} \quad (14.5)$$

provided this limit exists. For each outcome ω, $X(\omega)$ is just a number and $f(\theta) = e^{\theta X(\omega)}$ is a differentiable function of θ with derivative $f'(\theta) = X(\omega) e^{\theta X(\omega)}$, so

$$X(\omega) = f'(0) = \lim_{h\to 0} \frac{f(h) - f(0)}{h} = \lim_{h\to 0} \frac{e^{hX(\omega)} - 1}{h}.$$

It follows that, in the almost sure sense,

$$X^n e^{\theta X} \frac{e^{hX} - 1}{h} \to X^{n+1} e^{\theta X} \quad \text{as } h \to 0.$$

Presently we produce a random variable Y with $EY < \infty$ and

$$\left| X^n e^{\theta X} \frac{e^{hX} - 1}{h} \right| \le Y \quad \text{provided } |h| \le \alpha.$$

The Dominated Convergence Theorem then gives (see exercises) that

$$E\left[X^n e^{\theta X} \frac{e^{hX} - 1}{h}\right] \to E\big[X^{n+1} e^{\theta X}\big] \quad \text{as } h \to 0,$$

establishing that the limit in (14.5) exists and that (14.3) holds.

We conclude by finding an appropriate Y. Recall that $|\theta| < c$, and $|\theta| + 2\alpha = c$. If $|h| \le \alpha$,

$$\begin{aligned} \left| \frac{e^{hX} - 1}{h} \right| &= \left| \frac{hX + \frac{1}{2!} h^2 X^2 + \frac{1}{3!} h^3 X^3 + \cdots}{h} \right| \\ &\le |X| + \frac{1}{2!} |h| |X^2| + \frac{1}{3!} |h|^2 |X|^3 + \cdots \\ &\le \alpha^{-1} \big[\alpha|X| + \frac{1}{2!} \alpha^2 |X^2| + \frac{1}{3!} \alpha^3 |X|^3 + \cdots\big] \\ &\le \alpha^{-1} e^{\alpha|X|}. \end{aligned}$$

Combining this with (14.4), if $|h| \leq \alpha$ we have

$$\left| X^n \, e^{\theta X} \, \frac{e^{hX} - 1}{h} \right| \leq \alpha^{-n} n! e^{(\alpha + |\theta|)|X|} \alpha^{-1} e^{\alpha |X|} = \alpha^{-(n+1)} n! e^{c|X|}.$$

We therefore take $Y = \alpha^{-(n+1)} n! e^{c|X|}$; $EY < \infty$ because $E\big[e^{c|X|}\big] < \infty$. □

EXERCISES

1. Suppose X_h is a family of random variables indexed by a real-valued parameter h such that, for each $\omega \in \Omega$, $X_h(\omega) \to X(\omega)$ as $h \to 0$. If Y is a random variable with $EY < \infty$ and $|X_h| \leq Y$ for all h, show that $E[X_h] \to EX$ as $h \to 0$. Hint: consider sequences $h_n \to 0$.

CHAPTER III

MARKOV CHAINS WITH FINITELY MANY STATES

Markov chains describe the random evolution of systems over time. Here, time is discretized into *periods* and in any given period the system will be in one of finitely many *states*. The defining feature of a Markov chain is that the probability distribution of the state of the system in period $n+1$ depends on the system's state in period n, but not on the system's previous history.

§15. The Two-State System

Let us consider a simple system with only two states.

EXAMPLE 15.1. In the town of Binaryville it either rains or shines every day. These are the two states. If it rains today, tomorrow it will rain with probability 0.6 and shine with probability 0.4. If it shines today, tomorrow it will rain with probability 0.2 and shine with probability 0.8. The weather yesterday is irrelevant. Our Markov chains are *homogeneous* in time in the sense that these probabilities do not change as time elapses. These simple rules describe how the system evolves over time — we will refer to them as the Markov chain's *dynamics*. It is convenient to record these *transition probabilities* (here, the numbers $0.6, 0.4, 0.2, 0.8$) in a *probability transition matrix* (PTM for short). For Binaryville we have

$$\mathbf{P} = \begin{pmatrix} p_{rr} & p_{rs} \\ p_{sr} & p_{ss} \end{pmatrix} = \begin{pmatrix} 0.6 & 0.4 \\ 0.2 & 0.8 \end{pmatrix}.$$

Each row of the matrix corresponds to a state of the system today; each column corresponds to a possible state of the system tomorrow. For example, the entry p_{rs} ("r" means rain, "s" means shine) gives the probability of being in state s tomorrow if we're in state r today.

Note that the entries of each row of the PTM sum to 1. This is always true, since from any given state one must transition to *some* state one period hence. This system may be generated on $\mathbf{\Omega}$ as follows. Using the iid sequence $U_0, U_1, U_2, \ldots$:

Step 1: Set $X_0 = r$ (here we start with a rainy day). Put $n = 1$.

Step 2: If $X_{n-1} = r$ put $X_n = \begin{cases} r & \text{if } U_n \le p_{rr} \\ s & \text{otherwise,} \end{cases}$

if $X_{n-1} = s$ put $X_n = \begin{cases} r & \text{if } U_n \le p_{sr} \\ s & \text{otherwise.} \end{cases}$

Step 3: Increase n by 1 and go back to step 2.

This produces a sequence $X_0, X_1, X_2, \ldots$ of states that might look like this:

$$r\ r\ r\ s\ r\ s\ s\ r\ s\ s\ r\ r\ s\ s\ s\ s\ r\ r\ r\ s\ r\ s\ s\ r\ r\ r\ s\ s\ s\ s\ s\ s\ s\ \ldots, \tag{15.2}$$

where the initial r comes from Step 1 above. In this section we rigorously study the long-run behavior of such two-state systems.

Some Restrictions on P. Before getting started we must impose a simple restriction on the PTM $\mathbf{P}$: specifically, we require that the four entries of the matrix $\mathbf{P}^2$ are all strictly positive. It is not difficult to see (see exercises) that this is equivalent to

$$\mathbf{P} \ne \begin{pmatrix} 1 & 0 \\ p & q \end{pmatrix}, \begin{pmatrix} p & q \\ 0 & 1 \end{pmatrix}, \text{ or } \begin{pmatrix} 0 & 1 \\ 1 & 0 \end{pmatrix}. \tag{15.3}$$

Here '$p\ \ q$' are any two probabilities that sum to 1; for example $\begin{pmatrix} 1 & 0 \\ .8 & .2 \end{pmatrix}$ is precluded by the first form in (15.3). These PTMs are not very interesting in any case. For example, if $p_{rr} = 1$ (first form) then once it rains it rains every day from that day forth. In the second form, once it is sunny it is forever sunny. If the PTM is given by the third form, the weather toggles back and forth non-randomly between rain and sun. As we develop the mathematics, these restrictions will play a role. If $\mathbf{P}^2$ has all strictly positive entries (equivalently, if (15.3) holds), the resulting two state Markov chain is said to be *ergodic*. Binaryville's PTM yields an ergodic Markov chain.

The Invariant Distribution. We will use a row vector like $\boldsymbol{\nu} = (a, b)$, where $a, b \ge 0$ and $a + b = 1$, to represent a probability distribution on the two states r and s. For example, if we write $X_n \sim \boldsymbol{\nu}$ we mean that with probability a it is rainy in period n and with probability b it is sunny. Presently we address the following question: Suppose we know that the probability distribution of X_n is given by $\boldsymbol{\nu} = (a, b)$; what is the probability distribution of X_{n+1}?

We can have $X_{n+1} = r$ in one of two ways: (i) it can be rainy in period n followed by rain in period $n + 1$, or (ii) it can be sunny in period n followed by rain in period $n + 1$. Possibility (i) happens with probability $a \cdot p_{rr}$ while possibility (ii) happens with probability $b \cdot p_{sr}$. Hence $P[X_{n+1} = r] = ap_{rr} + bp_{sr}$. Similarly we have that $P[X_{n+1} = s] = ap_{rs} + bp_{ss}$ — see Figure 15.4 below.

Figure 15.4. Computing the distribution of X_{n+1} if $X_n \sim \boldsymbol{\nu} = (a, b)$.

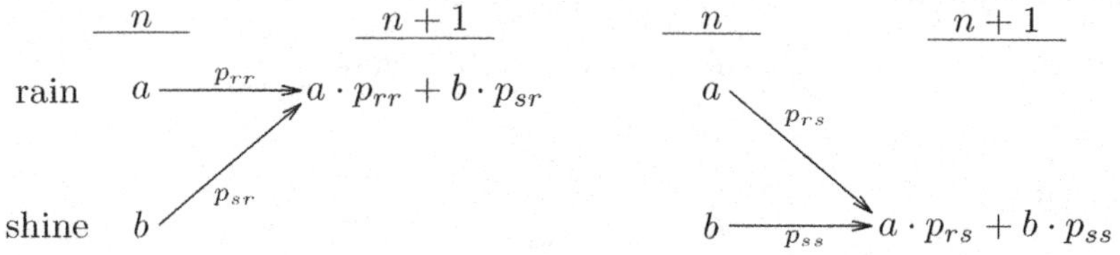

Here is where the matrix notation is most useful. By direct computation we see that

$$\boldsymbol{\nu}\mathbf{P} = (\,a \quad b\,) \cdot \begin{pmatrix} p_{rr} & p_{rs} \\ p_{sr} & p_{ss} \end{pmatrix} = (\,ap_{rr} + bp_{sr} \quad ap_{rs} + bp_{ss}\,)\,. \tag{15.5}$$

In other words, if $X_n \sim \boldsymbol{\nu}$ then $X_{n+1} \sim \boldsymbol{\nu}\mathbf{P}$. Additional applications give that $X_{n+2} \sim (\boldsymbol{\nu}\mathbf{P})\mathbf{P} = \boldsymbol{\nu}\mathbf{P}^2$, $X_{n+3} \sim \boldsymbol{\nu}\mathbf{P}^3$, and so forth.

REMARK. When the vector $\boldsymbol{\nu}$ is viewed as a matrix with one row, as in (15.5), it is customary to omit the comma. We will use both notations, (a, b) and $(a \;\; b)$, depending on the context.

A distribution $\boldsymbol{\mu}$ is *invariant* if we have $\boldsymbol{\mu} = \boldsymbol{\mu}\mathbf{P}$. (We reserve the symbol $\boldsymbol{\mu}$ for the invariant distribution which, as we shall see, exists and is unique. The symbol $\boldsymbol{\nu}$ will be used for *any* distribution.) If $\boldsymbol{\mu}$ is invariant, then:

$$\begin{aligned} X_0 \sim \boldsymbol{\mu} \quad &\Longrightarrow \quad X_1 \sim \boldsymbol{\mu}\mathbf{P} = \boldsymbol{\mu} \\ &\Longrightarrow \quad X_2 \sim \boldsymbol{\mu}\mathbf{P} = \boldsymbol{\mu} \\ &\Longrightarrow \quad X_3 \sim \boldsymbol{\mu} \\ &\Longrightarrow \quad \ldots \end{aligned}$$

In other words, if the initial state of the Markov chain is chosen randomly according to the invariant distribution, i.e., $X_0 \sim \boldsymbol{\mu}$, then the probability distribution of all future states remains the same distribution: all $X_n \sim \boldsymbol{\mu}$. Hence the term 'invariant'.

If $\boldsymbol{\mu} = (a, b)$ is to be invariant, by (15.5) we must have that

$$\begin{aligned} ap_{rr} + bp_{sr} &= a, \quad \text{and} \\ ap_{rs} + bp_{ss} &= b. \end{aligned} \tag{15.6}$$

Unfortunately we cannot solve this system for a and b as these equations are redundant (to see this, add the two equations and get that $a + b = a + b$ using that $p_{rr} + p_{rs} = 1 = p_{sr} + p_{ss}$). However we also have that $a + b = 1$. Since

$p_{rr} - 1 = -p_{rs}$, the first equation in (15.6) is equivalent to $-ap_{rs} + bp_{sr} = 0$. Solving the system

$$\begin{aligned} -ap_{rs} + bp_{sr} &= 0, \quad \text{and} \\ a + b &= 1 \end{aligned}$$

yields that

$$a = \frac{p_{sr}}{p_{rs} + p_{sr}} \quad \text{and} \quad b = \frac{p_{rs}}{p_{rs} + p_{sr}}. \tag{15.7}$$

For Binaryville, for example, we have that $\boldsymbol{\mu} = (\frac{1}{3}, \frac{2}{3})$.

Of course, we will run in to difficulty if $p_{rs}+p_{sr} = 0$ — but this cannot happen as $p_{rs} = 0$ is precluded by virtue of (15.3), first form. We have established that the invariance of $\boldsymbol{\mu} = (a, b)$ demands that (15.7) hold. To summarize, if we preclude the PTMs of (15.3), the invariant distribution exists and is unique and is given by (15.7).

Properties of $\boldsymbol{\mu}$. Here we prove two key facts about the invariant distribution. They establish two different ways in which the Markov chain "converges" to the invariant distribution as time elapses. We have:

FACT 1. In the long run, the fraction of time that the Markov chain spends in each state is given by the invariant distribution.

FACT 2. Regardless of the probability distribution of the initial state X_0, the probability distribution of X_n converges to the invariant distribution as $n \to \infty$.

PROOF OF FACT 1. Looking at (15.2), we see that the state of the system toggles randomly back and forth between r and s. Let R_i denote the number of rainy days in the ith block of rain. For example, in (15.2) we have $R_1 = 3$, $R_2 = R_3 = 1$, $R_4 = 2$, and so forth. Similarly, let S_i denote the number of sunny days in the ith block of sun. By the Markov property, each R_i is independent of the previous R_j's. For example, R_3 is independent of R_1 and R_2 (the past is irrelevant). Thus $R_1, R_2, \ldots$ is a sequence of independent random variables. Similarly $S_1, S_2, \ldots$ are independent. The distribution of the R_i's is easy to deduce. Note that $R_i = 1$ if the first rainy day in the ith block is immediately followed by a sunny day. This happens with probability p_{rs}, so $P[R_i = 1] = p_{rs}$. Similarly, $R_i = 2$ if the first rainy day is followed by a rainy day and then a sunny day. This happens with probability $p_{rr} \cdot p_{rs}$, so $P[R_i = 2] = p_{rr} \cdot p_{rs}$. In general, $R_i = k$ if the first rainy day is followed by $k - 1$ rainy days and then a sunny day. This happens with probability $p_{rr}^{k-1} \cdot p_{rs}$, so $P[R_i = k] = p_{rr}^{k-1} \cdot p_{rs}$. Since $p_{rr}+p_{rs} = 1$, we see that $R_i \sim \text{Geometric}\,(p_{rs})$. Thus the R_i's are iid with mean $1/p_{rs}$. A similar argument shows that the S_i's are iid with mean $1/p_{sr}$.

After n blocks of rain and sun, we have seen $R_1 + \cdots + R_n$ rainy days and $S_1 + \cdots + S_n$ sunny days. Hence, after n such blocks,

$$\textit{fraction of rainy days} = \frac{R_1 + \cdots + R_n}{R_1 + \cdots + R_n \ + \ S_1 + \cdots + S_n} = \frac{\overline{R}_n}{\overline{R}_n + \overline{S}_n},$$

where

$$\overline{R}_n = \frac{R_1 + \cdots + R_n}{n} \quad \text{and} \quad \overline{S}_n = \frac{S_1 + \cdots + S_n}{n}.$$

Now the Strong Law of Large numbers assures that $\overline{R}_n \to ER = 1/p_{rs}$ and $\overline{S}_n \to ES = 1/p_{sr}$ as $n \to \infty$, giving that

$$\textit{fraction of rainy days} \to \frac{\frac{1}{p_{rs}}}{\frac{1}{p_{rs}} + \frac{1}{p_{sr}}} = \frac{p_{sr}}{p_{sr} + p_{rs}} = a, \text{ from (15.7)}.$$

Similarly,

$$\textit{fraction of sunny days} \to b.$$

These calculations are valid provided $p_{sr} \neq 0$ and $p_{rs} \neq 0$, both of which possibilities are precluded by restriction (15.3). □

PROOF OF FACT 2. Here, again, we let $\boldsymbol{\mu} = (a, b)$ denote the invariant distribution. Suppose $X_0 \sim \boldsymbol{\nu} = (c, d) \neq (a, b)$. Then

$$\boldsymbol{\nu} = (c, d) = (a, b) + (c - a, d - b) = \boldsymbol{\mu} + (c - a, d - b).$$

Since $c + d = 1$ and $a + b = 1$ (they are probability distributions), we must have $c - a = b - d$. Letting $\delta = c - a$ we have

$$\boldsymbol{\nu} = \boldsymbol{\mu} + (\delta, -\delta) = \boldsymbol{\mu} + \delta \cdot (1, -1).$$

Then

$$\begin{aligned} X_1 \sim \boldsymbol{\nu}\mathbf{P} &= [\boldsymbol{\mu} + \delta \cdot (1 \quad -1)]\mathbf{P} \\ &= \boldsymbol{\mu}\mathbf{P} + \delta \cdot (1 \quad -1)\mathbf{P} \\ &= \boldsymbol{\mu} + \delta \cdot (1 \quad -1) \begin{pmatrix} p_{rr} & p_{rs} \\ p_{sr} & p_{ss} \end{pmatrix} \\ &= \boldsymbol{\mu} + \delta \cdot (p_{rr} - p_{sr} \quad p_{rs} - p_{ss}). \end{aligned}$$

Now $p_{rr} + p_{rs} = 1$ and $p_{sr} + p_{ss} = 1$ so $p_{rr} - p_{sr} = p_{ss} - p_{rs}$. Letting $\rho = p_{rr} - p_{sr}$, we have

$$X_1 \sim \boldsymbol{\mu} + \delta \cdot (\rho \quad -\rho) = \boldsymbol{\mu} + \delta\rho \cdot (1 \quad -1).$$

Since this works for any number δ, we have also that

$$X_2 \sim [\boldsymbol{\mu} + \delta\rho \cdot (1 \quad -1)]\mathbf{P} = \boldsymbol{\mu} + \delta\rho^2 \cdot (1 \quad -1),$$

and in general (by induction, to be formal),

$$X_n \sim \boldsymbol{\mu} + \delta\rho^n \cdot (1 \quad -1). \tag{15.8}$$

Now $|\rho| = |p_{rr} - p_{sr}|$ and p_{rr} and p_{sr} are probabilities, so we must have $|\rho| \leq 1$. Furthermore, $|\rho| = 1$ only if either (i) $p_{rr} = 1$ and $p_{sr} = 0$ or (ii) $p_{rr} = 0$ and $p_{sr} = 1$. In case (i), we have $\mathbf{P} = \begin{pmatrix} 1 & 0 \\ 0 & 1 \end{pmatrix}$, while in case (ii) we have $\mathbf{P} = \begin{pmatrix} 0 & 1 \\ 1 & 0 \end{pmatrix}$, both of which are precluded by condition (15.3). Hence, subject to (15.3), we must have $|\rho| < 1$. It follows that $\rho^n \to 0$ as $n \to \infty$ so the vector $\delta\rho^n \cdot (1 \quad -1) \to (0 \quad 0)$. From this and (15.8) we see that the distribution of X_n converges to $\boldsymbol{\mu}$ as $n \to \infty$. □

We have seen that for two-state ergodic Markov chains, the invariant distribution exists and is unique and Facts 1 and 2 hold. This continues to be true for ergodic Markov chains with more than two states. In this situation, we must formulate a more general notion of ergodicity and the underlying mathematics is more difficult. In the next several sections we turn our attention to this.

EXERCISES

1. Suppose p and q are fixed positive numbers that sum to 1. Find an invariant distribution for $\mathbf{P} = \begin{pmatrix} p & q \\ 0 & 1 \end{pmatrix}$. How many such invariant distributions are there? Do Facts 1 and 2 hold in this case?

2. Same for $\mathbf{P} = \begin{pmatrix} 1 & 0 \\ p & q \end{pmatrix}$.

3. Same for $\mathbf{P} = \begin{pmatrix} 1 & 0 \\ 0 & 1 \end{pmatrix}$.

4. Same for $\mathbf{P} = \begin{pmatrix} 0 & 1 \\ 1 & 0 \end{pmatrix}$.

5. For general PTM $\mathbf{P}$ on $S = \{1, 2\}$ show that all the entries of $\mathbf{P}^2$ are positive if and only if $\mathbf{P}$ satisfies the condition (15.3).

6. Modify Step 1 of the construction of the Binaryville process on $\boldsymbol{\Omega}$ so that the weather on day 0 is chosen randomly with $P[X_0 = r] = p$.

§16. Markov Chains with More Than Two States

Suppose $X_0, X_1, X_2, \ldots$ is a sequence of random variables taking values in $S = \{1, 2, 3, \ldots, s\}$ where the value of X_n specifies the state of the system at time n. For example, if $X_6 = 4$, then at time 6 the system is in state 4. Exactly what state 4 means will depend on the specific problem at hand. The sequence $X_0, X_1, X_2, \ldots$ is a Markov chain if, for any sequence of states $i_0, i_1, \ldots$ we have

$$\begin{aligned} P[X_{n+1} = i_{n+1} | X_0 = i_0, X_1 = i_1, \ldots, X_n = i_n] \\ = P[X_{n+1} = i_{n+1} | X_n = i_n]. \end{aligned} \tag{16.1}$$

With

$$\begin{aligned} A &= \{X_0 = i_0, X_1 = i_1, \ldots, X_{n-1} = i_{n-1}\}, \\ B &= \{X_n = i_n\}, \text{ and} \\ C &= \{X_{n+1} = i_{n+1}\}. \end{aligned}$$

the Markov condition says that $P[C \mid A \cap B] = P[C \mid B]$, which is equivalent to $P[A \cap C \mid B] = P[A \mid B] \cdot P[C \mid B]$. Thinking of A as the past, B as the present, and C as the future (all relative to period n), this latter equality says that, conditioned on the present, the past and the future are independent.

The probabilities on the right side of (16.1) are called the *transition probabilities*. If the transition probabilities do not change with time (don't depend on n), the Markov chain is called *homogeneous*. Henceforth we will study only homogeneous Markov chains. As with the two-state model, it is convenient to hold the transition probabilities in a *probability transition matrix* $\mathbf{P} = (\mathbf{P}_{ij})$, where

$$\mathbf{P}_{ij} = P[X_{n+1} = j | X_n = i]. \tag{16.2}$$

EXAMPLE 16.3. Suppose Binaryville's weather dynamics are the same as in Example 15.1, but with the exception that it never rains three days in a row. In the two state framework this process is not Markovian. To see this observe that $P[X_2 = r \mid X_0 = r, X_1 = r] = 0$ (it never rains 3 days in a row) but we still have $P[X_2 = r \mid X_0 = s, X_1 = r] = 0.6$ (the regular dynamics). Hence history is important. Sometimes a system that is not Markovian when described in the most natural way can be made Markovian when described differently. To accomplish this here we will use three states: $\{r_1, r_2, s\}$. State r_1 means that it is rainy today but sunny yesterday; state r_2 means that it is rainy today and was also rainy yesterday; state s means that it is sunny today. With this more elaborate description of the state space the process is Markovian and we have:

$$\mathbf{P} = \begin{pmatrix} \mathbf{P}_{r_1 r_1} & \mathbf{P}_{r_1 r_2} & \mathbf{P}_{r_1 s} \\ \mathbf{P}_{r_2 r_1} & \mathbf{P}_{r_2 r_2} & \mathbf{P}_{r_2 s} \\ \mathbf{P}_{s r_1} & \mathbf{P}_{s r_2} & \mathbf{P}_{ss} \end{pmatrix} = \begin{pmatrix} 0 & 0.6 & 0.4 \\ 0 & 0 & 1 \\ 0.2 & 0 & 0.8 \end{pmatrix}.$$

Notice that for this three state example the sum of the entries of each row of the probability transition matrix is 1. This is always the case. With $S = \{1, 2, \ldots, s\}$ we have

$$\sum_{j=1}^{s} \mathbf{P}_{ij} = \sum_{j=1}^{s} P[X_{n+1} = j | X_n = i] = 1.$$

Indeed, the events $\{X_{n+1} = j\}$ for $j = 1, 2, \ldots, s$ form a partition of the outcome space so their probabilities sum to 1 under any fixed probability measure.

NOTATIONAL REMARK. Sometimes we descriptively name the states of a Markov chain, as in $S = \{r_1, r_2, s\}$. For a 'generic' Markov chain, we would number the states, as in $S = \{1, 2, 3\}$. With 'descriptive' notation, since the states correspond to the rows and columns of the PTM we list them in the order in which they appear in the PTM. For example, state s is row 3 and column 3 of the matrix above.

The Chapman–Kolmogorov Equations. Suppose we know the probability distribution of X_n and we are interested in computing the distribution of X_{n+1}. Specifically, suppose $P[X_n = i] = \nu_i$ for $i = 1, 2, \ldots, s$, i.e., $X_n \sim \boldsymbol{\nu}$ where $\boldsymbol{\nu} = (\nu_1, \ldots, \nu_s)$. Then, by the partition theorem,

$$\begin{aligned} P[X_{n+1} = j] &= \sum_{i=1}^{s} P[X_{n+1} = j | X_n = i]\, P[X_n = i] \\ &= \sum_{i=1}^{s} \nu_i \mathbf{P}_{ij} = (\boldsymbol{\nu}\mathbf{P})_j. \end{aligned}$$

Here, $\boldsymbol{\nu}\mathbf{P}$ is the matrix multiplication of the $1 \times s$ row vector $\boldsymbol{\nu}$ by the $s \times s$ probability transition matrix $\mathbf{P}$, and $(\boldsymbol{\nu}\mathbf{P})_j$ is the jth entry of the resulting $1 \times s$ row vector as shown in Figure 16.4 below.

Figure 16.4. If $X_n \sim \boldsymbol{\nu}$, then $X_{n+1} \sim \boldsymbol{\nu}\mathbf{P}$.

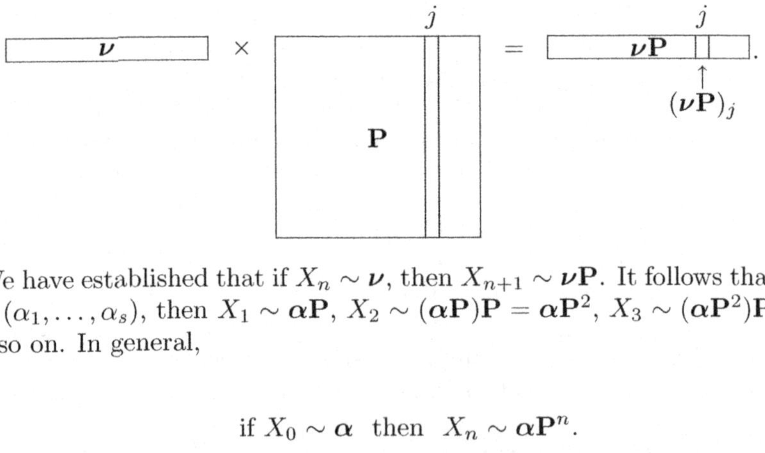

We have established that if $X_n \sim \boldsymbol{\nu}$, then $X_{n+1} \sim \boldsymbol{\nu}\mathbf{P}$. It follows that if $X_0 \sim \boldsymbol{\alpha} = (\alpha_1, \ldots, \alpha_s)$, then $X_1 \sim \boldsymbol{\alpha}\mathbf{P}$, $X_2 \sim (\boldsymbol{\alpha}\mathbf{P})\mathbf{P} = \boldsymbol{\alpha}\mathbf{P}^2$, $X_3 \sim (\boldsymbol{\alpha}\mathbf{P}^2)\mathbf{P} = \boldsymbol{\alpha}\mathbf{P}^3$, and so on. In general,

$$\text{if } X_0 \sim \boldsymbol{\alpha} \text{ then } X_n \sim \boldsymbol{\alpha}\mathbf{P}^n. \tag{16.5}$$

EXAMPLE 16.6. Suppose, referring to Example 16.3, that $\boldsymbol{\alpha} = (\frac{1}{4}, \frac{1}{4}, \frac{1}{2})$. What is the probability that it is sunny on day 3? With $\mathbf{P} = \begin{pmatrix} 0 & 0.6 & 0.4 \\ 0 & 0 & 1 \\ 0.2 & 0 & 0.8 \end{pmatrix}$, the reader will confirm that the distribution of Binaryville's weather state after three days is given by

$$X_3 \sim \boldsymbol{\alpha}\mathbf{P}^3 = \begin{pmatrix}\frac{1}{4} & \frac{1}{4} & \frac{1}{2}\end{pmatrix}\begin{pmatrix}0 & 0.6 & 0.4\\ 0 & 0 & 1\\ 0.2 & 0 & 0.8\end{pmatrix}^3$$

$$= \begin{pmatrix}\frac{1}{4} & \frac{1}{4} & \frac{1}{2}\end{pmatrix}\begin{pmatrix}.184 & .048 & .768\\ .16 & .12 & .72\\ .144 & .096 & .76\end{pmatrix} = (\,.1 \quad .15 \quad .75\,).$$

So, if $X_0 \sim (\frac{1}{4}, \frac{1}{4}, \frac{1}{2})$, $P[X_3 = s] = .75$. Suppose we are told that it is sunny on day 0. Conditioned on this, what is the probability that it is rainy on both day 2 and day 3? Here, the distribution of X_0 is given by $\boldsymbol{\alpha} = (0\ 0\ 1)$. Given this,

$$X_3 \sim (0 \quad 0 \quad 1)\begin{pmatrix}.184 & .048 & .768\\ .16 & .12 & .72\\ .144 & .096 & .76\end{pmatrix} = (\,.144 \quad .096 \quad .76\,).$$

We are interested in the probability of being in state r_2 on day 3, which is $P[X_3 = r_2 \,|\, X_0 = s] = 0.096$. Note that this is just the 3-2 entry of the matrix $\mathbf{P}^3$.

EXERCISES

1. The Markov process $X_0, X_1, X_2, \ldots$ has PTM

$$\begin{pmatrix}1/2 & 1/2 & 0\\ 1/4 & 0 & 3/4\\ 1/4 & 1/2 & 1/4\end{pmatrix}$$

(call the states 1, 2, and 3). Calculate $P[X_8 = 1|X_6 = 2]$.

2. A homogeneous Markov chain on $S = \{1, 2, 3, 4\}$ works as follows. From state 1, it branches to 1, 2, 3, or 4 with equal probability. From state i, with $i \neq 1$, it branches to state $i - 1$ with probability 1. (a) What is this Markov chain's PTM? (b) If $X_0 \sim \text{Uniform}\{1, 2, 3, 4\}$ what is the distribution of X_1?

3. Initially three green balls are placed in urn A and 1 green and two red balls are placed in urn B. At each step, we simultaneously draw a ball randomly from each urn and place the ball drawn from urn A into urn B and place the ball drawn from urn B into urn A. (So after each step both urns still contain 3 balls.) Let X_n denote the number of red balls in urn A after n steps (so $X_0 = 0$). Compute the PTM for the Markov chain $(X_n : n \geq 0)$. Compute the probability that urn A contains 3 green marbles after two steps.

4. Each day is either sunny or rainy. A statistician has discovered that the probability of sun on any particular day depends on the weather over the previous two days as follows, Let r denote the number of days of rain over the previous two days. Then

$$P[\text{sun that particular day}] = \begin{cases} \frac{1}{6} & \text{if } r = 2 \\ \frac{2}{6} & \text{if } r = 1 \\ \frac{3}{6} & \text{if } r = 0. \end{cases}$$

Set this up as a Markov chain (you will need 4 states). What is the PTM?

5. Label the vertices of a triangle 1, 2, and 3, going clockwise. An ant walks from vertex to vertex on the triangle going clockwise with probability: $\frac{1}{2}$ if it's at vertex 1; $\frac{3}{4}$ if it's at vertex 2; and 1 if it's at vertex 3. Write down the PTM for this Markov chain. If the ant starts at vertex 1, what is the probability that it's back at vertex 1 after 3 steps?

6. The unit cube has 8 vertices whose (x, y, z) coordinates are

$$S = \{(0,0,0),(1,0,0),(0,1,0),(1,1,0),(0,0,1),(1,0,1),(0,1,1),(1,1,1)\}.$$

An ant walks from vertex to vertex each step choosing one of the three neighboring vertices randomly and independently of the walk's history. For example, from vertex $(0,1,0)$ it will walk to one of $(1,1,0)$, $(0,0,0)$, or $(0,1,1)$ — each with probability $\frac{1}{3}$. Labeling the vertices 1 through 8 in the order presented above, what is the PTM corresponding to this Markov chain?

7. When tossed, a coin produces heads with probability p. Zoe tosses the coin repeatedly. Let S_n denote the number of heads among the first n tosses and let R_n denote the remainder when S_n is divided by 3. Show that $(R_n : n \geq 0)$ is a Markov chain and compute its PTM.

8. A Markov chain X_n on $S = \{1, 2, 3\}$ has initial distribution $\boldsymbol{\alpha} = (\frac{1}{4}, \frac{1}{2}, \frac{1}{4})$ and PTM

$$\mathbf{P} = \begin{pmatrix} 1/3 & 1/3 & 1/3 \\ 2/5 & 0 & 3/5 \\ 0 & 1 & 0 \end{pmatrix}.$$

Describe an algorithm to use independent Uniform $(0,1)$'s $U_0, U_1, U_2, \ldots$ to generate $X_0, X_1, X_2, \ldots$ on $\boldsymbol{\Omega}$. Hint: Modify the two-state Binaryville construction from §15.

9. Describe the construction of a Markov chain $X_0, X_1, \ldots$ on $\boldsymbol{\Omega}$ with states $S = \{1, 2, \ldots, s\}$, $s \times s$ PTM $\mathbf{P}$, and with $X_0 \sim \boldsymbol{\alpha}$. Use the sequence $U_0, U_1, \ldots$ to generate the X_n's.

§17. Irreducible and Ergodic Markov Chains

From this discussion we know that, given $X_0 = i$, the conditional distribution of X_n is given by

$$X_n \sim (0, \cdots, 0, 1, 0, \ldots, 0)\mathbf{P}^n,$$

where the sole 1 in X_0's distribution is in the ith location. The vector on the right side above is the ith row of the matrix $\mathbf{P}^n$, so

$$P[X_n = j | X_0 = i] \ = \ ((0,\dots,0,1,0,\dots,0)\mathbf{P}^n)_j \ = \ \mathbf{P}^n_{ij}.$$

Similarly, for any k, $P[X_{k+n} = j | X_k = i] = \mathbf{P}^n_{ij}$. In words, $\mathbf{P}^n_{ij}$ is the probability of being in state j, n periods hence, given that the system is currently in state i.

NOTATIONAL COMMENT. $\mathbf{P}^n_{ij}$ denotes the i-j^{th} entry of the matrix $\mathbf{P}^n$, NOT the i-j^{th} entry of $\mathbf{P}$, raised to the power n. This is more streamlined notation than the unambiguous $(\mathbf{P}^n)_{ij}$.

If i, j, and k are any three states and $n, m \geq 0$ then $\mathbf{P}^{n+m}_{ik} \geq \mathbf{P}^n_{ij}\mathbf{P}^m_{jk}$. The left side is the probability of going from state i to state k in $n+m$ steps; the right side is the probability of going from state i to state k in $n+m$ steps by passing through state j after the first n steps. More formally we have

$$\mathbf{P}^{n+m}_{ik} \ = \ (\mathbf{P}^n\mathbf{P}^m)_{ik} \ = \ \sum_{h=1}^{s} \mathbf{P}^n_{ih}\mathbf{P}^m_{hk} \ \geq \ \mathbf{P}^n_{ij}\mathbf{P}^m_{jk}.$$

Similarly, if i, j, h, and k are any four states and $n, m, r \geq 0$, then $\mathbf{P}^{n+m+r}_{ik} \geq \mathbf{P}^n_{ij}\mathbf{P}^m_{jh}\mathbf{P}^r_{hk}$. (Interpret the right and left sides of this latter inequality.) We will make use of these inequalities repeatedly without reference.

If $\mathbf{P}^n_{ij} > 0$ for some $n \geq 0$ we say that state j is accessible from state i and write $i \to j$. This holds if and only if there is a sequence of states $i_0, i_1, \dots, i_n$ with $i_0 = i$, $i_n = j$, and each $\mathbf{P}_{i_k, i_{k+1}} > 0$. We will call such a sequence a *walk* of *length* n from i to j. Of course, if $i \to j$, there may be more than one such walk. Note that for each state i, $i \to i$, as $\mathbf{P}^0_{ii} = \mathbf{I}_{ii} = 1$.

The Markov chain with PTM $\mathbf{P}$ is said to be *irreducible* if $i \to j$ for any two states i and j. The following preliminary result says that, for an irreducible Markov chain, all states are visited quickly as time elapses.

THEOREM 17.1. *Suppose a Markov chain with probability transition matrix* $\mathbf{P}$ *on the finite state space* $S = \{1, 2, \dots, s\}$ *is irreducible. Let* A_n *denote the event that there is some state that has not been visited by time* n. *Then there are positive numbers* c *and* α *such that* $P[A_n] \leq ce^{-\alpha n}$. *This inequality holds regardless of the distribution of* X_0.

PROOF. Since $\mathbf{P}$ is irreducible, we have that $1 \to 2, 2 \to 3, \dots, s-1 \to s$ and $s \to 1$. Hence there is a cycle of states $(i_0, i_1, \dots, i_m)$ with (i) $i_0 = i_m = 1$; (ii) each $\mathbf{P}_{i_k i_{k+1}} > 0$; and (iii) each of the states $\{1, 2, \dots, s\}$ present among the i_k. [For example, for the PTM $\mathbf{P}$ of Example 16.3 the sequence $(1, 2, 3, 1)$ accomplishes this.] The probability of observing this cycle over the next m steps of the Markov chain given that the current state is i_0 is $\rho = \mathbf{P}_{i_0 i_1}\mathbf{P}_{i_1 i_2}\cdots\mathbf{P}_{i_{m-1} i_m} > 0$. Additionally, we may start this cycle at any one of the s states. [To illustrate, starting the cycle of Example 16.3 at 2 produces $(2, 3, 1, 2)$.] Regardless of which state the Markov chain is currently in, the conditional probability of observing this particular cycle over the next m steps is ρ. The computation of this probability simply re-orders the multiplication of the transition probabilities. [For example, $\mathbf{P}_{12}\mathbf{P}_{23}\mathbf{P}_{31} = \mathbf{P}_{23}\mathbf{P}_{31}\mathbf{P}_{12}$.] We conclude that regardless of what state the chain is in, every state is visited over the next m steps with probability at least ρ.

By time km, the system has had k such opportunities to visit every state. With each such opportunity, regardless of the previous history (using the Markov property), the system will visit every state with probability at least ρ. Refer to Figure 17.2.

Figure 17.2.

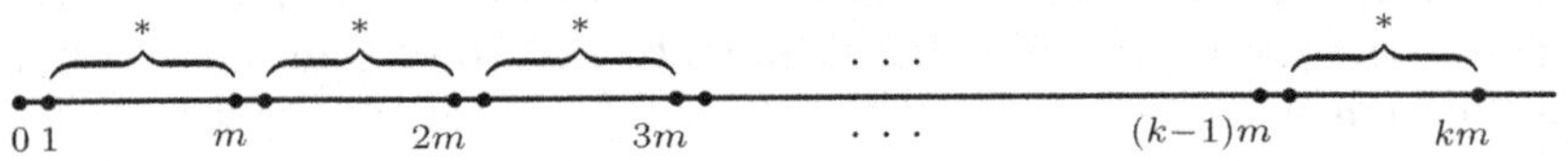

$*$ The probability that the chain visits every state in each of these blocks of periods is at least ρ, regardless of previous history.

Let T denote the first time by which every state has been visited. If $T > km$, then the process has *failed to visit every state* for k consecutive blocks of periods. It follows that $P[T > km] \leq (1-\rho)^k$. Fix any $n > 0$. Then for some positive integer k, we will have $km \leq n < (k+1)m$. Now $km \leq n \implies \{T > n\} \subset \{T > km\}$ and $n < (k+1)m) \implies k > \frac{n}{m} - 1$, so

$$P[T > n] \leq P[T > km] \leq (1-\rho)^k \leq (1-\rho)^{\frac{n}{m}-1} = cb^n,$$

where $c = (1-\rho)^{-1}$ and $b = (1-\rho)^{1/m}$. Here, $b < 1$ since $\rho > 0$. With $\alpha = -\log b > 0$, we have $P[A_n] = P[T > n] \leq ce^{-\alpha n}$. □

This theorem has the following important corollary:

COROLLARY 17.3. *If a Markov chain with finite state space is irreducible, then, with probability one, the Markov chain visits every state infinitely often.*

PROOF. See exercises. □

Invariant Measures. In this section we generalize the notion of an invariant measure to Markov chains with finitely many states s, where $2 \leq s < \infty$.

EXAMPLE 17.4. Let us compute the invariant distribution for the PTM of Example 16.3. The condition $\boldsymbol{\mu}\mathbf{P} = \boldsymbol{\mu}$ gives us

$$(\mu_1 \quad \mu_2 \quad \mu_3)\begin{pmatrix} 0 & 0.6 & 0.4 \\ 0 & 0 & 1 \\ 0.2 & 0 & 0.8 \end{pmatrix} = (\mu_1 \quad \mu_2 \quad \mu_3),$$

or $\mu_1 = .2\mu_3$, $\mu_2 = .6\mu_1$, and $\mu_3 = .4\mu_1 + \mu_2 + .8\mu_3$. Here we have three equations in three unknowns, but only the first two are linearly independent. However $\mu_1 + \mu_2 + \mu_3 = 1$ provides us with the needed third equation (analogous to the two state case). The first two equations imply that $\mu_2 = .12\mu_3$, so $\boldsymbol{\mu} = (\,.2\mu_3 \quad .12\mu_3 \quad \mu_3\,)$. Since $\mu_1 + \mu_2 + \mu_3 = 1$, we get that $\mu_3 = \frac{1}{1.32} \approx .7576$ and then $\mu_2 \approx .0909$, and $\mu_1 \approx .1515$.

In the two-state case, we found that the invariant measure gave the long-term fraction of time that the chain spends in each of the two states. Let's proceed on the assumption that this continues to hold in the general case (which it does). If a state, say state 1, has a positive representation in the invariant distribution (i.e., $\mu_1 > 0$), then state 1 should appear with positive long-term density μ_1 in the sequence $X_0, X_1, \ldots$. It should also be true (and it is surely intuitive to expect), that the long-term average length of time between visits to state 1 should be $\frac{1}{\mu_1}$. For example, if state 1 is present with density $\mu_1 = 0.1$ then, on average, one period out of 10 the Markov chain is in state 1 and on average there should be 10 steps between successive visits to state 1. The next example illustrates this for the PTM of Example 16.3.

EXAMPLE 17.5. In the three state Example 16.3 it is not difficult to compute the expected length of time between visits to any particular state. Let us continue with this example and focus on state 's'. Refer to Figure 17.6, which diagrams the Markov chain and displays the transition probabilities. Let x denote the expected number of steps until state s is visited starting from state s. Formally, if $T = \min\{n > 0 : X_n = s\}$ then $x = E[T \mid X_0 = s]$. Similarly, let $y = E[T \mid X_0 = r_1]$ and $z = E[T \mid X_0 = r_2]$.

Figure 17.6. A transition diagram for **P** *from Example 16.3.*

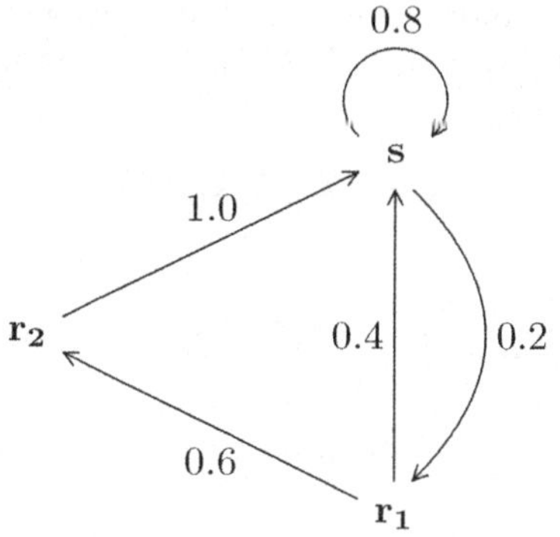

Then x is the expected time between visits to state s, which is the quantity we are interested in. By conditioning on the first step away from the starting state, we see that $x = 1 + .8 \cdot 0 + .2y$, $y = 1 + .4 \cdot 0 + .6z$, and $z = 1 + 1.0 \cdot 0$. Solving, we get $z = 1$, then $y = 1.6$, then $x = 1.32$. We have confirmed for this three state example by direct computation that μ_3 is the reciprocal of the expected time between visits to state s.

The General Construction. Let's condition our Markov chain on $S = \{1, \ldots, s\}$ with irreducible PTM $\mathbf{P}$ to start in state 1, i.e. condition on $X_0 = 1$. The chain will wander away from state 1 but return to it infinitely often. In other words, the chain takes repeated excursions that begin at state 1, as illustrated hypothetically in Figure 17.7. In this diagram we take T to be the first period in which the chain returns to state 1, i.e., $T = \min\{n > 0 : X_n = 1\}$.

Figure 17.7. Illustrative excursions from state 1.

Excursion 1 Excursion 2 Excursion 3

X_n 1 2 5 ⋯ 3 1 3 6 ⋯⋯⋯⋯ 2 1 5 s ⋯⋯ 7 1 ⋯

n 0 1 ⋯ $T-1$ T ⋯

For each state i, let N_i be the number of times that state i is visited by the chain before time T. If $i \neq 1$, N_i is a random quantity; but, by the definition of an excursion, we must have $N_1 = 1$ (since only $X_0 = 1$ before time T). We must also have

$$\sum_{i=1}^{s} N_i = T,$$

since in each of the T time periods $0, 1, \ldots, T-1$ the Markov chain is in exactly one of the s states. By the Markov property, the excursions (which are vectors with a random number of components) are independent and identically distributed. It follows that the first excursion, distributionally, is representative of the excursions in the chain as a whole. In the two-state Markov chain, μ_1 and μ_2 were equal to the long-run fraction of time the chain spent in states 1 and 2, respectively. If this continues to hold for Markov chains with more than two states (which under prescribed conditions it does), we would expect to have $E[N_i|X_0 = 1] = \mu_i \cdot E[T|X_0 = 1]$. Reflecting this intuition, let $\boldsymbol{\nu} = (\nu_1, \nu_2, \ldots, \nu_s)$ where

$$\nu_i = E[N_i|X_0 = 1],$$

and put

$$\tau = E[T|X_0 = 1] = \sum_{i=1}^{s} E[N_i \,|\, X_0 = 1] = \sum_{i=1}^{s} \nu_i,$$

so τ is the expected number of steps between successive visits to state 1. Now let

$$\boldsymbol{\mu} = (\mu_1, \ldots, \mu_s) = \frac{1}{\tau}\boldsymbol{\nu} = \left(\frac{\nu_1}{\tau}, \ldots, \frac{\nu_s}{\tau}\right).$$

Then $\sum_{i=1}^{s} \mu_i = 1$ and each $\mu_i \geq 0$, so $\boldsymbol{\mu}$ is a probability distribution on the states of the Markov chain. Note, in particular, that

$$\nu_1 = E[N_1|X_0 = 1] = E[1|X_0 = 1] = 1,$$

so $\mu_1 = \frac{1}{T}$ is the reciprocal of the expected number of steps between visits to state 1.

Presently we show that $\boldsymbol{\mu}$ is an invariant distribution. Since $\boldsymbol{\nu}$ and $\boldsymbol{\mu}$ differ by a scalar multiple, $\boldsymbol{\mu}$ will be invariant if $\boldsymbol{\nu}P = \boldsymbol{\nu}$, for then

$$\boldsymbol{\mu}P = \left(\frac{1}{T}\boldsymbol{\nu}\right)\mathbf{P} = \frac{1}{T}(\boldsymbol{\nu}\mathbf{P}) = \frac{1}{T}\boldsymbol{\nu} = \boldsymbol{\mu}. \tag{17.8}$$

Since N_i counts how often the chain is in state i before time T, it follows that

$$N_i = \sum_{n=0}^{\infty} I_{\{X_n=i,\ n<T\}}.$$

(Recall, the indicator random variable I_A is 1 if A occurs and 0 otherwise. Also, the notation $\{X_n = i,\ n < T\}$ means the event $\{X_n = i\} \cap \{n < T\}$.) This gives that

$$\begin{aligned} \nu_i = E[N_i|X_0 = 1] &= \sum_{n=0}^{\infty} E\left[I_{\{X_n=i,\ n<T\}} \,|\, X_0 = 1\right] \\ &= \sum_{n=0}^{\infty} P[X_n = i,\ n < T \,|\, X_0 = 1]. \end{aligned} \tag{17.9}$$

The astute reader will notice that the interchange of the infinite summation with the expected value operator in the second equality above must be justified. Since the summands are all non-negative, Corollary 12.4 of the Monotone Convergence Theorem assures that this is correct. Computing the jth component of the vector $\boldsymbol{\nu}\mathbf{P}$, we get

$$\begin{aligned} (\boldsymbol{\nu}\mathbf{P})_j &= \sum_{i=1}^{s} \nu_i \mathbf{P}_{ij} \\ &= \sum_{i=1}^{s} \Big(\sum_{n=0}^{\infty} P[X_n = i,\ n < T \,|\, X_0 = 1] \Big) \mathbf{P}_{ij} \\ &= \sum_{n=0}^{\infty} \Big(\sum_{i=1}^{s} \mathbf{P}_{ij} \cdot P[X_n = i,\ n < T \,|\, X_0 = 1] \Big). \end{aligned} \tag{17.10}$$

Since

$$\{n < T,\ X_0 = 1\} = \{X_0 = 1, X_1 \neq 1, X_2 \neq 1, \ldots, X_n \neq 1\},$$

the event $\{n < T,\ X_0 = 1\}$ depends on the history of the Markov chain up to and including time n, and it follows from the Markov property that

$$\begin{aligned} \mathbf{P}_{ij} &= P[X_{n+1} = j \,|\, X_n = i] \\ &= P[X_{n+1} = j \,|\, X_n = i,\ n < T,\ X_0 = 1]. \end{aligned}$$

Working on the inner summation of (17.10), we have

$$\begin{aligned}
&\sum_{i=1}^{s} \mathbf{P}_{ij} \cdot P[X_n = i,\ n < T \,|\, X_0 = 1] \\
&\quad = \sum_{i=1}^{s} P[X_{n+1} = j \,|\, X_n = i,\ n < T,\ X_0 = 1] \cdot P[X_n = i,\ n < T \,|\, X_0 = 1] \\
&\quad = \sum_{i=1}^{s} P[X_{n+1} = j,\ X_n = i,\ n < T \,|\, X_0 = 1] \\
&\quad = P[X_{n+1} = j,\ n < T \,|\, X_0 = 1]. \qquad (17.11)
\end{aligned}$$

The second equality in (17.11) uses that $P[A|B \cap C] \cdot P[B|C] = P[A \cap B|C]$ and the last equality holds because the events $\{X_n = i\}$ partition the outcome space as i ranges over all the states $1 \le i \le s$.

Now

$$\{X_{n+1} = j,\ n < T\} = \begin{cases} \{T = n+1\} & \text{if } j = 1 \\ \{X_{n+1} = j,\ n+1 < T\} & \text{if } 1 < j \le s, \end{cases}$$

and setting this into (17.11) gives

$$\begin{aligned}
&\sum_{i=1}^{s} \mathbf{P}_{ij} \cdot P[X_n = i,\ n < T \,|\, X_0 = 1] \\
&\quad = \begin{cases} P[T = n+1 \,|\, X_0 = 1] & \text{if } j = 1 \\ P[X_{n+1} = j,\ n+1 < T \,|\, X_0 = 1] & \text{if } 1 < j \le s. \end{cases} \qquad (17.12)
\end{aligned}$$

Setting (17.12) into (17.10), first for the case $j = 1$, gives

$$\begin{aligned}
(\boldsymbol{\nu}\mathbf{P})_1 &= \sum_{n=0}^{\infty} P[T = n+1 \,|\, X_0 = 1] \\
&= P[1 \le T < \infty \,|\, X_0 = 1] = 1 = \nu_1. \qquad (17.13)
\end{aligned}$$

For the case $1 < j \le s$, we get

$$\begin{aligned}
(\boldsymbol{\nu}\mathbf{P})_j &= \sum_{n=0}^{\infty} P[X_{n+1} = j,\ n+1 < T \,|\, X_0 = 1] \\
&= \sum_{n=0}^{\infty} P[X_n = j,\ n < T \,|\, X_0 = 1] = \nu_j. \qquad (17.14)
\end{aligned}$$

The second equality holds because the latter summation includes exactly one additional term, namely $P[X_0 = j,\ 0 < T \,|\, X_0 = 1]$, which must be 0 if $j > 1$ (conditioned on $X_0 = 1$ we cannot have $X_0 = j > 1$). The last equality is just (17.9).

We have shown in (17.13) and (17.14) that $\boldsymbol{\nu}\mathbf{P} = \boldsymbol{\nu}$ and hence, by (17.8), that $\boldsymbol{\mu}\mathbf{P} = \boldsymbol{\mu}$. That is, $\boldsymbol{\mu}$ is an invariant distribution for $\mathbf{P}$. We have proven the following:

THEOREM 17.15. *If a Markov chain with PTM* $\mathbf{P}$ *is irreducible on the finite state space* S, *then there is an invariant measure* $\boldsymbol{\mu}$ *on* S *given by the recipe above.*

We shall see that invariant measures are not always unique, although they are for irreducible Markov chains.

The Ergodic Case. The Markov chain (or its PTM $\mathbf{P}$) is said to be *ergodic* if, for some $n^* \geq 0$, $\mathbf{P}_{ij}^{n^*} > 0$ for all pairs of states i and j (including $j = i$). Ergodicity is a stronger condition than irreducibility, because the same walk length n^* must work for all pairs of states. First we establish:

LEMMA 17.16. *If* $\mathbf{P}_{ij}^{n^*} > 0$ *for all pairs of states* i *and* j *and* $n \geq n^*$ *then also* $\mathbf{P}_{ij}^{n} > 0$ *for all pairs of states* i *and* j.

PROOF. We will show that $\mathbf{P}_{ij}^{n} > 0$ for all $i, j \implies \mathbf{P}_{ij}^{n+1} > 0$ for all i, j. The result follows from a repeated application of this. Let $p = \min\{\mathbf{P}_{ij}^{n} : i, j \in S\}$, so $p > 0$. Then

$$\begin{aligned} \mathbf{P}_{ij}^{n+1} &= (\mathbf{P} \cdot \mathbf{P}^{n})_{ij} \\ &= \sum_{k=1}^{s} \mathbf{P}_{ik}\mathbf{P}_{kj}^{n} \geq p \sum_{k=1}^{s} \mathbf{P}_{ik} = p > 0. \quad \square \end{aligned}$$

Additionally we have the following sufficient, but not necessary, condition for ergodicity:

THEOREM 17.17 (CRITERIA FOR ERGODICITY). *If a Markov chain on* S *with PTM* $\mathbf{P}$ *is irreducible and there is a state* i^* *with* $\mathbf{P}_{i^*i^*} > 0$ *then it is ergodic.*

PROOF. We must show that, for some n^*, $\mathbf{P}_{ij}^{n^*} > 0$ for all states i and j. Since our Markov chain is irreducible, as in the proof of Theorem 17.1 we have a cycle of some length m that visits all the states in S. We claim that $n^* = 2m$ will work. The existence of our cycle insures that, for any $i, j \in S$, we have $\mathbf{P}_{ii^*}^{n} > 0$ and $\mathbf{P}_{i^*j}^{\tilde{n}} > 0$ for some numbers n and $\tilde{n}$ less than or equal to m. If $n + \tilde{n} = 2m$, we have

$$\mathbf{P}_{ij}^{n^*} = \mathbf{P}_{ij}^{n+\tilde{n}} \geq \mathbf{P}_{ii^*}^{n}\mathbf{P}_{i^*j}^{\tilde{n}} > 0.$$

If $n + \tilde{n} < 2m$, put $k = 2m - (n + \tilde{n})$ so $k > 0$. Then

$$\mathbf{P}_{ij}^{n^*} = \mathbf{P}_{ij}^{n+k+\tilde{n}} \geq \mathbf{P}_{ii^*}^{n}\mathbf{P}_{i^*i^*}^{k}\mathbf{P}_{i^*j}^{\tilde{n}} > 0.$$

Here we have used that $\mathbf{P}_{i^*i^*} > 0 \implies \mathbf{P}_{i^*i^*}^{k} > 0$ (see exercises). $\square$

For ergodic Markov chains we have:

THEOREM 17.18. *Ergodic Markov chains have a unique invariant measure. If $\boldsymbol{\mu}$ is the unique invariant measure, then all the rows of $\mathbf{P}^n$ converge to $\boldsymbol{\mu}$ as $n \to \infty$. If $\boldsymbol{\nu}$ is any measure on S, then $\boldsymbol{\nu}\mathbf{P}^n$ converges to $\boldsymbol{\mu}$ as $n \to \infty$.*

PROOF. Theorem 17.15 guarantees the existence of at least one invariant measure. We will show that if $\boldsymbol{\mu}$ is *any* invariant measure, then all the rows of $\mathbf{P}^n$ converge to $\boldsymbol{\mu}$ as $n \to \infty$. That is, for each $i, j \in S$, $\mathbf{P}^n_{ij} \to \mu_j$ as $n \to \infty$. Since $\mathbf{P}^n_{ij}$ can have at most one such limit as $n \to \infty$, $\boldsymbol{\mu}$ must be unique.

Let X_n and Y_n be two independent Markov chains on S, each with probability transition matrix $\mathbf{P}$, where we start the two processes with $X_0 = i$ (any fixed deterministic state) and $Y_0 \sim \boldsymbol{\mu}$. Let

$$T \;=\; \min\{n > 0 : X_n = Y_n = 1\},$$

so T is the first time at which both the X and the Y processes are in state 1. This will eventually happen, in fact we presently show that

$$P[T > n] \;\leq\; ce^{-\alpha n} \tag{17.19}$$

for some positive numbers c and α. Since $\mathbf{P}$ is ergodic, for some number m we have that all the entries of $\mathbf{P}^m$ are positive. Let $p = \min\{\mathbf{P}^m_{ab} : a, b \in S\}$, so $p > 0$ and, in particular, $\mathbf{P}^m_{a1} \geq p$ for every state a. This means that whatever state the Markov chain may be in, after m additional steps it will be in state 1 with probability at least p. Since X and Y evolve independently we have, for instance, that

$$P[X_m = 1, Y_m = 1] \;=\; P[X_m = 1] \cdot P[Y_m = 1] \;\geq\; p^2.$$

If, after m steps, X is in state a and Y is in state b (so $H = \{X_m = a, Y_m = b\}$ has occurred), then we also have

$$P[X_{2m} = 1, Y_{2m} = 1 \,|\, H] \;\geq\; p^2.$$

In general, for any $k \geq 1$ and any realization H of the two Markov chains up to period $(k-1)m$, after m additional steps we have

$$P[X_{km} = 1, Y_{km} = 1 \,|\, H] \;\geq\; p^2.$$

That is, after every block of m steps the X and Y processes are both in state 1 with probability at least p^2 regardless of previous history. It follows from this that

$$P[T > km] \;\leq\; (1 - p^2)^k.$$

We obtain (17.19) from this just as in Theorem 17.1 (where T has a different meaning).

Now X_n and Y_n are independent Markov chains both evolving according to the probability transition matrix $\mathbf{P}$. They have different distributions, however, as X_n starts off in deterministic state i ($X_0 = i$) while Y_0 is randomly distributed according to $\boldsymbol{\mu}$. We now define a third Markov chain, putting

$$Z_n = \begin{cases} Y_n & \text{if } n < T \\ X_n & \text{if } n \geq T, \end{cases} \tag{17.20}$$

so Z_n follows the Y_n process until both X_n and Y_n are in state 1 at which time it "jumps over to" the X_n process. (See Figure 17.21.)

Figure 17.21. The Z_n process (indicated by the •'s).

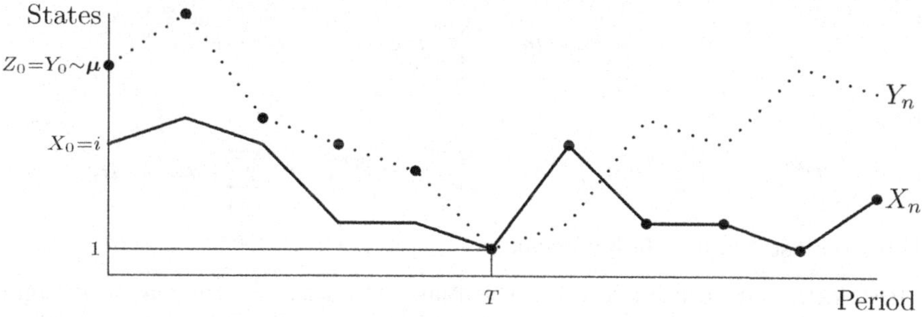

Then the Z process: (i) has $Z_0 = Y_0 \sim \boldsymbol{\mu}$; and (ii) has PTM $\mathbf{P}$. Statement (ii) holds because at any given time the Z process is either following X or Y, both of which have PTM $\mathbf{P}$. Since $\boldsymbol{\mu}$ is $\mathbf{P}$-invariant, we get that each $Z_n \sim \boldsymbol{\mu}$. For each state j we therefore have

$$P[X_n = j] = \mathbf{P}_{ij}^n \text{ and } P[Z_n = j] = \mu_j, \tag{17.22}$$

where the first equality uses that $X_0 = i$. Now

$$\{X_n = j\} \subset \{Z_n = j\} \cup \{X_n \neq Z_n\}$$

(if $X_n = j$ then either Z_n is also j or $Z_n \neq X_n$), and hence

$$P[X_n = j] \leq P[Z_n = j] + P[X_n \neq Z_n].$$

In exactly the same fashion we obtain

$$P[Z_n = j] \leq P[X_n = j] + P[X_n \neq Z_n],$$

which yields $|P[X_n = j] - P[Z_n = j]| \leq P[X_n \neq Z_n]$ or, from (17.22), that

$$|\mathbf{P}_{ij}^n - \mu_j| \leq P[X_n \neq Z_n].$$

Now $X_n \neq Z_n \implies n < T$ — refer back to (17.20), so

$$|\mathbf{P}^n_{ij} - \mu_j| \;\le\; P[X_n \neq Z_n] \;\le\; P[T > n] \;\le\; ce^{-\alpha n}.$$

Since this holds for each state j, this means that the ith row of $\mathbf{P}^n$ converges to the vector $\boldsymbol{\mu}$ as $n \to \infty$, and exponentially fast.

Next, we establish that this implies that $\boldsymbol{\nu}\mathbf{P}^n \to \boldsymbol{\mu}$ for any measure $\boldsymbol{\nu}$ on S. For each $i \in S$, let

$$\mathbf{e}_i \;=\; (0\,0\;\ldots\;0\,1\,0\,0\;\ldots\;0),$$

where the 1 is in the ith position. Then $\mathbf{e}_i\mathbf{P}^n$ is the ith row of $\mathbf{P}^n$, so $\mathbf{e}_i\mathbf{P}^n \to \boldsymbol{\mu}$ as $n \to \infty$. If $\boldsymbol{\nu} = (\nu_1, \ldots, \nu_s)$, then we may express $\boldsymbol{\nu}$ as a linear combination of the $\mathbf{e}_i$ as follows:

$$\boldsymbol{\nu} \;=\; \sum_{i=1}^{s} \nu_i \mathbf{e}_i.$$

Then

$$\boldsymbol{\nu}\mathbf{P}^n \;=\; \left(\sum_{i=1}^{s} \nu_i \mathbf{e}_i\right)\mathbf{P}^n \;=\; \sum_{i=1}^{s} \nu_i(\mathbf{e}_i\mathbf{P}^n) \;\to\; \sum_{i=1}^{s} \nu_i \boldsymbol{\mu} \;=\; \boldsymbol{\mu},$$

where the last equality holds because $\sum_i \nu_i = 1$. □

REMARK. This establishes that the analog of Fact 2 (for the two-state ergodic Markov chain of §15) holds for general Markov chains with s states. We will establish Fact 1's analog in §19.

EXERCISES

1. For general PTM $\mathbf{P}$ on $S = \{1, 2\}$ show that if all the entries of $\mathbf{P}^{n^*}$ are positive for some $n^* \ge 1$ then all the entries of $\mathbf{P}^2$ are positive. (The converse is trivial.) Hints: (i) Use that $\mathbf{P}^n_{ij}$ is the probability of going from state i to state j in n steps to show that $\mathbf{P}_{12} > 0$ and $\mathbf{P}_{21} > 0$; (ii) show that $\mathbf{P} = \begin{pmatrix} 0 & 1 \\ 1 & 0 \end{pmatrix}$ is not possible; (iii) refer to problem 5, §15.

2. Conclude from problem 1 that the definition of ergodicity presented in this section is consistent with the definition presented for the two-state system in §15.

3. Draw transition diagrams for the Markov chains of problems 1–6 in §16.

4. Show that these Markov chains are irreducible by displaying a cycle through all the states.

5. By solving the system of equations $\boldsymbol{\mu}\mathbf{P} = \boldsymbol{\mu}$ together with $\sum_i \mu_i = 1$, compute the invariant distribution for each of these Markov chains.

6. Which of these Markov chains are ergodic?

7. Referring to the Markov chain of problem 5, §16, use the methodology of Example 17.5 above to compute the expected number of steps required for the ant to return to vertex 1 if it starts at vertex 1. Square this with the invariant distribution you obtained in problem 5 above.

8. Prove Corollary 17.3.

9. Prove that if $\mathbf{P}_{ii} > 0$ for some state i then $\mathbf{P}^k_{ii} > 0$ for all $k \geq 1$.

10. Consider the Markov chain $(R_n : n \geq 0)$ of problem 7, §16. For what values of p $(0 \leq p \leq 1)$ is this Markov chain irreducible?

11. For those p for which R_n is irreducible, what is the invariant distribution?

12.* A Markov chain on the *infinite* state space $S = \{0, 1, 2, 3, \ldots\}$ works as follows: $\mathbf{P}_{01} = 1$ and, for $n > 0$, $\mathbf{P}_{n,n+1} = \frac{1}{n+1}$ and $\mathbf{P}_{n,n-1} = \frac{n}{n+1}$. Find the invariant distribution for this Markov chain.

§18. Example: An Interacting Particle System

We did a lot of work in the last section — here we have some fun. The following beautiful example is called an *interacting particle system*. We start with a very simple case to illustrate the idea. Consider a grid with three cells and two indistinguishable particles. Each cell can be occupied by only one particle – which we'll call the 'single-occupancy rule'. A state (we'll also call it a *configuration*) consists of any way to allocate the two particles to the three cells. There are only three such configurations as shown in Figure 18.1, where they are labeled x, y, and z.

Figure 18.1. A three-state interacting particle system.

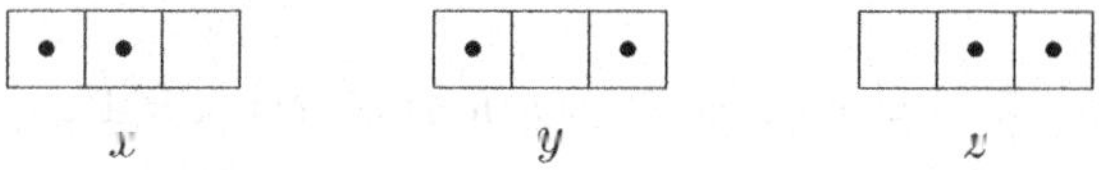

The transition rule is as follows. At each step of the Markov chain, pick a particle at random. (Here 'at random' means each possibility with equal likelihood and independently of any other choices.) Then pick a direction (east or west) at random. If the chosen particle can be moved one cell in the chosen direction, move it. Otherwise the configuration remains unchanged until (possibly) the next step. Suppose the current configuration is x and the rightmost particle is selected. If the direction is 'east', the transition is to configuration y; if the direction is 'west', the configuration remains unchanged due to the single-occupancy rule. If the leftmost particle is selected, then either direction leaves the configuration unchanged. A move to the west puts it off of the grid (not permitted); a move to the east violates single-occupancy. This shows that $\mathbf{P}_{xy} = \frac{1}{2} \cdot \frac{1}{2} = 1/4$ — the first $\frac{1}{2}$ is for picking the correct particle, the second $\frac{1}{2}$ is for picking the correct direction. Clearly $\mathbf{P}_{xz} = 0$, as z is not obtainable from

x by moving one particle one cell. The reader should verify that the PTM for this Markov chain is given by

$$\mathbf{P} = \begin{pmatrix} \mathbf{P}_{xx} & \mathbf{P}_{xy} & \mathbf{P}_{xz} \\ \mathbf{P}_{yx} & \mathbf{P}_{yy} & \mathbf{P}_{yz} \\ \mathbf{P}_{zx} & \mathbf{P}_{zy} & \mathbf{P}_{zz} \end{pmatrix} = \begin{pmatrix} 3/4 & 1/4 & 0 \\ 1/4 & 1/2 & 1/4 \\ 0 & 1/4 & 3/4 \end{pmatrix}.$$

Ergodicity is easily established as $\mathbf{P}^2$ has all positive entries.

Note that this PTM is *symmetric* in the sense that $\mathbf{P}_{ij} = \mathbf{P}_{ji}$ for any two states i and j. Since the rows of a PTM sum to 1 (always), if the PTM is symmetric the columns do as well. This makes the unique invariant distribution particularly easy to compute: $\boldsymbol{\mu} = (\frac{1}{3}, \frac{1}{3}, \frac{1}{3})$. To verify that the second component of $\boldsymbol{\mu}\mathbf{P}$ is $\frac{1}{3}$, for example, we compute

$$(\boldsymbol{\mu}\mathbf{P})_2 = \frac{1}{3}\cdot\frac{1}{4} + \frac{1}{3}\cdot\frac{1}{2} + \frac{1}{3}\cdot\frac{1}{4} = \frac{1}{3}\cdot\left(\frac{1}{4} + \frac{1}{2} + \frac{1}{4}\right) = \frac{1}{3}. \checkmark$$

This holds in general. If the columns of the $s \times s$ PTM $\mathbf{P} = (\mathbf{P}_{ij})$ sum to one and $\boldsymbol{\mu} = (\frac{1}{s} \;\; \frac{1}{s} \;\; \cdots \;\; \frac{1}{s})$, then necessarily $\boldsymbol{\mu}\mathbf{P} = \boldsymbol{\mu}$ (see exercises).

We may generalize this interacting particle system to arbitrary dimension, grid, and number of particles. For example, consider a 4×4 2-dimensional grid with 4 particles. Again we impose a single-occupancy rule for the cells so there are $s = \binom{16}{4} = 1820$ possible particle configurations. We will consider any two configurations x and y to be *neighbors* if y can be obtained from x by moving one particle exactly one cell to the north, east, south, or west. Clearly this is a symmetric relation, as x can be obtained from y by reversing the move. If x and y are neighbors we write $x \sim y$. Two such neighboring configurations are shown in Figure 18.2 below.

Figure 18.2. Two neighboring configurations on the 4×4 2-dimensional grid.

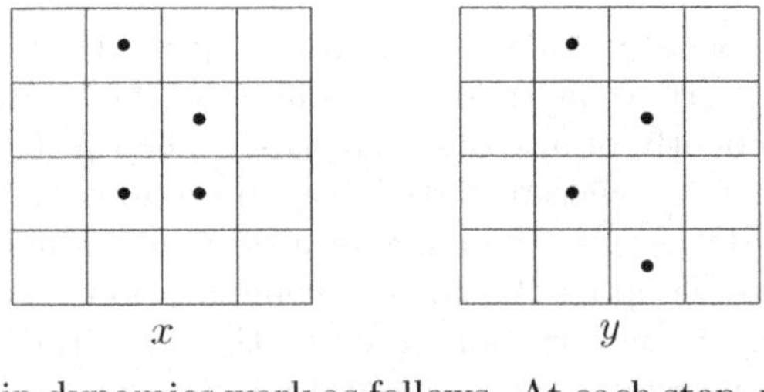

The Markov chain dynamics work as follows. At each step, pick one of the four particles randomly and one of the four directions (north, east, south, or west) randomly. If it is legitimate, move the chosen particle one cell in the chosen direction. If this move is not legitimate, leave the configuration unchanged. As before, a move is not legitimate if it would move the particle off of the grid or to an already occupied cell.

It is easy to see that this Markov chain satisfies the criteria for ergodicity of Theorem 17.17. Clearly we may proceed from any configuration to any other configuration by a sequence of single-particle moves, establishing irreducibility. (Although rigorously spelling out an algorithm to do this is harder than you might think.) Also, any configuration z with two adjacent particles has $\mathbf{P}_{zz} > 0$. For example, in Figure 18.2 we have $\mathbf{P}_{xx} > 0$ for this reason. (We also have $\mathbf{P}_{yy} > 0$. Why? — it has no adjacent particles.)

The PTM $\mathbf{P}$ for this process is a 1820×1820 matrix. Clearly we cannot write it down, let alone solve the equations $\boldsymbol{\mu}\mathbf{P} = \boldsymbol{\mu}$ to determine the invariant distribution! However we can deduce its invariant distribution. If $y \neq x$ and y is not a neighbor of x, we must have $\mathbf{P}_{xy} = 0$ by virtue of the system's dynamics. On the other hand, if $x \sim y$, to transition from x to y we must: (i) select the correct particle to move; and (ii) select the correct direction to move it. Hence $\mathbf{P}_{xy} = \frac{1}{4} \cdot \frac{1}{4} = \frac{1}{16}$ in this case. Summarizing we have:

$$\text{for } y \neq x, \quad \mathbf{P}_{xy} \;=\; \begin{cases} \frac{1}{16} & \text{if } x \sim y \\ 0 & \text{otherwise.} \end{cases} \tag{18.3}$$

Of course $\mathbf{P}_{xx}$ must get the balance of the probability:

$$\mathbf{P}_{xx} \;=\; 1 - \sum_{y:y\neq x} \mathbf{P}_{xy}, \tag{18.4}$$

since the rows of the PTM sum to 1. Equations (18.3) and (18.4) give formulas for the entries of the PTM $\mathbf{P}$ — equation (18.4) is for the diagonal entries of $\mathbf{P}$ and equation (18.3) is for the off-diagonal entries. The crucial point here is that $\mathbf{P}$ is a symmetric matrix, just as in the 1-dimensional 2-particle model. To show this we must show that if $x \neq y$ then $\mathbf{P}_{xy} = \mathbf{P}_{yx}$. This works because the neighbor relation $x \sim y$ is symmetric. If $x \sim y$ then $y \sim x$ and $\mathbf{P}_{xy} = \mathbf{P}_{yx} = \frac{1}{16}$. On the other hand, if $x \not\sim y$ then $y \not\sim x$ and $\mathbf{P}_{xy} = \mathbf{P}_{yx} = 0$. Since the PTM is symmetric, we know immediately that its invariant distribution is uniformly distributed on all 1820 configurations.

A Larger Version of the Same Model. Now for the fun part. Here we consider a 200×200 2-dimensional grid with 2500 particles. Originally they are confined to a 50×50 'container' as shown in the first frame of Figure 18.5. (The actual grid in Figure 18.5 cannot be shown as the entire picture would be black.) Then the top of the container is removed, but the sides are left in place. This is accomplished by imposing two vertical columns (1×50 each) of permanently occupied grid cells — these are discernible in the final configuration. The particles in these cells are not among the 2500 and are not subject to being moved. The Markov chain dynamics are exactly the same. At each step pick one of the 2500 movable particles and a direction at random. If the chosen particle can legitimately be moved one cell in the chosen direction, do so. Otherwise leave the configuration unchanged. (Here moving a particle to one of the 100 permanently occupied cells is illegitimate — so particles cannot pass through the walls of the container.)

Here the number of configurations is $\binom{40000-100}{2500} \approx 3.07 \times 10^{4056}$, which is a truly super-astronomical number. Furthermore the PTM $\mathbf{P}$ is a $(3.07 \times 10^{4056}) \times (3.07 \times 10^{4056})$ matrix! Yet we know from the reasoning above (which applies here as well) that $\mathbf{P}$ is a symmetric matrix. Thus the invariant distribution is uniformly distributed on all 3.07×10^{4056} states. Stated differently, in the long run the particles are uniformly distributed on the grid. A simulated realization of this Markov chain, shown at various stages in Figure 18.5, beautifully illustrates the convergence to the invariant distribution.

Figure 18.5. Simulated gas escaping from a container in a vacuum.

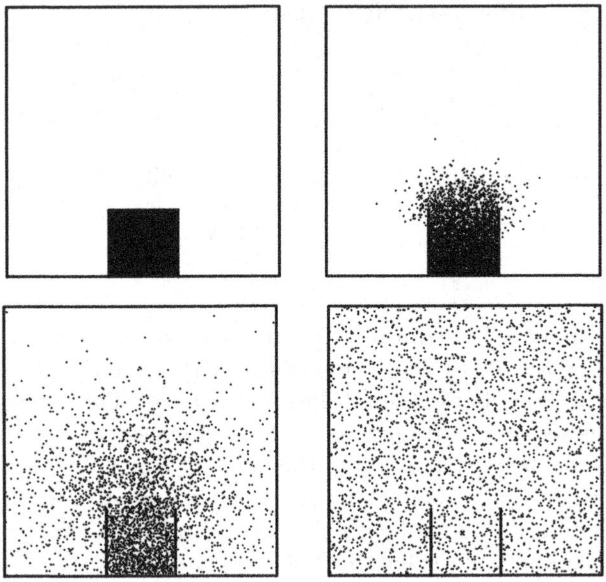

EXERCISES

1. Show that if the columns of the $s \times s$ PTM $\mathbf{P}$ sum to one and $\boldsymbol{\mu} = \left(\frac{1}{s} \ \ \frac{1}{s} \ \ \cdots \ \ \frac{1}{s}\right)$, then $\boldsymbol{\mu}\mathbf{P} = \boldsymbol{\mu}$. (A PTM where the columns also sum to one is called *doubly stochastic.*

2. Use the result in problem 1 above to compute the invariant distribution for R_n of problem 7, §16. Compare with your answer to problem 11, §17.

3. Jim does a 'random walk' on the face of a clock (a real clock, not a digital clock). Starting at 12:00, he repeatedly steps clockwise to the next hour with probability $\frac{2}{3}$ and counterclockwise to the previous hour with probability $\frac{1}{3}$.

Show that the invariant distribution is uniformly distributed on the twelve hours of the clock.

§19. A Strong Law for Ergodic Markov Chains

In this section we prove a strong law of large numbers for ergodic Markov chains. Suppose the the Markov chain $(X_0, X_1, X_2, \dots)$ is ergodic on the finite state space $\{1, 2, \dots, s\}$. We know from Theorem 17.18 that the Markov chain has a unique invariant distribution $\boldsymbol{\mu}$ and that, for each state i, $P[X_k = i] \to \mu_i$ as $k \to \infty$. This holds regardless of the distribution of X_0 (of course if $X_0 \sim \boldsymbol{\mu}$, then $P[X_k = i] = \mu_i$ for all k).

Let $N_n(i)$ denote the number of times the Markov chain (X_k) is in state i for $1 \le k \le n$. More formally, with

$$I_{\{X_k=i\}} = \begin{cases} 1 & \text{if the chain is in state } i \text{ in period } k \\ 0 & \text{otherwise,} \end{cases}$$

we have that $N_n(i) = \sum_{k=1}^n I_{\{X_k=i\}}$. Then we have:

THEOREM 19.1. *With the above setup, for every state i, $\frac{N_n(i)}{n} \to \mu_i$ almost surely as $n \to \infty$. This holds regardless of the distribution of X_0.*

REMARK. $\frac{N_n(i)}{n}$ is the fraction of the time that the Markov chain has spent in state i during the first n steps. Theorem 19.1 says that this converges with probability one to state i's weight in the invariant distribution. This is the analog of the Fact 1 statement for the two-state Markov chain of §15.

Before proving this theorem, we develop an easy corollary. Let $f : S \to \mathbb{R}$ be any real-valued function defined on the state space and put

$$\bar{f} = \mu_1 f(1) + \cdots + \mu_s f(s).$$

In words, $\bar{f}$ is the average value of $f(x)$ on the state space with weights given by the invariant distribution.

COROLLARY 19.2. *With the above setup, $\frac{1}{n}\sum_{k=1}^n f(X_k) \to \bar{f}$ almost surely as $n \to \infty$.*

PROOF. First note that

$$\sum_{k=1}^n f(X_k) = N_n(1)f(1) + \cdots + N_n(s)f(s), \tag{19.3}$$

and so

$$\frac{1}{n}\sum_{k=1}^n f(X_k) = \frac{N_n(1)}{n} f(1) + \cdots + \frac{N_n(s)}{n} f(s).$$

Equality (19.3) holds because exactly $N_n(1)$ of the summands on the left have their $X_k = 1$, with a similar statement holding for each of the s states. Theorem 19.1 tells us that each $\frac{N_n(i)}{n} \to \mu_i$ with probability one as $n \to \infty$. The corollary follows from this. $\square$

PROOF OF THEOREM 19.1. Without any loss of generality, we prove this for state $i = 1$. Referring back to Figure 17.7, let T_m denote the number of periods in the mth excursion of the chain away from state 1. (Figure 17.7 shows the chain beginning in state $X_0 = 1$, but this need not be true here. Also, T_1 here is called T in that figure.) Then the mth visit to state 1 (after period 0) occurs at period $T_1 + \cdots + T_m$. Now $N_n(1) = \max\{m : T_1 + \cdots + T_m \le n\}$, so

$$T_1 + \cdots + T_{N_n(1)} \le n < T_1 + \cdots + T_{N_n(1)+1},$$

and

$$\frac{T_1 + \cdots + T_{N_n(1)}}{N_n(1)} \le \frac{n}{N_n(1)} < \frac{T_1 + \cdots + T_{N_n(1)+1}}{N_n(1)+1} \cdot \frac{N_n(1)+1}{N_n(1)}. \tag{19.4}$$

By the Markov property, $T_1, T_2, T_3, \ldots$ are independent; in fact, $T_2, T_3, \ldots$ are iid. (Note that T_1 may be drawn from a different distribution if we do not have $P[X_0 = 1] = 1$). Referring back to the defining properties of the invariant distribution, recall that

$$\mu_1 = \frac{1}{E[T_1 \mid X_0 = 1]}.$$

Since each excursion except possibly the first starts at state 1, we see that, for $m > 1$, $ET_m = E[T_1 \mid X_0 = 1] = \frac{1}{\mu_1}$. From Theorem 17.1 we know that the T_i have finite exponential moment (see exercises below) and so the Strong Law of Large Numbers applies to the sequence $T_2, T_3, \ldots$. Therefore

$$\begin{aligned} \frac{T_1 + \cdots + T_m}{m} &= \frac{T_1}{m} + \frac{T_2 + \cdots + T_m}{m-1} \cdot \frac{m-1}{m} \\ & \qquad \downarrow \qquad\qquad \downarrow \qquad\qquad \downarrow \\ & \qquad 0 \quad + \quad \frac{1}{\mu_1} \quad \cdot \quad 1 \\ &= \frac{1}{\mu_1} \quad \text{as } m \to \infty. \end{aligned} \tag{19.5}$$

Looking at (19.4), we would like to conclude that

$$\frac{T_1 + \cdots + T_{N_n(1)}}{N_n(1)} \to \frac{1}{\mu_1} \quad \text{as } n \to \infty. \tag{19.6}$$

The problem here is that $N_n(1)$ is *random*, so (19.5) does not directly apply. The following lemma comes to the rescue:

LEMMA 19.7. *Suppose* $(Z_n : n \geq 1)$ *is a sequence of random variables that converges almost surely to the limit* c *as* $n \to \infty$. *Suppose further that* $(N_n : n \geq 1)$ *is a sequence of positive integer-valued random variables with* $P[N_n \to \infty] = 1$. *Then* $Z_{N_n} \to c$ *almost surely as* $n \to \infty$.

REMARK. It is crucial to understand what Z_{N_n} means. Suppose, for some outcome ω, we have $N_n(\omega) = 23$ and $Z_{23}(\omega) = 6$. Then $Z_{N_n}(\omega) = Z_{N_n(\omega)}(\omega) = Z_{23}(\omega) = 6$.

PROOF OF LEMMA. The lemma follows from the fact that

$$\{\omega : N_n(\omega) \to \infty\} \cap \{\omega : Z_n(\omega) \to c\} \subset \{\omega : Z_{N_n}(\omega) \to c\}, \tag{19.8}$$

where in each of these three events convergence is as $n \to \infty$. Since the two events on the left have probability one by hypotheses, so does the event on the right. To see (19.8) assume an outcome ω^* belongs to the left hand side: we must show that ω^* also belongs to the right hand side. Choose any $\epsilon > 0$. Since $\omega^* \in \{\omega : Z_n(\omega) \to c\}$, there is a number m_0 such that $|Z_m(\omega^*) - c| < \epsilon$ provided $m > m_0$. Since $\omega^* \in \{\omega : N_n(\omega) \to \infty\}$, there is a number n_0 such that $N_n(\omega^*) > m_0$ provided $n > n_0$. But then

$$n > n_0 \implies N_n(\omega^*) > m_0 \implies |Z_{N_n(\omega^*)}(\omega^*) - c| < \epsilon.$$

Since this holds for arbitrary $\epsilon > 0$, we see that $Z_{N_n}(\omega^*) \to c$ so ω^* indeed belongs to the right-hand-side of (19.8). □

Returning to the proof of Theorem 19.1, we apply Lemma 19.7 to $Z_n = \frac{T_1 + \cdots + T_n}{n}$. We see that (19.6) will hold provided $P[N_n(1) \to \infty] = 1$. But this latter condition must hold since the Markov chain visits state 1 infinitely often. Looking back at (19.4), we see that the expressions on both the left and right side converge to $\frac{1}{\mu_1}$ as $n \to \infty$. (Here we use that $\frac{N_n(1)+1}{N_n(1)} \to 1$, which holds since $N_n(1) \to \infty$.) It follows that $\frac{n}{N_n(1)} \to \frac{1}{\mu_1}$, and so $\frac{N_n(1)}{n} \to \mu_1$. □

EXERCISES

1. In the proof of Theorem 19.1, show that Theorem 17.1 implies that the T_i have finite exponential moment.

2. I walk back and forth between home and work. I have two umbrellas, which I keep at either location. If it is sunny when I do the walk, I leave the umbrellas where they are. If it is rainy: if there is an umbrella available I take it and don't get wet; if there is no umbrella available I get wet. If it rains with probability 1/4 (independently with each walk), in the long run what fraction of the time do I get wet?

§20. Communication Classes and Their Properties

Let i and j be any two states in S. If $i \to j$ and $j \to i$, then states i and j are said to *communicate*; we denote this by $i \leftrightarrow j$. Communication of states has the following properties: (i) for any state i, $i \leftrightarrow i$ (communication is *reflexive*); (ii) if $i \leftrightarrow j$ then $j \leftrightarrow i$ (communication is *symmetric*); and (iii) if $i \leftrightarrow j$ and $j \leftrightarrow k$ then $i \leftrightarrow k$ (communication is *transitive*). Communication is reflexive because $\mathbf{P}^0_{ii} = P[X_0 = i | X_0 = i] = 1$. It is symmetric because the definition of communication is symmetric in i and j. Now suppose $i \leftrightarrow j$ and $j \leftrightarrow k$. Then $\mathbf{P}^n_{ij} > 0$ and $\mathbf{P}^m_{jk} > 0$ for some n and m, so

$$\mathbf{P}^{n+m}_{ik} \geq \mathbf{P}^n_{ij}\mathbf{P}^m_{jk} > 0.$$

This shows that $i \to k$. Similarly $k \to i$, so $i \leftrightarrow k$ establishing transitivity.

Any relation that is reflexive, symmetric, and transitive is called an *equivalence relation*. The equivalence relation of communication partitions the Markov chain states into *equivalence classes* which, in this setting, are called *communication classes*. Specifically, for any state $i \in S$, let

$$C(i) = \{j \in S : i \leftrightarrow j\};$$

$C(i)$ is called the *communication class* or, simply, the *class* of state i. Notice that automatically $i \in C(i)$, because communication is reflexive. Furthermore, any two classes are either disjoint or identical. To see this, suppose that $C(i)$ and $C(j)$ are not disjoint. We seek to show that $C(i) = C(j)$. Since they are not disjoint, there is some state k with $k \in C(i)$ and $k \in C(j)$ and we have $i \leftrightarrow k$ and $j \leftrightarrow k$. But then $k \leftrightarrow j$ (the relation is symmetric) and $i \leftrightarrow j$ (the relation is transitive). Now suppose $h \in C(j)$. Then $j \leftrightarrow h$ so also $i \leftrightarrow h$ (transitivity again) and $h \in C(i)$. Thus $C(j) \subset C(i)$. Similarly $C(i) \subset C(j)$ proving that $C(i) = C(j)$. In this terminology, a Markov chain is irreducible if it has only one communication class, i.e., if $C(i) = C(j)$ for all states i and j.

EXAMPLE 20.1. The Markov chain in Example 15.1 is irreducible: $C(1) = C(2) = \{1, 2\} = S$.

EXAMPLE 20.2. Suppose a Markov chain on states $S = \{1, 2, 3, 4\}$ has probability transition matrix:

$$\mathbf{P} = \begin{pmatrix} .5 & .5 & 0 & 0 \\ .5 & .5 & 0 & 0 \\ .25 & .25 & .25 & .25 \\ 0 & 0 & 0 & 1 \end{pmatrix}.$$

Then $1 \to 2$ and $2 \to 1$ so states 1 and 2 are in the same communication class. Now $3 \to 1$ and $3 \to 2$, but state 3 is inaccessible from either state 1 or 2. Similarly $3 \to 4$ but state 3 is inaccessible from state 4. The communication classes are therefore $\{1,2\}$, $\{3\}$, and $\{4\}$. Specifically, $C(1) = C(2) = \{1,2\}$, $C(3) = \{3\}$, and $C(4) = \{4\}$. Note that once the chain enters state 4, it remains there forever. Such a state (where $P_{ii} = 1$) is called *absorbing*, e.g., here 4 is an absorbing state.

Recurrence/Transience. A Markov chain state is called *recurrent* if, upon entering that state, with probability 1 the system will eventually return to that state. More precisely, state i is recurrent if

$$\rho \;=\; P[X_n = i \text{ for some } n > 0 \,|\, X_0 = i] \;=\; 1. \tag{20.3}$$

Note that in (20.3) we can replace "0" (both occurrences) with an arbitrary period "k" without changing the value of ρ. If $\rho < 1$, the state is called *transient*.

EXAMPLE 20.4. In Example 20.2, states 1 and 2 are recurrent, state 3 is transient (why?), and state 4 is recurrent (an absorbing state is obviously recurrent).

Let N denote the number of times over the long run that the system enters state i: $N = \#\{n \geq 0 : X_n = i\}$. If state i is recurrent and $X_0 = i$ then necessarily $N = \infty$, for each time the system enters state i it is destined to return to it at some future time with probability one. Hence, if i is recurrent $E[N|X_0 = i] = \infty$. If state i is transient then $\rho < 1$ and, conditioned on $X_0 = i$, N is a geometric random variable with parameter $1 - \rho$. (A geometric random variable with parameter p is the number of trials of an experiment it takes to obtain a success when each trial is successful independently with probability p. Think of "not returning to state i" as a success.) It follows that

$$E[N|X_0 = i] \;=\; \frac{1}{1-\rho} \;<\; \infty.$$

To summarize, state i is recurrent if and only if $E[N \,|\, X_0 = i] = \infty$.

Next we seek to relate this to the Markov chain's probability transition matrix. Let

$$I_n \;=\; \begin{cases} 1 & \text{if } X_n = i \\ 0 & \text{otherwise,} \end{cases}$$

so $N = \sum_{n=0}^{\infty} I_n$ and $E[I_n \,|\, X_0 = i] = P[X_n = i \,|\, X_0 = i] = \mathbf{P}^n_{ii}$. Therefore

$$\begin{aligned} E[N|X_0 = i] \;&=\; E\left[\sum_{n=0}^{\infty} I_n \,\middle|\, X_0 = i\right] \\ &=\; \sum_{n=0}^{\infty} E[I_n \,|\, X_0 = i] \\ &=\; \sum_{n=0}^{\infty} \mathbf{P}^n_{ii}, \end{aligned}$$

where the Monotone Convergence Theorem justifies the second equality above. We therefore have the following:

THEOREM 20.5. *State i of a Markov chain with PTM $\mathbf{P}$ is recurrent if and only if $\sum_{n=0}^{\infty} \mathbf{P}_{ii}^n = \infty$.*

A consequence of this is:

COROLLARY 20.6. *If state i is recurrent and $i \leftrightarrow j$, then state j is recurrent, so recurrence is a property inherent to entire communication classes.*

PROOF OF COROLLARY. To prove the corollary, pick q and r so that $\mathbf{P}_{ij}^q$ and $\mathbf{P}_{ji}^r$ are greater than zero. This is possible as $i \leftrightarrow j$. Then $\mathbf{P}_{jj}^{r+n+q} \geq \mathbf{P}_{ji}^r \mathbf{P}_{ii}^n \mathbf{P}_{ij}^q$, so

$$\begin{aligned} \sum_{n=0}^{\infty} \mathbf{P}_{jj}^n \geq \sum_{n=r+q}^{\infty} \mathbf{P}_{jj}^n &= \sum_{n=0}^{\infty} \mathbf{P}_{jj}^{r+n+q} \\ &\geq \sum_{n=0}^{\infty} \mathbf{P}_{ji}^r \mathbf{P}_{ii}^n \mathbf{P}_{ij}^q = \mathbf{P}_{ji}^r \mathbf{P}_{ij}^q \sum_{n=0}^{\infty} \mathbf{P}_{ii}^n = \infty. \end{aligned}$$

The first inequality holds since $r+q$ nonnegative terms are dropped going from the left to the right side of the inequality. The final equality holds since $\mathbf{P}_{ji}^r \mathbf{P}_{ij}^q > 0$ and $\sum_{n=0}^{\infty} \mathbf{P}_{ii}^n = \infty$ (state i is recurrent). Hence $\sum_{n=0}^{\infty} \mathbf{P}_{jj}^n = \infty$ and the corollary follows from Theorem 20.5. □

Another way to see this is the following less formal reasoning. If the system is ever in state j, then with probability 1 it will eventually visit state i. (If not, then starting from state i it would be possible to go to state j (as $i \leftrightarrow j$) and then never return to state i. This contradicts the recurrence of state i.) Also, starting from state i, with positive probability the system will visit state j before returning to state i. (Otherwise we would not have $i \rightarrow j$.) Putting this together we have that starting from state j the system will eventually visit state i. Since i is recurrent it will visit i infinitely often. With each new visit to i, with positive probability (independently of history) it will visit state j before the next return to i. It follows that, starting from state j, with probability one the system will eventually revisit state j.

Periodicity. Another important idea is that of a recurrent state's *period.* (There is potential for confusion here, because we also use the word "period" as in "time period". These are two distinct usages which must be distinguished by context.) For any recurrent state i, let

$$I_i = \{n \geq 0 : \mathbf{P}_{ii}^n > 0\}.$$

Then the period of state i, d_i, is the greatest common divisor of all the elements of I_i:

$$d_i = \gcd(I_i) = \gcd\{n \geq 0 : \mathbf{P}_{ii}^n > 0\}. \tag{20.7}$$

If $d_i = 1$, state i is said to be *aperiodic.*

Since d_i divides every element of I_i, if n is not a multiple of d_i, then necessarily $n \notin I_i$ and $\mathbf{P}_{ii}^n = 0$. The converse is not true. However, we do have:

THEOREM 20.8. *For recurrent state i, all sufficiently large multiples of d_i are in I_i. That is, there is a number m_0 such that $md_i \in I_i$ whenever $m \geq m_0$.*

EXAMPLE 20.9. Suppose a Markov chain on states $\{1, 2, 3, 4\}$ has PTM

$$\mathbf{P} = \begin{pmatrix} 0 & .5 & 0 & .5 \\ 0 & 0 & 1 & 0 \\ 0 & 0 & 0 & 1 \\ 1 & 0 & 0 & 0 \end{pmatrix},$$

as diagrammed in Figure 20.10. First note that, since we can cycle through the states ($1 \to 2 \to 3 \to 4 \to 1$, e.g.), $\mathbf{P}$ is irreducible. Additionally, we can see from the diagram that, with each new time period, the parity (odd vs. even) of the state number changes. It follows that each d_i is a multiple of 2. Clearly, $I_1 = \{0, 2, 4, \dots\}$, since the sequence of states $1, 4, 1, 4, 1, \dots$ is possible. Hence the period of state 1 is 2. For the same reason, we have $I_4 = \{0, 2, 4, \dots\}$ and $d_4 = 2$. I_2 is more interesting. It takes a minimum of 4 steps to return to state 2, i.e., via the sequence $2, 3, 4, 1, 2$. The chain can also go from state 2 to state 2 in six steps: $2, 3, 4, 1, 4, 1, 2$. By pasting these together appropriately, we see that $\mathbf{P}^n_{22} > 0$ for any even number exceeding 2. Hence $I_2 = \{0, 4, 6, 8, \dots\}$. Again, $d_2 = \gcd(I_2) = 2$; but here $2 \notin I_2$. Regarding Theorem 20.8, this example illustrates that it is possible for I_i *not* to contain all multiples of the period. What is I_3?

Figure 20.10. A Transition Diagram for $\mathbf{P}$.

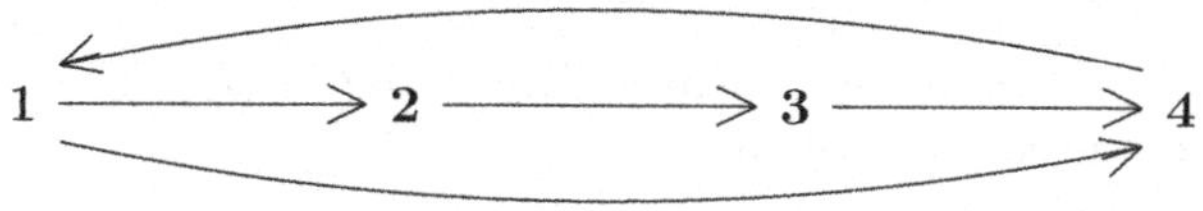

PROOF OF THEOREM 20.8. We prove this in four steps.

Step 1. First note that $0 \in I_i$ because $\mathbf{P}^0_{ii} = 1$. Now suppose $m, n \in I_i$ so $\mathbf{P}^m_{ii} > 0$ and $\mathbf{P}^n_{ii} > 0$. Then $\mathbf{P}^{m+n}_{ii} \geq \mathbf{P}^m_{ii}\mathbf{P}^n_{ii} > 0$, so also $m + n \in I_i$. It follows from repeated application of this that

$$am + bn \in I_i \text{ for all non-negative integers } a \text{ and } b. \tag{20.11}$$

For example, $2m + 3n \in I_i$ because: $2m = m + m$ so $2m \in I_i$; similarly $2n \in I_i$; then $3n = n + 2n$ so $3n \in I_i$; finally $2m + 3n \in I_1$. Fact (20.11), as it turns out, is all we use about I_i.

Step 2. In steps 2 and 3 we handle the case where $d_i = 1$ — step 4 deals with the general situation. Suppose, then, that $d_i = 1$. In this step we prove that for some number n, both n and $n + 1$ belong to I_i; that is, there are two *consecutive* integers in I_i. Suppose $n \in I_i$ and $n + k \in I_i$. (Since i is recurrent, I_i has infinitely many elements.) If $k = 1$, we are done. If $k > 1$, there is a number $m \in I_i$ that is not divisible by k (since $\gcd(I_i) = 1$). Then $m = qk + r$ for some quotient q and remainder r where

$$0 < r < k. \tag{20.12}$$

Since n, $n + k$, and m belong to I_i, step 1 implies that both $(q+1)n + m$ and $(q+1)(n+k)$ are in I_i. But the difference between these two numbers is

$$\begin{aligned} \tilde{k} &= (q+1)(n+k) - ((q+1)n + m) \\ &= (q+1)k - m \\ &= (q+1)k - (qk + r) \\ &= k - r, \end{aligned}$$

which, due to (20.12), satisfies $0 < \tilde{k} < k$. That is, we have found two distinct numbers in I_i that are *closer* than n and $n+k$. Repeat this process as necessary until the two numbers are consecutive.

Step 3. Suppose, using step 2, that n and $n+1$ are in I_i. In this step we show that $m \in I_i$ provided $m \geq n^2$. This will prove the theorem for the case $d_i = 1$ by taking $m_0 = n^2$. Suppose, then, that $m \geq n^2$. Then $m = nq + r$ for some quotient $q \geq n$ and remainder r with $0 \leq r < n$. Then $m = (q - r)n + r(n+1)$, where $q - r \geq n - r > 0$ and $r \geq 0$. It follows from step 1 that m is in I_i.

Step 4. In this final step we prove the theorem in the case $d_i = \gcd(I_i) > 1$. Let

$$\tilde{I}_i = \left\{ \frac{n}{d_i} : n \in I_i \right\}.$$

Then $\gcd(\tilde{I}_i) = 1$ and $m \in \tilde{I}_i$ if and only if $md_i \in I_i$. That $\tilde{I}_i$ satisfies the condition (20.11) is easily verified:

$$\begin{aligned} m, n \in \tilde{I}_i &\implies md_i, nd_i \in I_i \\ &\implies amd_i + bnd_i \in I_i \\ &\implies (am + bn)d_i \in I_i \\ &\implies am + bn \in \tilde{I}_i. \end{aligned}$$

Since $\gcd(\tilde{I}_i) = 1$ and $\tilde{I}_i$ satisfies (20.11) we get from steps 1–3 that, for some m_0, $m \in \tilde{I}_i$ provided $m \geq m_0$. But then $md_i \in I_i$ provided $m \geq m_0$. □

We also have the following, as illustrated by the earlier Example 20.9:

THEOREM 20.13. *If states i and j are in the same communication class, then they have the same period: $i \leftrightarrow j \implies d_i = d_j$.*

PROOF. Assume that $i \leftrightarrow j$ and choose r and q so that $\mathbf{P}^r_{ij} > 0$ and $\mathbf{P}^q_{ji} > 0$. Then $\mathbf{P}^{r+q}_{ii} \geq \mathbf{P}^r_{ij}\mathbf{P}^q_{ji} > 0$, so $r + q \in I_i$ and is therefore a multiple of d_i. Now suppose $n \in I_j$ so $\mathbf{P}^n_{jj} > 0$. Then $\mathbf{P}^{r+n+q}_{ii} \geq \mathbf{P}^r_{ij}\mathbf{P}^n_{jj}\mathbf{P}^q_{ji} > 0$, so $r+n+q \in I_i$ and is a multiple of d_i. Since $r+q$ and $r+n+q$ are both multiples of d_i, it follows that n is also a multiple of d_i. We have established that every element of I_j is divisible by d_i. Since d_j is the greatest common divisor of the elements of I_j it must be that $d_i \leq d_j$. Now the relation $i \leftrightarrow j$ is symmetric in i and j so we also have $d_j \leq d_i$, yielding $d_i = d_j$. □

Ergodicity Revisited. We conclude this section with another characterization of ergodicity.

THEOREM 20.14. *A Markov chain on $S = \{1, \dots, s\}$ is ergodic if and only if it is irreducible and aperiodic.*

PROOF. *Step 1:* $\Longrightarrow$. Suppose a Markov chain with PTM $\mathbf{P}$ on S is ergodic. We have already seen that ergodic Markov chains are irreducible. We must show that they are aperiodic. By definition, for some number n^* we have all $\mathbf{P}^{n^*}_{ij} > 0$. By Lemma 17.16, we also have all $\mathbf{P}^{n^*+1}_{ij} > 0$. In particular (take $j = i$), we have both $n^* \in I_i$ and $n^*+1 \in I_i$. Hence $d_i = \gcd I_i = 1$.

Step 2: $\Longleftarrow$. Now suppose the Markov chain is irreducible and aperiodic. We must show that all $\mathbf{P}^{n^*}_{ij} > 0$ for some number n^*. Since the Markov chain is irreducible, for any two states i and j there is a number n with $\mathbf{P}^n_{ij} > 0$. Let $n(i,j) = \min\{n : \mathbf{P}^n_{ij} > 0\}$. In words, $n(i,j)$ is the fewest number of steps it can take the Markov chain to go from state i to state j. Now let $n' = \max\{n(i,j) : i,j \in S\}$. Then for all states i and j, $n(i,j) \leq n'$ and $\mathbf{P}^{n(i,j)}_{ij} > 0$. Since each state j is aperiodic, by Theorem 20.8 there is a number n_j such that $n \geq n_j \implies n \in I_j$. Let $n'' = \max\{n_1, \dots, n_s\}$. Then for *any* state j, $n \geq n'' \implies n \geq n_j \implies n \in I_j$, i.e., $\mathbf{P}^n_{jj} > 0$.

Take $n^* = n' + n''$. For any states i and j we have

$$n^* - n(i,j) \;=\; n' + n'' - n(i,j) \;\geq\; n'',$$

so $\mathbf{P}^{n^*}_{ij} \geq \mathbf{P}^{n(i,j)}_{ij}\mathbf{P}^{n^*-n(i,j)}_{jj} > 0$. □

REMARKS. This theorem sheds new light on Theorem 17.17 (criteria for ergodicity). We see that the condition $\mathbf{P}_{i^*i^*} > 0$ implies that state i^*, and hence the entire irreducible Markov chain, has period 1. Theorem 20.14 above then gives ergodicity. Also, referring back to the prohibited PTM forms of (15.3) for the two-state Markov chain, we see that the first two forms have absorbing states and are therefore reducible while the third form has period 2.

EXERCISES

1. For probability transition matrices shown below, diagram the states and:
   - determine the communication classes;
   - determine which states are transient, recurrent, and absorbing;
   - compute $I_i = \{n \geq 0 : \mathbf{P}^n_{ii} > 0\}$;

- determine the period of each state;
- find the invariant distribution.

$$\begin{pmatrix} 2/3 & 1/3 \\ 0 & 1 \end{pmatrix}, \quad \begin{pmatrix} 1/3 & 2/3 \\ 2/3 & 1/3 \end{pmatrix}, \quad \begin{pmatrix} 1/2 & 1/2 & 0 \\ 0 & 1/2 & 1/2 \\ 0 & 1/2 & 1/2 \end{pmatrix},$$

$$\begin{pmatrix} 0 & 1/2 & 0 & 1/2 & 0 \\ 0 & 0 & 1 & 0 & 0 \\ 1/3 & 0 & 0 & 0 & 2/3 \\ 0 & 0 & 1 & 0 & 0 \\ 0 & 2/3 & 0 & 1/3 & 0 \end{pmatrix}.$$

2. Prove that once a Markov chain enters a recurrent communication class C, it never leaves that class. That is, $X_n \in C \implies X_m \in C$ for all $m \geq n$.

3. If a Markov chain on s states is irreducible, show that for any states i and j there is a walk from i to j of length less than or equal to s. (Hence $\mathbf{P}^n_{ij} > 0$ for some $n \leq s$.) If $i \neq j$ show that there is a walk from i to j of length strictly less than s.

4. Consider the Markov chain $(R_n : n \geq 0)$ of problem 7, §16. For what values of p $(0 \leq p \leq 1)$ is this Markov chain aperiodic? How about ergodic?

§21. When the Markov Chain is Periodic

In this section we continue to assume that the state space S is finite and that the Markov chain is irreducible but we relax the condition that it is aperiodic, which we have seen is required for ergodicity. Once again by Theorem 17.15, there exists at least one invariant measure on S. Letting d denote the period of each state in S (they all have the same period as S is irreducible), we relax the assumption that $d = 1$. We will show that this Markov chain also has a unique invariant measure and show how to compute it.

Recall that for any state j, $\mathbf{P}^k_{jj} > 0 \implies d|k$. Fix some state $i \in S$ and suppose that

$$\mathbf{P}^{n_1 d + r_1}_{1i} > 0 \quad \text{and} \quad \mathbf{P}^{n_2 d + r_2}_{1i} > 0$$

for non-negative integers n_1 and n_2 and remainders r_1, r_2 (we use the term "remainder" here to mean an integer r with $0 \leq r < d$). Here we show that necessarily $r_1 = r_2$. Since S is irreducible we know that $\mathbf{P}^k_{i1} > 0$ for some non-negative integer k. Then

$$\mathbf{P}^{n_1 d + r_1 + k}_{11} \geq \mathbf{P}^{n_1 d + r_1}_{1i} \cdot \mathbf{P}^k_{i1} > 0,$$

so $d \,|\, (n_1 d + r_1 + k)$. Similarly, $d \,|\, (n_2 d + r_2 + k)$. But then d divides their difference:

$$(n_1 d + r_1 + k) - (n_2 d + r_2 + k) = (n_1 - n_2)d + r_1 - r_2.$$

It follows that d divides $|r_1 - r_2|$ and, since $0 \le |r_1 - r_2| < d$, that $|r_1 - r_2| = 0$, i.e., that $r_1 = r_2$.

For each state $i \in S$, $\mathbf{P}^k_{1i} > 0$ for some non-negative integer k (in fact, for infinitely many k), again since S is irreducible. We may write $k = qd + r$ for some quotient q and remainder r. By the previous paragraph, the remainder r will not depend on our choice of k and we may unambiguously define $r(i)$ to be the unique such remainder associated with the state i. Then, for $0 \le r < d$, set

$$D(r) \;=\; \{i \in S : r(i) = r\},$$

so $D(r)$ consists of all the states whose associated remainder is r. For example, since $\mathbf{P}^0_{11} = 1 > 0$ we have $r(1) = 0$ and $1 \in D(0)$. Since every state has a unique remainder, the subsets $D(r)$, as r ranges from 0 to $d-1$, partition the state space S.

EXAMPLE 21.1. Suppose $S = \{1,2,3,4,5,6\}$ and

$$\mathbf{P} \;=\; \begin{pmatrix} 0 & .25 & .75 & 0 & 0 & 0 \\ 0 & 0 & 0 & .5 & .5 & 0 \\ 0 & 0 & 0 & .5 & .5 & 0 \\ 0 & 0 & 0 & 0 & 0 & 1 \\ 0 & 0 & 0 & 0 & 0 & 1 \\ 1 & 0 & 0 & 0 & 0 & 0 \end{pmatrix}.$$

Refer to Figure 21.2 for a diagram. Here $d = 4$ and $D(0) = \{1\}$, $D(1) = \{2,3\}$, $D(2) = \{4,5\}$, and $D(3) = \{6\}$.

Figure 21.2. Diagram of $\mathbf{P}$.

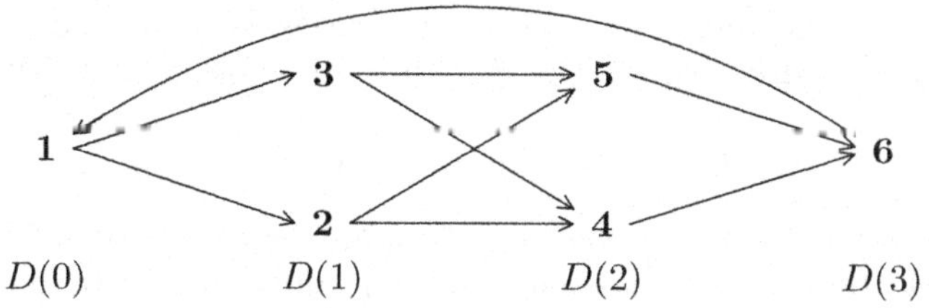

For any two remainders r and $\tilde{r}$, let the operation '$\oplus$' (addition modulo d) be defined as follows:

$$r \oplus \tilde{r} \;=\; \begin{cases} r + \tilde{r} & \text{if } r + \tilde{r} < d \\ r + \tilde{r} - d & \text{if } r + \tilde{r} \ge d, \end{cases}$$

so, for example, $r \oplus 1 = r + 1$ if $0 \le r < d - 1$ while $(d-1) \oplus 1 = 0$. The reader will verify that

$$\text{if } X_n \in D(r) \text{ then } X_{n+\tilde{r}} \in D(r \oplus \tilde{r}). \tag{21.3}$$

EXAMPLE 21.4. In Example 21.1 suppose $X_7 \in D(1)$, so here $n = 7$ and $r = 1$. Letting $\tilde{r} = 3$, We have $X_{n+\tilde{r}} = X_{10}$ and $r \oplus \tilde{r} = 1 + 3 - 4 = 0$. We conclude that $X_{10} \in D(0)$, i.e., $X_{10} = 1$.

Next we show that the probability transition matrix $\mathbf{Q} = \mathbf{P}^d$ defines a Markov chain on S where $D(0), D(1), \ldots, D(d-1)$ are each irreducible communication classes with period 1. If we condition the chain to start in state 1 ($X_0 = 1$) then, since state 1 is in $D(0)$, for $0 \leq r < d$ we must have $X_r \in D(r)$ showing that none of the $D(r)$'s are empty. Since $\mathbf{Q}$ is a stochastic matrix (the entries are non-negative and the rows sum to one) it defines a Markov chain on S. Now fix any remainder r. If state $i \in D(r)$ then $\mathbf{P}_{1i}^{nd+r} > 0$ for some non-negative integer n. If $\mathbf{Q}_{ij} > 0$, then

$$\mathbf{P}_{1j}^{(n+1)d+r} \geq \mathbf{P}_{1i}^{nd+r} \cdot \mathbf{P}_{ij}^{d} = \mathbf{P}_{1i}^{nd+r} \cdot \mathbf{Q}_{ij} > 0,$$

so we must also have $j \in D(r)$. (This follows since r is the remainder when $(n+1)d + r$ is divided by d.) We have shown that

$$i \in D(r), \ \mathbf{Q}_{ij} > 0 \implies j \in D(r),$$

which further implies that states in $D(r)$ can only communicate with other states in $D(r)$ under the process induced by $\mathbf{Q}$.

Now pick any two states $i, j \in D(r)$. Then $\mathbf{P}_{ij}^k > 0$ for some integer k as the process induced by $\mathbf{P}$ is irreducible. Recalling that $\mathbf{P}_{1i}^{nd+r} > 0$, we obtain

$$\mathbf{P}_{1j}^{nd+r+k} \geq \mathbf{P}_{1i}^{nd+r} \cdot \mathbf{P}_{ij}^{k} > 0.$$

Since $j \in D(r)$, the remainder when $nd + r + k$ is divided by d must be r, implying that k is divisible by d so $k = dm$ for some non-negative integer m. But then

$$\mathbf{Q}_{ij}^m = ((\mathbf{P}^d)^m)_{ij} = \mathbf{P}_{ij}^k > 0,$$

i.e., j is accessible from i under the Markov chain induced by $\mathbf{Q}$. Similarly i is accessible from j and therefore i and j communicate under the Markov chain induced by $\mathbf{Q}$. This proves that the subsets $D(r) \subset S$ are irreducible communication classes under $\mathbf{Q}$.

EXAMPLE 21.5. In our current example, by brute force we could obtain that

$$\mathbf{Q} = \mathbf{P}^d = \mathbf{P}^4 = \begin{pmatrix} 1 & 0 & 0 & 0 & 0 & 0 \\ 0 & .25 & .75 & 0 & 0 & 0 \\ 0 & .25 & .75 & 0 & 0 & 0 \\ 0 & 0 & 0 & .5 & .5 & 0 \\ 0 & 0 & 0 & .5 & .5 & 0 \\ 0 & 0 & 0 & 0 & 0 & 1 \end{pmatrix}.$$

However, simple reasoning also yields this. From state 1, in four steps the chain must be in state 1 again (refer to Figure 21.2). The same holds for state 6. This explains rows 1 and 6 of $\mathbf{Q}$. From either state in $D(1) = \{2, 3\}$, the $\mathbf{P}$ Markov chain must pass through state 1 just prior to returning to $D(1)$ in four steps. A look at $\mathbf{P}$ shows that from state 1, we transition (under $\mathbf{P}$) to state 2 with probability .25 and to state 3 with probability .75. This explains rows 2 and 3 of $\mathbf{Q}$. For either state in $D(2) = \{4, 5\}$, the $\mathbf{P}$ Markov chain is in $D(1)$ after 3 steps. It could be at state 2 or 3, but in either case it transitions to states 4 and 5 each with probability .5 (see rows 2 and 3 of $\mathbf{P}$). This explains lines 4 and 5 of $\mathbf{Q}$.

Finally, we show that each state i — and therefore each class $D(r)$ — has period 1 under $\mathbf{Q}$. Since i has period d under $\mathbf{P}$, we know that, for some n_0, $\mathbf{P}_{ii}^{nd} > 0$ provided $n \geq n_0$ (by Theorem 20.8). But then $\mathbf{Q}_{ii}^n = (\mathbf{P}^d)_{ii}^n = \mathbf{P}_{ii}^{nd} > 0$ provided $n \geq n_0$ and therefore $\gcd\{n \geq 0 : \mathbf{Q}_{ii}^n > 0\} = 1$.

Under $\mathbf{Q} = \mathbf{P}^d$, as advertised, $D(0), D(1), \ldots, D(d-1)$ are each finite irreducible communication classes with period 1. For each remainder r, we may view $\mathbf{Q}$ as inducing a Markov chain on the states only in $D(r)$, ignoring the rest of S. It follows from the aforementioned observations that this Markov chain is ergodic and therefore has a unique invariant measure, call it $\boldsymbol{\mu}^{D(r)}$, on the states in $D(r)$. It will prove convenient to think of $\boldsymbol{\mu}^{D(r)}$ as a measure on all the states in S, where $\boldsymbol{\mu}_i^{D(r)} = 0$ for states $i \notin D(r)$.

EXAMPLE 21.6. Clearly $\boldsymbol{\mu}^{D(0)} = (1\ 0\ 0\ 0\ 0\ 0)$ and $\boldsymbol{\mu}^{D(3)} = (0\ 0\ 0\ 0\ 0\ 1)$. Since $D(1)$ and $D(2)$ both contain 2 states, we may use (15.7) to compute their invariant measures. This gives:

$$\boldsymbol{\mu}^{D(1)} = (0\ \frac{.25}{.75+.25}\ \frac{.75}{.75+.25}\ 0\ 0\ 0) = (0\ .25\ .75\ 0\ 0\ 0),$$

and

$$\boldsymbol{\mu}^{D(2)} = (0\ 0\ 0\ \frac{.5}{.5+.5}\ \frac{.5}{.5+.5}\ 0) = (0\ 0\ 0\ .5\ .5\ 0).$$

We prove the following:

THEOREM 21.7. *If the Markov chain on S with probability transition matrix $\mathbf{P}$ is irreducible with period d, it has a unique invariant measure $\boldsymbol{\mu}$ defined by*

$$\boldsymbol{\mu} = \frac{1}{d}\left[\boldsymbol{\mu}^{D(0)} + \boldsymbol{\mu}^{D(1)} + \cdots + \boldsymbol{\mu}^{D(d-1)}\right]. \tag{21.8}$$

Furthermore, all the rows of the matrix

$$\frac{1}{d}\left[\mathbf{I} + \mathbf{P} + \mathbf{P}^2 + \cdots + \mathbf{P}^{d-1}\right]\mathbf{P}^{dn} \tag{21.9}$$

converge to $\boldsymbol{\mu}$ as $n \to \infty$.

EXAMPLE 21.10. In Example 21.6,

$$\begin{aligned}\boldsymbol{\mu} &= \frac{1}{4}(1\ 0\ 0\ 0\ 0\ 0) + \frac{1}{4}(\ 0\ .25\ .75\ 0\ 0\ 0) \\ &\quad + \frac{1}{4}(0\ 0\ 0\ .5\ .5\ 0) + \frac{1}{4}(0\ 0\ 0\ 0\ 0\ 0\ 1) \\ &= \left(\frac{1}{4}\ \frac{1}{16}\ \frac{3}{16}\ \frac{1}{8}\ \frac{1}{8}\ \frac{1}{4}\right),\end{aligned}$$

which we presently confirm by computation is invariant. Clearly it suffices to confirm that $16\boldsymbol{\mu}$ is invariant — a computation that is simpler:

$$(4\ 1\ 3\ 2\ 2\ 4) \cdot \begin{pmatrix} 0 & .25 & .75 & 0 & 0 & 0 \\ 0 & 0 & 0 & .5 & .5 & 0 \\ 0 & 0 & 0 & .5 & .5 & 0 \\ 0 & 0 & 0 & 0 & 0 & 1 \\ 0 & 0 & 0 & 0 & 0 & 1 \\ 1 & 0 & 0 & 0 & 0 & 0 \end{pmatrix} = (4\ 1\ 3\ 2\ 2\ 4).\ \checkmark$$

REMARK. Theorem 21.7 is consistent with the ergodic case (Theorem 17.18). When $d = 1$, there is only one remainder, namely 0. In this case $D(0)$ is all of S. Equation (21.8) says that $\boldsymbol{\mu} = \boldsymbol{\mu}^{D(0)}$. Additionally, the expression in (21.9) is just $\mathbf{P}^n$ when $d = 1$.

PROOF OF THEOREM 21.7. Let $\boldsymbol{\nu}$ be *any* measure on S that is invariant under $\mathbf{P}$ and suppose $X_0 \sim \boldsymbol{\nu}$. (Theorem 17.15 guarantees the existence of at least one such measure.) Fixing any state i, we have $P[X_m = i] = \nu_i$ for all $m = 0, 1, 2, \ldots$. Suppose $0 \le k < d$ and $n \ge 0$. The Partition Theorem gives that

$$\begin{aligned}\nu_i &= P[X_{nd+k} = i] \\ &= \sum_{r=0}^{d-1} P[X_{nd+k} = i \,|\, X_k \in D(r)] \cdot P[X_k \in D(r)].\end{aligned} \tag{21.11}$$

There are exactly nd steps in the $\mathbf{P}$ Markov chain between time k and time $nd + k$. This corresponds to n steps in the $\mathbf{Q}$ Markov chain, so, by Theorem 17.18,

$$P[X_{nd+k} = i \,|\, X_k \in D(r)] \to \boldsymbol{\mu}_i^{D(r)} \quad \text{as} \quad n \to \infty. \tag{21.12}$$

Equations (21.11) and (21.12) imply that

$$\nu_i = \sum_{r=0}^{d-1} \boldsymbol{\mu}_i^{D(r)} P[X_k \in D(r)]. \tag{21.13}$$

Since (21.13) holds for $0 \le k < d$, we have

$$\nu_i = \frac{1}{d}\sum_{k=0}^{d-1}\left(\sum_{r=0}^{d-1}\boldsymbol{\mu}_i^{D(r)} P[X_k \in D(r)]\right),$$

as the outer summation consists of d summands each of value ν_i. Switching the order of summation and noting that $\boldsymbol{\mu}_i^{D(r)}$ does not depend on the index k gives

$$\nu_i = \frac{1}{d}\sum_{r=0}^{d-1}\boldsymbol{\mu}_i^{D(r)}\left(\sum_{k=0}^{d-1} P[X_k \in D(r)]\right). \qquad (21.14)$$

For any fixed remainder r, precisely one of the states $X_0, X_1, \ldots, X_{d-1}$ lies in $D(r)$. Hence the events $\{X_0 \in D(r)\}, \ldots, \{X_{d-1} \in D(r)\}$ partition the outcome space and

$$\sum_{k=0}^{d-1} P[X_k \in D(r)] = 1,$$

which together with (21.14) yield

$$\nu_i = \frac{1}{d}\sum_{r=0}^{d-1}\boldsymbol{\mu}_i^{D(r)}.$$

Since this holds for each coordinate ν_i, we have $\boldsymbol{\nu} = \boldsymbol{\mu}$ as defined in (21.8), proving the first part of the theorem.

To complete the proof of Theorem 21.7, we show that each row of (21.9) converges to $\boldsymbol{\mu}$ as $n \to \infty$. To prove this for the general row i, recall that $\mathbf{e}_i$ is the measure placing probability one on state $i \in S$ and $r(i)$ is state i's associated remainder. For each remainder r, $\mathbf{e}_i\mathbf{P}^r$ is the conditional distribution of X_r given that $X_0 = i$. Since $i \in D(r(i))$, it follows from (21.3) that $X_r \in D(r(i) \oplus r)$. Therefore $\mathbf{e}_i\mathbf{P}^r$ is a measure supported by the states in $D(r(i) \oplus r)$. It follows from Theorem 17.18 that

$$(\mathbf{e}_i\mathbf{P}^r)\mathbf{P}^{dn} = (\mathbf{e}_i\mathbf{P}^r)\mathbf{Q}^n \to \boldsymbol{\mu}^{D(r(i)\oplus r)} \quad \text{as} \quad n \to \infty. \qquad (21.15)$$

Now, as r ranges over the numbers

$$\{0, 1, \ldots, d-1\},$$

$r(i) \oplus r$ ranges over the numbers

$$\{r(i), r(i)+1, \ldots, d-1, 0, 1, \ldots, r(i)-1\}.$$

That is, $r(i) \oplus r$ ranges over the same set of numbers in a different order. From this fact and from (21.15) we obtain that

$$\frac{1}{d}\sum_{r=0}^{d-1}(\mathbf{e}_i\mathbf{P}^r)\mathbf{P}^{dn} \to \frac{1}{d}\sum_{r=0}^{d-1}\boldsymbol{\mu}^{D(r(i)\oplus r)} = \frac{1}{d}\sum_{r=0}^{d-1}\boldsymbol{\mu}^{D(r)},$$

as $n \to \infty$, or

$$\mathbf{e}_i\left(\frac{1}{d}[\mathbf{I}+\mathbf{P}+\mathbf{P}^2+\cdots+\mathbf{P}^{d-1}]\mathbf{P}^{dn}\right) \to \boldsymbol{\mu} \quad \text{as} \quad n \to \infty.$$

Recall that if $\mathbf{M}$ is any $s \times s$ matrix, then $\mathbf{e}_i\mathbf{M}$ is the ith row of $\mathbf{M}$. This establishes that the arbitrary row i of (21.9) converges to the unique invariant measure $\boldsymbol{\mu}$ as $n \to \infty$ completing the proof of Theorem 21.7. □

EXERCISES

1. Write down an irreducible PTM on state space $S = \{1, 2, 3, 4\}$ that has period three. Hint: first diagram your Markov chain.

2. If PTM $\mathbf{P}$ on state space $S = \{1, 2, \ldots, s\}$ is irreducible show that the period is at most s.

3. Write down an irreducible PTM on $S = \{1, 2, 3, 4, 5, 6\}$ with period 1. Do the same for periods 2, 3, 4, 5, and 6.

4. How many irreducible PTM's of period 6 are there on $S = \{1, 2, 3, 4, 5, 6\}$?

5. Show that the PTM's of problem 4 all have the same invariant distribution and compute what it is.

• For problems 6 – 13, the PTM $\mathbf{P} = \begin{pmatrix} 0 & 0 & .5 & .3 & .2 \\ 0 & 0 & .5 & 0 & .5 \\ 0 & 1 & 0 & 0 & 0 \\ 0 & 1 & 0 & 0 & 0 \\ 1 & 0 & 0 & 0 & 0 \end{pmatrix}$ defines a Markov chain on state space $S = \{1, 2, 3, 4, 5\}$.

6. Show that this Markov chain is irreducible by displaying a cycle through all the states.

7. Show that the period is $d = 2$.

8. If $1 \in D(0)$, what are $D(0)$ and $D(1)$?

9. Compute the two PTMs for the "**Q**" Markov chains restricted to $D(0)$ and $D(1)$. (You should get a 2×2 and a 3×3 matrix.)

10. Compute the invariant distributions for these two matrices. What are $\boldsymbol{\mu}^{D(0)}$ and $\boldsymbol{\mu}^{D(1)}$, viewed as measures on all of S? (Express your answer as fractions with a common denominator. Hint: the common denominator is 26.)

11. Compute the $\mathbf{P}$-invariant distribution $\boldsymbol{\mu} = \frac{1}{2}\left(\boldsymbol{\mu}^{D(0)} + \boldsymbol{\mu}^{D(1)}\right)$. (Now the denominator is 52.)

12. Verify by computing $\boldsymbol{\mu}\mathbf{P}$ that $\boldsymbol{\mu}$ is $\mathbf{P}$ invariant. (Hint: easier to compute $(52\boldsymbol{\mu}) \cdot \mathbf{P}$.)

13. Compute the invariant distribution directly by solving the system of equations $\boldsymbol{\mu}\mathbf{P} = \boldsymbol{\mu}$ together with $\mu_1 + \cdots + \mu_5 = 1$.

§22. Ergodic Decomposition

In this final section of the chapter we relax the assumption of irreducibility and completely characterize all invariant distributions on a Markov chain with finite state space. Our initial endeavor is to show that such invariant distributions assign zero probability collectively to the transient states. Throughout this section the sequence X_n is a Markov chain on a finite state space S with PTM $\mathbf{P}$. We let $R_1, R_2, \ldots, R_k$ denote the recurrent communication classes of S and let T denote all transient states of S. Then the sets $R_1, R_2, \ldots, R_k, T$ partition S. For any realization $X_0, X_1, \ldots$ of the Markov chain, one of two things happens. Either the Markov chain starts in a recurrent class R and never leaves that class, so all $X_n \in R$ (see Figure 22.1 below, where $R = R_k$).

Figure 22.1. Two possibilities for the Markov chain.

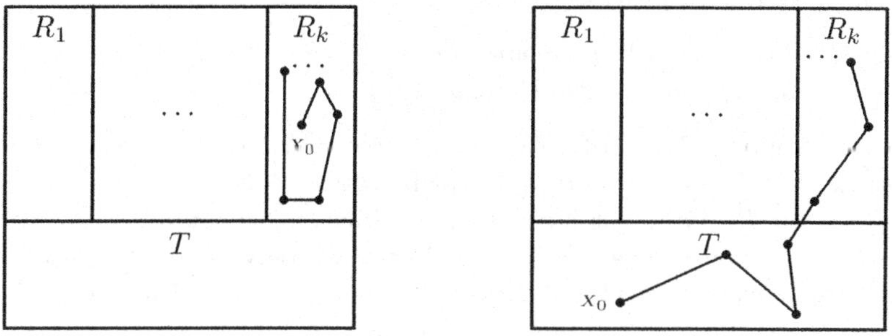

Or the Markov chain starts at a transient state, so $X_0 \in T$, and wanders about the transient states for a while, and then enters a recurrent class that it never subsequently leaves (see Figure 22.1 again). This second possibility uses that the number of transient states is finite: since a Markov chain visits a transient state only finitely often, it visits the finitely many transient states in T only finitely often — eventually it must leave T.

Letting $A = \{X_n \in T \text{ for infinitely many } n\}$, the above discussion shows that $PA = 0$. With $A_m = \{X_n \in T \text{ for some } n \geq m\}$, we have that $A_m \downarrow A$

as $m \to \infty$ and thus $PA_m \to 0$ by the continuity of probability measure. Now $\{X_m \in T\} \subset A_m$, so we also have that $P[X_m \in T] \to 0$. On the other hand, if $X_0 \sim \boldsymbol{\nu}$ where $\boldsymbol{\nu}$ is any invariant distribution and t is a transient state, then

$$\nu_t \;=\; P[X_0 = t] \;=\; P[X_m = t] \;\le\; P[X_m \in T].$$

Since the right side above can be made arbitrarily small by choosing m large we get that $\nu_t = 0$. If $\tilde{S} \subset S$ is any collection of states, let $\boldsymbol{\nu}(\tilde{S})$ denote the collective probability assigned by $\boldsymbol{\nu}$ to states in $\tilde{S}$, so $\boldsymbol{\nu}(\tilde{S}) = \sum_{i \in \tilde{S}} \nu_i$. Then we have proved:

LEMMA 22.2. *If $\boldsymbol{\nu}$ is $\mathbf{P}$-invariant distribution on S, then $\boldsymbol{\nu}(T) = 0$.*

Now suppose $R \subset S$ is any collection of states to which $\boldsymbol{\nu}$ assigns positive collective probability: $\boldsymbol{\nu}(R) > 0$. We then define a measure $\boldsymbol{\nu}$ restricted to R, denoted $\boldsymbol{\nu}|_R$, by

$$\boldsymbol{\nu}|_R(i) \;=\; \begin{cases} \frac{\nu(i)}{\nu(R)} & \text{if } i \in R \\ 0 & \text{if } i \notin R. \end{cases} \tag{22.3}$$

NOTATIONAL REMARK. If $\boldsymbol{\nu}$ is a measure on S we use the notation ν_i and $\nu(i)$ interchangeably. With the measure $\boldsymbol{\nu}|_R$, the first of these options is unwieldy.

The measure $\boldsymbol{\nu}|_R$ is a conditional measure: if $X_0 \sim \boldsymbol{\nu}$, then $\boldsymbol{\nu}|_R(i) = P[X_0 = i \,|\, X_0 \in R]$. In particular we always have $\boldsymbol{\nu}|_R(S) = \boldsymbol{\nu}|_R(R) = 1$.

EXAMPLE 22.4. Take $S = \{1, 2, 3, 4, 5, 6, 7, 8\}$, $\boldsymbol{\nu} = (.3, 0, .1, .2, .1, .1, .1, .1)$, and $R = \{4, 7, 8\}$. Then $\boldsymbol{\nu}(R) = .4$ and $\boldsymbol{\nu}|_R = (0, 0, 0, .5, 0, 0, .25, .25)$.

The next step is to prove:

LEMMA 22.5. *If $\boldsymbol{\nu}$ is $\mathbf{P}$-invariant and R is a recurrent communication class with $\boldsymbol{\nu}(R) > 0$, then $\boldsymbol{\nu}|_R$ is also $\mathbf{P}$-invariant.*

PROOF. Suppose $X_0 \sim \boldsymbol{\nu}|_R$. We must show that also $X_1 \sim \boldsymbol{\nu}|_R$. That is, we must show that $P[X_1 = j] = \boldsymbol{\nu}|_R(j)$ for all states $j \in S$.

Case 1: $j \notin R$. We must have $X_0 \in R$ since $\boldsymbol{\nu}|_R$ is supported by the states in R, so we must also have $X_1 \in R$. (Once a Markov chain enters a recurrent class it cannot leave that class.) Hence $P[X_1 = j] = 0$. But also $\boldsymbol{\nu}|_R(j) = 0$, as $j \notin R$. Hence $P[X_1 = j] = \boldsymbol{\nu}|_R(j)$.

Case 2: $j \in R$. By the Partition Theorem,

$$P[X_1 = j] \;=\; \sum_{i \in S} P[X_1 = j \,|\, X_0 = i] \cdot P[X_0 = i] \;=\; \sum_{i \in S} \boldsymbol{\nu}|_R(i) \cdot \mathbf{P}_{ij}.$$

If $i \in R$, then $\boldsymbol{\nu}|_R(i) = \frac{\nu(i)}{\nu(R)}$. If $i \notin R$, then $\boldsymbol{\nu}|_R(i) = 0 = \nu_i \mathbf{P}_{ij}$. (To see the latter equality, note that either i belongs to another recurrent class $\tilde{R}$, or $i \in T$. If $i \in \tilde{R}$, then $\mathbf{P}_{ij} = 0$ because i and j belong to two different recurrent classes.

If $i \in T$, then $\nu_i = 0$ by virtue of Lemma 22.2.) Either way ($i \in R$ or $i \notin R$) we have

$$\boldsymbol{\nu}|_R(i) \cdot \mathbf{P}_{ij} = \frac{1}{\boldsymbol{\nu}(R)} \nu_i \mathbf{P}_{ij},$$

giving that

$$P[X_1 = j] = \frac{1}{\boldsymbol{\nu}(R)} \sum_{i \in S} \nu_i \mathbf{P}_{ij} = \frac{1}{\boldsymbol{\nu}(R)} \nu_j = \boldsymbol{\nu}|_R(j).$$

The second equality above holds as $\boldsymbol{\nu}$ is **P**-invariant and the last by (22.3) as $j \in R$. □

From Theorems 17.18 and 21.7, we know that to each recurrent communication class R there corresponds a unique invariant measure $\boldsymbol{\mu}^R$ on S that assigns probability 1 collectively to the states in R, that is, with $\boldsymbol{\mu}^R(R) = 1$. It follows from Lemma 22.5 that if $\boldsymbol{\nu}$ is any **P**-invariant distribution we must have $\boldsymbol{\nu}|_R = \boldsymbol{\mu}^R$:

LEMMA 22.6. *If $\boldsymbol{\nu}$ is* **P***-invariant and R is a recurrent communication class with $\boldsymbol{\nu}(R) > 0$, then $\boldsymbol{\nu}|_R = \boldsymbol{\mu}^R$.*

We are now in a position to prove the main result of this section:

THEOREM 22.7. *Let $\boldsymbol{\nu}$ denote any measure on S that is* **P***-invariant. Then there are numbers $\lambda_1, \lambda_2, \ldots, \lambda_k$ with*

$$\boldsymbol{\nu} = \sum_{i=1}^{k} \lambda_i \cdot \boldsymbol{\mu}^{R_i}, \quad \text{where each } \lambda_i \geq 0 \text{ and } \sum_{i=1}^{k} \lambda_i = 1. \tag{22.8}$$

Furthermore, any measure $\boldsymbol{\nu}$ satisfying (22.8) *is invariant with respect to* **P**.

REMARK. Equation (22.8) means that $\nu_j = \sum_{i=1}^{k} \lambda_i \cdot \mu_j^{R_i}$ for all states $j \in S$. Any measure $\boldsymbol{\nu}$ satisfying (22.8) is said to be a *convex combination* of the $\boldsymbol{\mu}^{R_i}$'s.

This theorem completely characterizes all **P**-invariant measures on S.

PROOF. We dispense with the easier second statement of the theorem. Suppose $\boldsymbol{\nu}$ is of the form (22.8). Then

$$\boldsymbol{\nu}\mathbf{P} = \Big[\sum_{i=1}^{k} \lambda_i \cdot \boldsymbol{\mu}^{R_i}\Big]\mathbf{P} = \sum_{i=1}^{k} \lambda_i \cdot (\boldsymbol{\mu}^{R_i}\mathbf{P}) = \sum_{i=1}^{k} \lambda_i \cdot \boldsymbol{\mu}^{R_i} = \boldsymbol{\nu},$$

and $\boldsymbol{\nu}$ is **P**-invariant. In general, any linear combination of vectors that are invariant under right-multiplication by **P** is also invariant under right-multiplication by **P**.

Regarding the theorem's first statement, suppose $\boldsymbol{\nu}$ is **P**-invariant and $X_0 \sim \boldsymbol{\nu}$. The events $\{X_0 \in R_i\}$, $1 \leq i \leq k$, together with $\{X_0 \in T\}$ partition the outcome space. By Lemma 22.2, $P\{X_0 \in T\} = \boldsymbol{\nu}(T) = 0$. The measure $\boldsymbol{\nu}$ may assign zero probability to some of the recurrent classes, but it must assign non-zero probability to at least one recurrent class as $\boldsymbol{\nu}(S) = 1$. By the Partition Theorem (see exercises below) we have

$$\begin{aligned}\nu_j &= P[X_0 = j] = \sum_{i:\, P[X_0 \in R_i]>0} P[X_0 = j \mid X_0 \in R_i] \cdot P[X_0 \in R_i] \\ &= \sum_{i:\, \boldsymbol{\nu}(R_i)>0} \boldsymbol{\nu}|_{R_i}(j) \cdot \boldsymbol{\nu}(R_i). \end{aligned} \tag{22.9}$$

Now

$$\sum_{i:\, \boldsymbol{\nu}(R_i)>0} \boldsymbol{\nu}(R_i) = \sum_{i=1}^{k} \boldsymbol{\nu}(R_i) = 1,$$

where the first equality holds as the additional terms thrown in are all zero. Letting $\lambda_i = \boldsymbol{\nu}(R_i)$ for $1 \le i \le k$ we have each $\lambda_i \ge 0$ and $\sum_{i=1}^{k} \lambda_i = 1$. Using (22.9) and Lemma 22.6 we get that:

$$\nu_j = \sum_{i:\, \lambda_i>0} \lambda_i \cdot \boldsymbol{\nu}|_{R_i}(j) = \sum_{i:\, \lambda_i>0} \lambda_i \cdot \boldsymbol{\mu}_j^{R_i} = \sum_{i=1}^{k} \lambda_i \cdot \boldsymbol{\mu}_j^{R_i}.$$

Again, the last equality holds because the addition terms thrown in are all zero. □

EXAMPLE 22.10. Suppose a Markov chain on states $S = \{1, 2, \ldots, 9\}$ has PTM

$$\mathbf{P} = \begin{pmatrix} 0 & 1 & 0 & 0 & 0 & 0 & 0 & 0 & 0 \\ 0 & 0 & 1 & 0 & 0 & 0 & 0 & 0 & 0 \\ 1/2 & 0 & 1/2 & 0 & 0 & 0 & 0 & 0 & 0 \\ 0 & 1/4 & 0 & 1/2 & 1/4 & 0 & 0 & 0 & 0 \\ 0 & 0 & 0 & 0 & 0 & 1/2 & 1/2 & 0 & 0 \\ 0 & 0 & 0 & 0 & 0 & 0 & 0 & 2/3 & 1/3 \\ 0 & 0 & 0 & 0 & 0 & 0 & 0 & 1/2 & 1/2 \\ 0 & 0 & 0 & 0 & 1 & 0 & 0 & 0 & 0 \\ 0 & 0 & 0 & 0 & 1 & 0 & 0 & 0 & 0 \end{pmatrix},$$

a diagram of which is shown in Figure 22.11 below.

Figure 22.11. Diagram of **P**.

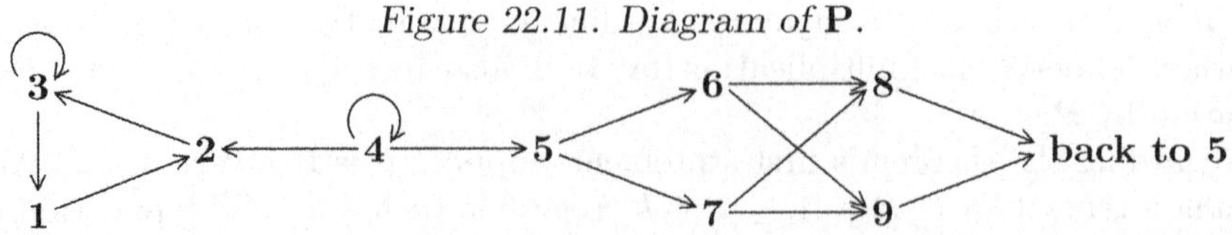

Clearly $R_1 = \{1,2,3\}$, $R_2 = \{5,6,7,8,9\}$ and $T = \{4\}$. The recurrent class R_1 is aperiodic and the reader will readily verify that

$$\boldsymbol{\mu}^{R_1} = \left(\tfrac{1}{4} \;\; \tfrac{1}{4} \;\; \tfrac{1}{2} \;\; 0 \;\; 0 \;\; 0 \;\; 0 \;\; 0 \;\; 0 \right).$$

Also, R_2 has period $d = 3$ and if we take $5 \in D(0)$ we have $D(0) = \{5\}$, $D(1) = \{6,7\}$ and $D(2) = \{8,9\}$. Here $\mathbf{Q} = \mathbf{P}^3$ and the reader should verify that the $\mathbf{Q}$-invariant measures on $D(0), D(1)$, and $D(2)$ are:

$$\begin{aligned}
\boldsymbol{\mu}^{D(0)} &= \left(0 \;\; 0 \;\; 0 \;\; 0 \;\; 1 \;\; 0 \;\; 0 \;\; 0 \;\; 0 \right), \\
\boldsymbol{\mu}^{D(1)} &= \left(0 \;\; 0 \;\; 0 \;\; 0 \;\; 0 \;\; \tfrac{1}{2} \;\; \tfrac{1}{2} \;\; 0 \;\; 0 \right), \text{ and} \\
\boldsymbol{\mu}^{D(2)} &= \left(0 \;\; 0 \;\; 0 \;\; 0 \;\; 0 \;\; 0 \;\; 0 \;\; \tfrac{7}{12} \;\; \tfrac{5}{12} \right),
\end{aligned}$$

producing

$$\boldsymbol{\mu}^{R_2} = \left(0 \;\; 0 \;\; 0 \;\; 0 \;\; \tfrac{1}{3} \;\; \tfrac{1}{6} \;\; \tfrac{1}{6} \;\; \tfrac{7}{36} \;\; \tfrac{5}{36} \right)$$

as the unique $\mathbf{P}$-invariant measure on R_2. The general $\mathbf{P}$-invariant distribution on S is therefore given by

$$\boldsymbol{\mu} = \left(\tfrac{\lambda_1}{4} \;\; \tfrac{\lambda_1}{4} \;\; \tfrac{\lambda_1}{2} \;\; 0 \;\; \tfrac{\lambda_2}{3} \;\; \tfrac{\lambda_2}{6} \;\; \tfrac{\lambda_2}{6} \;\; \tfrac{7\lambda_2}{36} \;\; \tfrac{5\lambda_2}{36} \right),$$

where $\lambda_1, \lambda_2 \geq 0$ and $\lambda_1 + \lambda_2 = 1$.

EXERCISES

1. Use the techniques developed in this section to analyze the two-state Markov chains with PTM's $\begin{pmatrix} p & q \\ 0 & 1 \end{pmatrix}$ and $\begin{pmatrix} 1 & 0 \\ p & q \end{pmatrix}$ where $p + q = 1$. Treat the case $\begin{pmatrix} 1 & 0 \\ 0 & 1 \end{pmatrix}$ separately.

• For problems 2 – 7, the PTM

$$\mathbf{P} = \begin{pmatrix}
0 & 1 & 0 & 0 & 0 & 0 & 0 \\
1 & 0 & 0 & 0 & 0 & 0 & 0 \\
.25 & 0 & 0 & .5 & 0 & .25 & 0 \\
0 & 0 & .5 & .5 & 0 & 0 & 0 \\
0 & 0 & 0 & 0 & .5 & 0 & .5 \\
0 & 0 & 0 & 0 & 1 & 0 & 0 \\
0 & 0 & 0 & 0 & 0 & 1 & 0
\end{pmatrix}$$

defines a Markov chain on state space $S = \{1,2,3,4,5,6,7\}$.

2. Make a diagram of this Markov chain.

3. What are the recurrent communication classes? What are the transient states T?

4. What are the periods of the recurrent classes?

5. Compute $\boldsymbol{\mu}^R$ for each recurrent communication class R.

6. What invariant distribution assigns probability $\frac{1}{4}$ to state 2?

7. Classify all invariant distributions for $\mathbf{P}$.

8. (Partition Theorem, revisited.) Suppose that events $B_1, B_2, \ldots$ partition an outcome space Ω, where possibly some of the B_i have $PB_i = 0$. If $A \subset \Omega$ is an event, show that

$$PA = \sum_{i:\, PB_i > 0} P[A \,|\, B_i]\, PB_i.$$

CHAPTER IV

RANDOM WALKS

§23. Preliminaries

In this chapter we study *random walks* in one, two, and three dimensions. These are examples of discretely indexed Markov chains where the state space S is countably infinite. Recall that both period and recurrence are class properties of Markov chains. Our proofs of these facts in Chapter III did not require the hypothesis of a finite state space (go back and check this!). In the case where S is finite, we can make the following statements, all of which were proved in Chapter III:

• Any recurrent communication class has an invariant distribution that is supported by that class (it assigns collective probability one to the states in that class).

• Every Markov chain has at least one recurrent communication class, so...

• Every Markov chain has an invariant distribution.

• If a Markov chain is irreducible every state in its sole recurrent communication class is visited exponentially fast in the sense of Theorem 17.1.

• The transient states collectively are visited by a Markov chain only finitely many times. Eventually the process will enter a recurrent class and remain there.

When S is infinite none of these statements hold. The random walks of this chapter will provide us with counterexamples. The following preliminary example illustrates various possibilities.

EXAMPLE 23.1. Suppose $S = \mathbb{N} = \{1, 2, 3, \dots\}$ and the transition probabilities are as shown in Figure 23.2. Here $0 < p < 1$ and $q = 1 - p$. First, observe that this Markov chain is irreducible: if m and n are any two states it is possible to go from state m to state n by taking $|m - n|$ consecutive steps to the right if $n > m$ or to the left if $n < m$. Second, this Markov chain is aperiodic. This is easy to see as the period of state 1 is clearly 1. If $p > \frac{1}{2} > q$, then the "drift" is to the right. In this case we will show in the next section that all states are transient. The Markov chain moves inexorably to the right eventually leaving any fixed state behind forever. If $p = \frac{1}{2} = q$, the Markov chain is recurrent but possesses no invariant distribution. We will treat this in the next section as well. If $p < \frac{1}{2} < q$, then the drift is to the left and, as we will see, the Markov chain will always eventually return to state 1. In fact, this will happen quickly enough so that there *is* an invariant distribution. The exercises below will lead you to the conclusion that there is a unique invariant distribution (a geometric distribution) in the case $p < \frac{1}{2} < q$ and no invariant distribution if $p \geq \frac{1}{2}$.

Figure 23.2. A transition diagram for Example 23.1.

EXERCISES

• In exercises 1 – 4, we show that the transition probabilities on $\mathbb{N}$ shown in Figure 23.2 have a Geometric $\left(1 - \frac{p}{q}\right)$ invariant distribution if $p < \frac{1}{2} < q$ and no invariant distribution if $p \geq \frac{1}{2}$. For these problems, let $\boldsymbol{\mu} = (\mu_1, \mu_2, \dots)$ be any probability distribution on $\mathbb{N}$.

1. Show that $\boldsymbol{\mu}$ is invariant under these dynamics if and only if: (i) $\mu_2 = \frac{p}{q}\mu_1$; and (ii) $\mu_n = \mu_{n-1}p + \mu_{n+1}q$ for $n \geq 2$.

2. Show that (i) and (ii) from problem 1 hold if and only if $\mu_n = \mu_1\left(\frac{p}{q}\right)^{n-1}$ for $n \geq 2$.

3. Suppose $0 < p < \frac{1}{2} < q < 1$. Show that $\boldsymbol{\mu}$ is invariant if and only if $\boldsymbol{\mu} \sim$ Geometric $\left(1 - \frac{p}{q}\right)$.

4. Now suppose $0 < q \leq \frac{1}{2} \leq p < 1$. Use your result from problem 2 to show that this Markov chain has no invariant distribution. Hint: separately rule out the two cases $\mu_1 = 0$ and $\mu_1 > 0$.

§24. Random Walks on $\mathbb{Z}$

A random walk can take place on any state space S. Associated with each state $i \in S$ is a collection of states $N(i) \subset S$ that are referred to as the *neighbors* or *nearest neighbors* of i. The walk starts at some state, either deterministic or random (always deterministic in our examples). At each "step" the walk moves to one of its neighbors selected randomly according to some fixed probability distribution and independently of the history of the walk. Suppose after n steps the walk is in state i, i.e., $W_n = i$. Then the state W_{n+1} is chosen randomly (often uniformly) from among the neighboring states $N(i)$. This is a Markov chain, so given W_n the selection of W_{n+1} must be independent of the history of the walk.

EXAMPLE 24.1. The Markov chain of Example 23.1 is a random walk. Here, $N(1) = \{1, 2\}$ and for $i > 1$, $N(i) = \{i-1, i+1\}$.

In this section, we study the situation where the state space is the set of integers, $\mathbb{Z}$, and the neighbors of state i are simply $i-1$ and $i+1$. We will take

$W_0 = 0$ with the following transition rule:

$$W_{n+1} = \begin{cases} W_n + 1 & \text{with probability } p \\ W_n - 1 & \text{with probability } q, \end{cases}$$

independent of the history up to period n. A diagram of this is shown in Figure 24.2.

Figure 24.2. A transition diagram for the random walk on $\mathbb{Z}$.

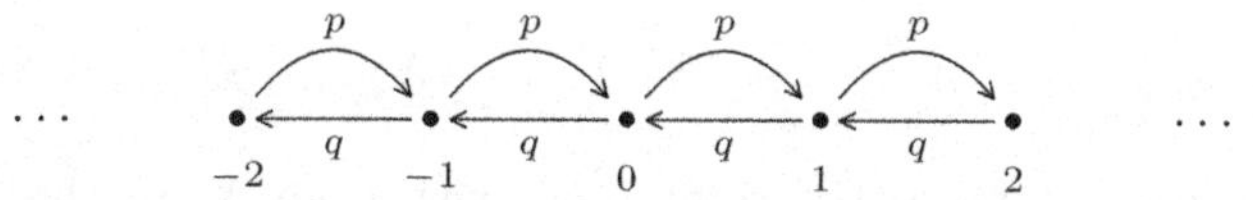

REMARK. You may ask "what distinguishes a random walk from a Markov chain?" This is a good question as the key point for both is that each 'step' is independent of history. Typically (but not always, Example 23.1, e.g.), a random walk model exhibits some kind of translation symmetry as in Figure 24.2.

As in Example 23.1, p is a parameter with $0 < p < 1$ and $q = 1 - p$. We construct this process on $\mathbf{\Omega}$ by letting $X_1, X_2, \ldots$ be iid with $P[X_i = +1] = p$ and $P[X_i = -1] = q$ and putting

$$W_n = 0 + X_1 + \cdots + X_n, \; n \geq 1,$$

where W_0, the starting point for the walk, is the origin (i.e., 0). In the special case where $p = q = \frac{1}{2}$ the process is called a *simple symmetric random walk* on $\mathbb{Z}$ ("SSRW on $\mathbb{Z}$" for short). Sometimes it will be convenient to start the walk at a location other than 0. We will let $W_n(a)$ denote a random walk starting at a, specifically,

$$W_n(a) = a + X_1 + \cdots + X_n, \; n \geq 1,$$

with $W_0(a) = a$. In particular, $W_n = W_n(0)$. Clearly this process is irreducible. It is not aperiodic, however — the period of this Markov chain is 2 because the parity (odd or even) of W_n changes with each step of the walk.

Gambler's Ruin. We begin with the following question. Fix integers a and b with $0 < a < b$. Suppose that a random walk starts at a. As the walk progresses, what is the probability that the walk hits 0 before it hits b? (In principle, the walk could remain between 0 and b *forever*, but we will show that this happens with probability 0. More on this below.) This is called the *gambler's ruin problem* because of the following amusing interpretation. Suppose a gambler has \$$a$ of resources. With each successive wager, he wins \$1 with probability p or loses \$1 with probability q. He plans to leave the casino a winner if his resources reach \$$b$. On the other hand, if his resources are reduced to 0 he must leave the casino in shame. What is the likelihood of shame?!

First we dispense with the issue raised in the previous paragraph. Fix a and for b with $-b < a < b$ let

$$E_b = \{-b < W_n(a) < b \text{ for all } n \geq 0\};$$

we will show that $PE_b = 0$. Observe that if the walk is *anywhere* between $-b$ and b and it then takes $2b$ consecutive steps to the right, it will hit b at some point. Such a sequence of steps happens with probability $p^{2b} > 0$. For E_b to occur, none of the m blocks of $2b$ consecutive steps

$$(X_1, \ldots, X_{2b}),\ (X_{2b+1}, \ldots, X_{4b}),\ \ldots,\ (X_{2(m-1)b+1}, \ldots, X_{2mb})$$

can be consecutive steps to the right (i.e., with each X_i in the block $= +1$). The probability of this occurring is $(1-p^{2b})^m$. Hence $PE_b \leq (1-p^{2b})^m$ — and this holds for any $m \geq 1$. But this latter quantity goes to 0 as $m \to \infty$ since $1 - p^{2b} < 1$. It follows that $PE_b = 0$. Pushing this a little further, let

$$E = \{\text{the walk } W_n(a) \text{ remains on } \textit{some} \text{ finite interval forever}\},$$

so $E_b \uparrow E$ as $b \to \infty$ (see exercises). By continuity of probability measure, we have that $PE = 0$. In other words, with probability one the walk starting at any point a eventually wanders out of every bounded interval. (It may, of course, subsequently return to that interval.) To avoid having to deal with those pesky exceptional outcomes $\omega \in \mathbf{\Omega}$ for which the walk does remain forever bounded, we henceforth work with the outcome space $\mathbf{\Omega} \setminus E$, which still has probability one despite removal of the exceptional outcomes. This will make life easier in the next few paragraphs.

Back to the problem at hand. Define the event

$$A(a) = \{W_n(a) \text{ hits } 0 \text{ before it hits } b\},$$

(we now know it eventually hits one of them) and let $s(a) = P[A(a)]$ (s is for shame!). For the moment we will think of b as a fixed number and a as a variable with $0 \leq a \leq b$. By definition, $s(0) = 1$ and $s(b) = 0$. Then for $0 < a < b$ we get an expression for $s(a)$ by conditioning on the first step of the walk:

$$\begin{aligned} s(a) &= P[A(a)] \\ &= P[A(a) \,|\, X_1 = +1] \cdot P[X_1 = +1] \,+\, P[A(a) \,|\, X_1 = -1] \cdot P[X_1 = -1] \\ &= P[A(a+1)] \cdot p + P[A(a-1)] \cdot q \;=\; s(a+1) \cdot p + s(a-1) \cdot q. \end{aligned} \tag{24.3}$$

To see that $P[A(a) \,|\, X_1 = +1] = P[A(a+1)]$, note that if a walk starting at a takes one step to the right it is at $a+1$ and, from that time forward, the process is a random walk starting at $a+1$. The same goes for the condition $X_1 = -1$. We treat the cases $p = \frac{1}{2}$ and $p \neq \frac{1}{2}$ separately.

If $p = q = \frac{1}{2}$, (24.3) says that

$$s(a) = \frac{s(a+1) + s(a-1)}{2},$$

which implies that

$$s(a) - s(a-1) \;=\; s(a+1) - s(a)$$

and the three points $(a-1, s(a-1))$, $(a, s(a))$, and $(a+1, s(a+1))$ are collinear in the x-y plane. If any three consecutive points on the graph of $s(a)$ are collinear, then *all* the points $(a, s(a))$, $0 \leq a \leq b$, are collinear, i.e., $s(a)$ is a linear function of a. Since $s(0) = 1$ and $s(b) = 0$ we must have

$$s(a) \;=\; 1 - \frac{a}{b}, \quad \text{when } p = q = \frac{1}{2}. \tag{24.4}$$

If $p \neq \frac{1}{2}$, the situation is a little harder. Re-writing (24.3) gives that

$$s(a+1) - \frac{1}{p}s(a) + \frac{q}{p}s(a-1) \;=\; 0. \tag{24.5}$$

Equation (24.5) is called a *difference equation*, or a *recurrence equation.* There is a substantial literature concerning the solution of difference equations. It turns out that the general solution to (24.5) is of the form $s(a) = \alpha + \beta\rho^a$, where $\rho = \frac{q}{p}$ and α and β are any two constants. The values of α and β will be pegged by the *boundary conditions* $s(0) = 1$ and $s(b) = 0$. We set this putative solution into (24.5) to verify that it is indeed a solution:

$$\alpha + \beta\rho^{a+1} - \frac{1}{p}(\alpha + \beta\rho^a) + \frac{q}{p}(\alpha + \beta\rho^{a-1}) \;\stackrel{?}{=}\; 0. \tag{24.6}$$

The terms involving α cancel, as $1 - \frac{1}{p} + \frac{q}{p} = 0$, and once these terms are removed we may divide through by β yielding

$$\rho^{a+1} - \frac{1}{p}\rho^a + \frac{q}{p}\rho^{a-1} \;\stackrel{?}{=}\; 0.$$

(We have assumed that $\beta \neq 0$, but if $\beta = 0$ both sides of (24.6) are 0.) Dividing through by ρ^{a-1} (which is also not 0 as $q > 0$), we get

$$\rho^2 - \frac{1}{p}\rho + \frac{q}{p} \;\stackrel{?}{=}\; 0. \tag{24.7}$$

The quadratic on the left side of (24.7) is called the characteristic polynomial of the difference equation, and the reader should verify that $\rho = \frac{q}{p}$ does indeed solve (24.7). (The other solution is $\rho = 1$, and the general solution to (24.5) is actually $s(a) = \alpha\rho_1^a + \beta\rho_2^a$, where ρ_1 and ρ_2 are the two solutions of (24.7).)

We now know that $s(a) = \alpha + \beta\rho^a$ solves (24.5) where $\rho = \frac{q}{p}$. To determine the appropriate values of α and β, we use the facts that $s(0) = 1$ and $s(b) = 0$, or

$$\begin{aligned} \alpha + \beta &= 1 \\ \alpha + \beta\rho^b &= 0. \end{aligned}$$

This yields $\alpha = \frac{-\rho^b}{1-\rho^b}$ and $\beta = \frac{1}{1-\rho^b}$, finally producing

$$s(a) = \frac{\rho^a - \rho^b}{1-\rho^b} = \frac{\left(\frac{q}{p}\right)^a - \left(\frac{q}{p}\right)^b}{1-\left(\frac{q}{p}\right)^b}, \quad \text{when } p \neq q. \tag{24.8}$$

REMARK. We cannot set $p = q = \frac{1}{2}$ into (24.8) for in this case $\rho = 1$ and the numerator and denominator are both 0. However, we can compute via l'Hôpital's rule

$$\lim_{\rho\to 1} \frac{\rho^a - \rho^b}{1-\rho^b} = \lim_{\rho\to 1} \frac{a\rho^{a-1} - b\rho^{b-1}}{-b\rho^{b-1}} = \frac{a-b}{-b} = 1 - \frac{a}{b},$$

reproducing (24.4). Nice.

REMARK. We have verified that (24.8) is a solution to (24.5) with $s(0) = 1$ and $s(b) = 0$. However, we have not established that this is the *only* solution. For this, see the exercises.

Recurrence/Transience of the Walk. Next we use the gambler's ruin solution to show that if $p = q = \frac{1}{2}$ the walk is recurrent while if $p \neq q$ the walk is transient. Define the event

$$A = \{W_n \text{ never returns to } 0\} = \{W_n \neq 0 \text{ for all } n > 0\}.$$

By conditioning on the first step of the walk, we see that

$$PA = P[A \mid X_1 = +1] \cdot p + P[A \mid X_1 = -1] \cdot q. \tag{24.9}$$

Consider the condition $X_1 = +1$. Here, the walk is at 1 after one step and continues its random walk from that location. Hence

$$P[A \mid X_1 = +1] = P[W_n(1) \text{ never hits } 0].$$

If a walk starting at 1 never hits 0 then it either: (i) hits *every* number $b \geq 2$ before it hits 0; or (ii) it wanders perpetually between 0 and *some* number $b \geq 2$. As we have deleted outcomes producing (ii), (i) must occur and with

$$B = \{W_n(1) \text{ hits every } b \geq 2 \text{ before it hits } 0\},$$

we have

$$\{W_n(1) \text{ never hits } 0\} \subset B.$$

On the other hand, any walk starting at 1 that hits every number $b \geq 2$ before it hits 0 can, in fact, never hit 0. We establish the contrapositive of this assertion. Suppose that $W(1)$ hits 0 after $2n+1$ steps, so $W_{2n+1} = 0$. The farthest to the right it could have walked prior to this is $n+1$ and hence it would not have hit $b = n+2$ before hitting 0. Therefore

$$B \subset \{W_n(1) \text{ never hits } 0\},$$

so

$$P[W_n(1) \text{ never hits } 0] = PB.$$

With

$$B_b = \{W_n(1) \text{ hits } b \text{ before it hits } 0\},$$

the reader will verify that $B_b \downarrow B$ as $b \to \infty$ and hence

$$P[W_n(1) \text{ never hits } 0] = PB = \lim_{b\to\infty} P[B_b] = \lim_{b\to\infty} (1 - s(1)). \tag{24.10}$$

Note that, although the notation "$s(1)$" suppresses it, this quantity depends on b — refer back to (24.4) and (24.8). In the symmetric case where $p = q = \frac{1}{2}$, (24.10) gives that

$$P[W_n(1) \text{ never hits } 0] = \lim_{b\to\infty} 1 - \left(1 - \frac{1}{b}\right) = 0,$$

hence $P[A \mid X_1 = +1] = 0$. By the symmetry of the walk ($p = q$) we also have $P[A \mid X_1 = -1] = 0$ which together with (24.9) yields that $PA = 0$. Hence, in the case $p = q = \frac{1}{2}$, the walk returns to 0 with probability 1, i.e., the SSRW on $\mathbb{Z}$ is recurrent.

Now suppose $p \neq q$. In this case (24.8) gives that $1 - s(1) = \frac{1-\rho}{1-\rho^b}$. Recalling that $\rho = \frac{q}{p}$, we see that as $b \to \infty$,

$$1 - s(1) \to \begin{cases} 0 & \text{if } q > p \ (\rho > 1) \\ 1 - \rho & \text{if } p > q \ (\rho < 1), \end{cases}$$

so

$$P[W_n(1) \text{ never hits } 0] = \begin{cases} 0 & \text{if } q > p \\ 1 - \frac{q}{p} & \text{if } p > q. \end{cases}$$

If $p > q$, then the random walk "drifts" toward the right and with positive probability $1-\rho$ it never returns to 0 from a starting point of 1. If $q > p$, the drift is to the left and from a starting point of 1, with probability one the walk will eventually be drawn back to 0. In the case where $p \neq q$, we do not have $P[A \mid X_1 = -1] = P[A \mid X_1 = +1]$ as we did for the SSRW. However, a different kind of symmetry is helpful here. First note that, as above for the condition $X_1 = +1$, we have

$$P[A \mid X_1 = -1] \;=\; P[W_n(-1) \text{ never hits } 0].$$

Next, observe that the random walk $W_n(-1)$ is distributed like the process $-W_n(1)$ with the roles of p and q switched, i.e., with $P[X_i = +1] = q$ and $P[X_i = -1] = p$. Hence

$$P[W_n(-1) \text{ never hits } 0] \;=\; \begin{cases} 0 & \text{if } p > q \\ 1 - \frac{p}{q} & \text{if } q > p. \end{cases}$$

In the case $p > q$, putting all this together with (24.9) yields

$$PA \;=\; \left(1 - \frac{q}{p}\right) \cdot p + 0 \cdot q \;=\; p - q.$$

The reader will verify that the case $q > p$ gives $PA = q - p$. Both cases may be combined by the formula $PA = |p - q|$. If $p \neq q$ we see that $PA > 0$ and the walk is transient. We summarize this with

THEOREM 24.11. *The SSRW on $\mathbb{Z}$ is recurrent. If $p \neq q$, the random walk on $\mathbb{Z}$ is transient and the probability that W_n never returns to 0 is $|p - q|$.*

Presently we give an alternative proof of the recurrence (transience) of the symmetric (asymmetric) random walks on $\mathbb{Z}$. We handle the asymmetric case first. Suppose $p > q$. Then $\mu = EX_i = 1 \cdot p + (-1) \cdot q = p - q > 0$ and the Strong Law of Large Numbers gives that

$$\frac{W_n}{n} \;=\; \frac{X_1 + \cdots + X_n}{n} \;=\; \overline{X}_n \;\to\; \mu$$

almost surely as $n \to \infty$. Now fix any number ℓ with $0 < \ell < \mu$, so

$$\begin{aligned} \{\overline{X}_n \to \mu \text{ as } n \to \infty\} &\subset \{\overline{X}_n > \ell \text{ for all sufficiently large } n\} \\ &= \{W_n > n\ell \text{ for all sufficiently large } n\} \\ &\subset \{W_n = 0 \text{ for only finitely many } n\}. \end{aligned}$$

The SLLN states that

$$P[\overline{X}_n \to \mu \text{ as } n \to \infty] \;=\; 1,$$

yielding also that

$$P[W_n = 0 \text{ for only finitely many } n] \;=\; 1,$$

so the asymmetric walk is transient. The walk eventually satisfies $W_n > n\ell$, wandering inexorably toward $+\infty$. A similar argument works for the case $q > p$, where the walk wanders inexorably toward $-\infty$. This argument, which follows directly from the SLLN, does not yield the quantity $|p - q|$ as the probability of not returning to 0.

For the symmetric case, our alternative approach is quite computational but will yield a result that we will use later. The heavy lifting is done by Stirling's formula, which says that

$$n! \;=\; c_n \left(\frac{n}{e}\right)^n \sqrt{n}, \tag{24.12}$$

where $c_n \to \sqrt{2\pi} \approx 2.5066$ as $n \to \infty$. In (24.12) the number c_n is the "plug" that makes the equation correct. For example, $c_{10} = 10!/\big[(\frac{10}{e})^{10}\sqrt{10}\big] \approx 2.5276$. We prove a weaker version of Stirling's formula at the end of this chapter. There it is proved that c_n converges to some number $c > 0$; that the limiting value is $\sqrt{2\pi}$ is left to the exercises. The crucial point for us is that there is a limiting value c that is finite but not zero.

We know that after an odd number of steps, the walk cannot be at 0. But $W_{2n} = 0$ if and only if exactly n steps are taken to the right and n to the left, in any order. Hence

$$P[W_{2n} = 0] \;=\; \binom{2n}{n}\left(\frac{1}{2}\right)^n\left(\frac{1}{2}\right)^n \;=\; \frac{(2n)!}{n!\,n!}\left(\frac{1}{2}\right)^{2n}.$$

Setting in Stirling's formula and executing a remarkable amount of cancellation produces

$$\begin{aligned} P[W_{2n} = 0] \;&=\; \frac{c_{2n}\left(\frac{2n}{e}\right)^{2n}\sqrt{2n}}{\left[c_n\left(\frac{n}{e}\right)^n\sqrt{n}\right]^2}\left(\frac{1}{2}\right)^{2n} \\ &=\; \frac{c_{2n}}{c_n^2}\sqrt{2}\cdot\frac{1}{\sqrt{n}} \;=\; b_n\cdot\frac{1}{\sqrt{n}}, \end{aligned} \tag{24.13}$$

where $b_n = \frac{c_{2n}}{c_n^2}\sqrt{2} \to \frac{c}{c^2}\sqrt{2} = \frac{\sqrt{2}}{c}$ as $n \to \infty$. Let us call this limiting value b — the actual limiting value is $\frac{1}{\sqrt{\pi}}$, but here we only need that $0 < b < \infty$. For sufficiently large values of n, we will have $P[W_{2n} = 0] > \frac{1}{2}b\frac{1}{\sqrt{n}}$. Since $\sum_n \frac{1}{\sqrt{n}}$ diverges, we see that

$$\sum_{n=0}^{\infty} P[W_{2n} = 0] \;=\; \infty,$$

giving from Theorem 20.5 that the SSRW is recurrent.

REMARK. In Theorem 20.5, $\mathbf{P}_{00}^{2n}$ corresponds to $P[W_{2n} = 0]$. Here the PTM $\mathbf{P}$ is infinite since there are an infinite number of states, and one must make sense of $\mathbf{P}^{2n}$. One may do this, but it is easier to recall what $\mathbf{P}_{00}^{2n}$ represents: it is the probability of being in state 0 after $2n$ steps of the Markov chain given that the process begins in state 0. The reader should confirm that the proof of Theorem 20.5 did not use the finiteness of the state space S.

Lack of Positive Recurrence. For a general Markov chain on state space S, fix any state $i \in S$ and let $T = \min\{n > 0 : X_n = i\}$, with $T = \infty$ if the Markov chain never enters state i. If i is recurrent we have $P[T < \infty | X_0 = i] = 1$. State i is said to be *positive recurrent* if, additionally, we have $E[T|X_0 = i] < \infty$. If S is finite, Theorem 17.1 guarantees that i is positive recurrent if it is recurrent. For infinite S this is not necessarily the case, as we shall see. Like recurrence, positive recurrence is also a class property: if i is positive recurrent and $i \leftrightarrow j$ then j is also positive recurrent. (We will not use this fact and we omit a proof.) In the setting of infinite S a communication class has an invariant distribution that is supported by that class if and only if that class is positive recurrent. A Markov chain on countably infinite S is ergodic if: (i) it is irreducible; (ii) it is aperiodic; and (iii) each state in S is positive recurrent. In this case we have:

THEOREM 24.14. *If a Markov chain $X_0, X_1, \ldots$ on S is ergodic then there is a unique invariant distribution $\boldsymbol{\mu}$. Regardless of the initial distribution of X_0, the distribution of X_n converges to $\boldsymbol{\mu}$ as $n \to \infty$ in the sense that $P[X_n = i] \to \mu_i$ for any state i.*

Again, we omit a proof of this fact. However, we show that the SSRW, while recurrent, is not positive recurrent — the expected return time to 0 is infinite. For this discussion, we fix $p = q = \frac{1}{2}$. For any two integers a and b let

$$T_{a,b} = \min\{n > 0 : W_n(a) = b\}.$$

In words, $T_{a,b}$ is the number of steps it takes for the SSRW starting at a to reach b. This definition is meaningful as the walk is irreducible and recurrent. Here we show that $ET_{0,0} = \infty$.

By conditioning on the first step of the walk we get that

$$\begin{aligned} ET_{0,0} &= E[T_{0,0}\,|X_1 = +1] \cdot \frac{1}{2} + E[T_{0,0}\,|X_1 = -1] \cdot \frac{1}{2} \\ &= (1 + ET_{1,0}) \cdot \frac{1}{2} + (1 + ET_{-1,0}) \cdot \frac{1}{2} \\ &= 1 + ET_{1,0}. \end{aligned} \tag{24.15}$$

$E[T_{0,0}\,|X_1 = +1] = 1 + ET_{1,0}$ holds because, conditioned on $X_1 = +1$, the walk first takes one step to the right and then begins a SSRW starting at 1.

$ET_{1,0} = ET_{-1,0}$ holds by virtue of the walk's symmetry. Next we apply a similar conditioning argument to $ET_{1,0}$. A walk beginning at 1 can either take one step to the left and hit 0 in one step, or it can take one step to the right and begin a SSRW starting at 2. Hence

$$ET_{1,0} = 1 \cdot \frac{1}{2} + (1 + ET_{2,0}) \cdot \frac{1}{2} = 1 + \frac{1}{2} ET_{2,0}.$$

Presently we will show that $ET_{2,0} = 2ET_{1,0}$, which will further produce $ET_{1,0} = 1 + ET_{1,0}$. Thus $ET_{1,0}$ cannot be a finite number and therefore $ET_{1,0} = \infty$. Referring back to (24.15), we see also that $ET_{0,0} = \infty$.

We conclude by establishing that $ET_{2,0} = 2ET_{1,0}$. Now $T_{2,1}$ is the number of steps required for the walk starting at 2 to first hit 1; clearly by translation symmetry we have $ET_{2,1} = ET_{1,0}$. Also, necessarily $T_{2,1} < T_{2,0}$ as a walk starting at 2 must hit 1 before it can hit 0. Put

$$\begin{aligned} T^* &= T_{2,0} - T_{2,1} \\ &= \begin{cases} \text{The number of additional steps, after the walk} \\ \quad \text{starting at 2 first hits 1, until it hits 0.} \end{cases} \end{aligned}$$

See Figure 24.16 below for a particular realization.

Figure 24.16. An illustrated realization of $T_{2,1}$, T^*, and $T_{2,0}$.

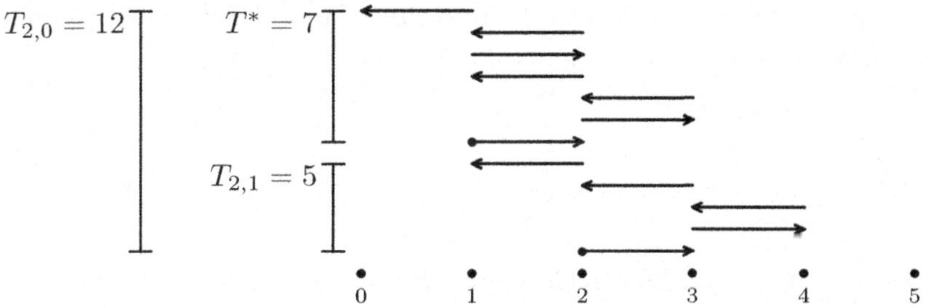

Clearly $T^* \sim T_{1,0}$, since after $T_{2,1}$ steps the walk (which started at 2) begins a SSRW starting at 1. Therefore $ET^* = ET_{1,0}$ and

$$ET_{2,0} = ET_{2,1} + ET^* = 2ET_{1,0},$$

and we have established:

THEOREM 24.17. *The SSRW on $\mathbb{Z}$ is not positive recurrent.*

The SSRW on $\mathbb{Z}$ has no Invariant Distribution. We continue our discussion of the SSRW on $\mathbb{Z}$ by showing that, while recurrent, it has no invariant distribution. (The issue is that it is not positive recurrent.) Suppose $\boldsymbol{\mu}$ is any distribution on $\mathbb{Z}$. If it were invariant and $W_0 \sim \boldsymbol{\mu}$ we would have also that $W_1 \sim \boldsymbol{\mu}$. Then for any state $i \in \mathbb{Z}$

$$\begin{aligned}\mu_i = P[W_1 = i] &= \sum_{j\in\mathbb{Z}} P[W_1 = i \mid W_0 = j] \cdot P[W_0 = j] \\ &= \frac{1}{2}\cdot\mu_{i-1} + \frac{1}{2}\cdot\mu_{i+1} \\ &= \frac{\mu_{i-1}+\mu_{i+1}}{2}. \end{aligned} \tag{24.18}$$

Here we have used that $P[W_1 = i \mid W_0 = j] = 0$ unless $j = i \pm 1$, while for $j = i \pm 1$, $P[W_1 = i \mid W_0 = j] = \frac{1}{2}$. Now (24.18) implies that the three points $(i-1, \mu_{i-1})$, (i, μ_i), and $(i+1, \mu_{i+1})$ are collinear in the x-y plane. Since this holds for every $i \in \mathbb{Z}$, it follows that *all* the points (i, μ_i), $i \in \mathbb{Z}$, are collinear and $\mu_i = mi + \mu_0$ for some slope m and "y-intercept" μ_0. If $m > 0$, we have that $\mu_i \to \infty$ as $i \to \infty$ which cannot be as we must always have $0 \le \mu_i \le 1$. Similarly, $m < 0$ produces $\mu_i \to -\infty$ as $i \to \infty$ — also an impossibility. Hence we must have $m = 0$ and each $\mu_i = \mu_0$, where $0 \le \mu_0 \le 1$. Now if $\mu_0 > 0$, $\sum_{i\in\mathbb{Z}} \mu_i = \infty$ while if $\mu_0 = 0$, $\sum_{i\in\mathbb{Z}} \mu_i = 0$. Yet we must have $\sum_{i\in\mathbb{Z}} \mu_i = 1$. Thus all possibilities are ruled out and there is no invariant distribution.

The Reflection Principle for the SSRW on $\mathbb{Z}$. Suppose a is any fixed positive integer and $(W_n : n \ge 0)$ is a SSRW on $\mathbb{Z}$ starting at $W_0 = 0$. Since this walk is recurrent, we know that W_n will eventually hit a (see exercises) — let $N = \min\{n : W_n = a\}$ denote the first time the walk hits a. We define a new stochastic process $(\widehat{W}_n : n \ge 0)$ as follows:

$$\widehat{W}_n = \begin{cases} W_n & \text{if } n < N \\ 2a - W_n & \text{if } n \ge N, \end{cases} \tag{24.19}$$

so, after the walk first hits a, $\widehat{W}$ is just W reflected about the horizontal line $w = a$ (see Figure 24.20, which shows this for a realization with $a = 3$ and $N = 5$).

Figure 24.20. A realization of W_n and $\widehat{W}_n$ for $0 \le n \le 10$ ($a = 3$, $N = 5$).

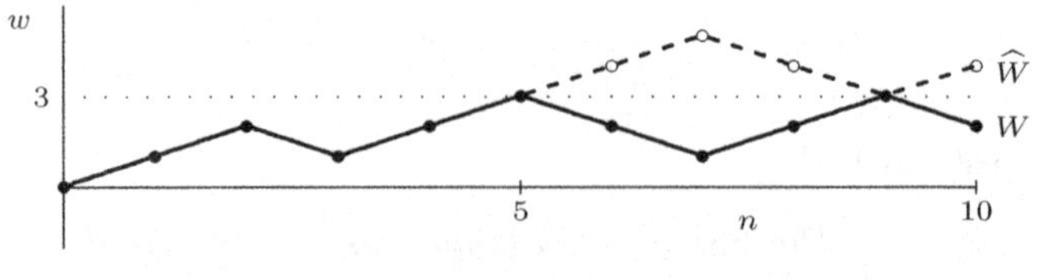

Then we have:

LEMMA 24.21 (REFLECTION PRINCIPLE). *The W and $\widehat{W}$ processes have the same distribution. That is, $(\widehat{W}_n : n \ge 0)$ is also a SSRW on $\mathbb{Z}$.*

PROOF. Note that if $n = N$, then both formulae in (24.19) produce the same value since $W_N = a$. For $n \le N$, the increments for $\widehat{W}$ are equal to those of W:

$$\widehat{X}_n = \widehat{W}_n - \widehat{W}_{n-1} = W_n - W_{n-1} = X_n.$$

On the other hand, for $n > N$ we have

$$\widehat{X}_n = \widehat{W}_n - \widehat{W}_{n-1} = (2a - W_n) - (2a - W_{n-1}) = W_{n-1} - W_n = -X_n.$$

Since $X_n = \pm 1$, each with probability $1/2$, we see that in either case the increments for the $\widehat{W}$ process are drawn from the same distribution as those of the W process.

To show that the $\widehat{X}_n$'s are independent, we generate the W and $\widehat{W}$ processes simultaneously on $\boldsymbol{\Omega}$ using our sequence U_n of iid Uniform $(0,1)$ random variables as follows: (i) initialize $n = W_0 = \widehat{W}_0 = I = 0$; (ii) increase n by 1 and, if $I = 0$ put

$$X_n = \begin{cases} +1 & \text{if } U_n \le 0.5 \\ -1 & \text{otherwise,} \end{cases} \quad \text{and} \quad \widehat{X}_n = \begin{cases} +1 & \text{if } U_n \le 0.5 \\ -1 & \text{otherwise,} \end{cases}$$

while if $I = 1$ put

$$X_n = \begin{cases} +1 & \text{if } U_n \le 0.5 \\ -1 & \text{otherwise,} \end{cases} \quad \text{and} \quad \widehat{X}_n = \begin{cases} -1 & \text{if } U_n \le 0.5 \\ +1 & \text{otherwise,} \end{cases}$$

(iii) put $W_n = W_{n-1} + X_n$ and $\widehat{W}_n = \widehat{W}_{n-1} + \widehat{X}_n$; (iv) if $W_n = a$, put $I = 1$; (v) go back to step (ii). It is apparent that, regardless of the history of the construction up to and including period $n - 1$, we have $\widehat{X}_n = \pm 1$ each with probability $1/2$. Since the conditional distribution of $\widehat{X}_n$ does not depend on the history, we conclude that $\widehat{X}_n$ is independent of the construction through period $n - 1$.

This establishes that the increments for the two processes, while different, are drawn from the same joint distribution and hence the $\widehat{W}$ process is also a SSRW on $\mathbb{Z}$. □

The reflection principle is useful in a variety of applications. For example, for $n \ge 0$, put $M_n = \max\{W_i : 0 \le i \le n\}$. Then:

THEOREM 24.22. *For any positive integer a,*

$$\begin{aligned} P[M_n \ge a] &= 2P[W_n \ge a] - P[W_n = a] \\ &= P[W_n \ge a] + P[W_n \ge a + 1]. \end{aligned}$$

PROOF. The second equality follows immediately from the first as

$$P[W_n \ge a] - P[W_n = a] = P[W_n \ge a + 1].$$

As for the first equality, we deduce from (24.19) that:

$$\begin{aligned} n < N &\implies \widehat{W}_n = W_n < a; \quad \text{while} \\ n \geq N &\implies W_n + \widehat{W}_n = 2a. \end{aligned} \tag{24.23}$$

First we argue that $\{M_n \geq a\} = \{W_n \geq a\} \cup \{\widehat{W}_n \geq a\}$. Call the union on the right A. If $M_n \geq a$, then the walk has already hit a so $n \geq N$. Referring back to (24.23 – second), we see that either $W_n \geq a$ or $\widehat{W}_n \geq a$ yielding $\{M_n \geq a\} \subset A$. Now suppose A happens. From (24.23 – first) we see that $n \geq N$ so the walk has hit a by time n, hence $M_n \geq a$ yielding $A \subset \{M_n \geq a\}$.

Next we show that $\{W_n \geq a\} \cap \{\widehat{W}_n \geq a\} = \{W_n = a\}$. Call the intersection on the left B. If $W_n = a$ then (24.23 – first) gives $n \geq N$ and (24.23 – second) then gives $\widehat{W}_n = a$, so B happens and $\{W_n = a\} \subset B$. If B happens then $n \geq N$ (24.23 – first) and also $W_n + \widehat{W}_n \geq 2a$. By (24.23 – second) we must have $W_n = \widehat{W}_n = a$, yielding that $B \subset \{W_n = a\}$.

We conclude the proof with a simple calculation using these two facts:

$$\begin{aligned} P[M_n \geq a] &= P[\{W_n \geq a\} \cup \{\widehat{W}_n \geq a\}] \\ &= P[W_n \geq a] + P[\widehat{W}_n \geq a] - P[W_n = a] \\ &= 2P[W_n \geq a] - P[W_n = a], \end{aligned}$$

where the last equality holds as the W and $\widehat{W}$ processes are both SSRWs on $\mathbb{Z}$. □

In the next two sections we examine the recurrence/transience of simple symmetric random walks in dimensions higher than 1.

EXERCISES

1. Suppose $0 < p < 1$, $q = 1 - p$, and b is a positive integer. Let $s(a)$ be defined for integers a with $0 \leq a \leq b$ satisfying the recursive relation

$$s(a) = s(a+1) \cdot p + s(a-1) \cdot q \quad \text{for } 0 < a < b,$$

and the boundary conditions $s(0) = u$ and $s(b) = v$. We know that such a function exists, show that it is unique.

Hints: Suppose both $s_1()$ and $s_2()$ both satisfy the recursion relation and the boundary conditions. Let $d(a) = s_1(a) - s_2(a)$. Show that $d()$ satisfies the recursion relation with boundary conditions $d(0) = d(b) = 0$. Let M be the maximum value that $d()$ attains. Show that $M > 0$ contradicts the recursive relation to conclude that $M = 0$. Similarly show that the minimum value attained by $d()$ is 0. Finish the uniqueness argument.

2. Suppose a random walk on $S = \{0, 1, 2, \ldots, L\}$ works as follows. If $1 \le k \le L-1$ and $X_n = k$, then X_{n+1} is $k \pm 1$ each with probability $1/2$. If $X_n = 0$, then $X_{n+1} = 0$ (the state 0 is absorbing). If $X_n = L$, then X_{n+1} is either L or $L-1$, each with probability $1/2$ (L is "retaining"). Let $T = \min\{n \ge 0 : X_n = 0\}$ denote the time that the walk gets absorbed at 0. For general k with $0 \le k \le L$, determine $\tau_k = E[T|X_0 = k]$.

3. Suppose $(W_n(a) : n \ge 0)$ is a SSRW starting at the integer a. Fix a positive integer b and for $|a| \le b$, let $T(a) = \min\{n \ge 0 : |W_n(a)| = b\}$, so $T(a)$ is the first time the walk starting at a hits either b or $-b$. Note that if $a = \pm b$, $T(a) = 0$. Compute $E[T(0)]$, the expected time required for the walk to travel a distance b in either direction, as follows:

(a) Let $e(a) = E[T(a)]$. For $|a| < b$, find a recurrence equation relating $e(a-1)$, $e(a)$, and $e(a+1)$.

(b) Show that $\alpha + \beta a - a^2$ satisfies this recursive relation for any choice of α and β. (These are the only solutions as you may verify as in exercise 1.)

(c) Use the symmetry of the walk and the appropriate boundary condition to deduce what β and α must be.

(d) Compute $E[T(0)]$.

4. Reconsider the setup of exercise 6 from §16. If the ant starts at the origin, what is the expected number of steps until the ant walks to the vertex $(1, 1, 1)$ of the cube?

5. Here $W_n(a)$ is a simple random walk on the integers starting at a with $1 > p > \frac{1}{2} > q = 1 - p > 0$. Compute $P[R(a)]$ for all integers a where $R(a)$ is the event that the walk returns to 0, i.e., $R(a) = \{W_n(a) = 0 \text{ for some } n > 0\}$.

6. With p and q as in exercise 5, let $T = \min\{n \ge 0 : W_n(1) = 0\}$ where $T = \infty$ if the walk never hits 0. We know that $P[T = \infty] > 0$ so $ET = \infty$. What is $E[T \mid R(1)]$? ($R(1)$ is as in exercise 5.)

7. Andrea is doing a random walk on the face of a clock. She starts at 4:00 and, at each step, moves one hour clockwise or one hour counterclockwise. The steps are independent and both directions are equally likely. Find the probability that Andrea visits all the hours 1:00, 2:00, ..., 11:00 before visiting the hour 12:00.

8. With the notation of the gambler's ruin problem, show that $E_b \uparrow E$ as $b \to \infty$.

9. Let W_n be a SSRW on $\mathbb{Z}$ and let a be any integer. Show that the recurrence of W_n implies that the walk will eventually hit a: $P[W_n = a \text{ for some } n] = 1$.

§25. Random Walks in Two Dimensions

Let $\mathbb{Z}^2$ denote all points in the Euclidean plane with integer coordinates. Specifically, $\mathbb{Z}^2 = \{(x, y) : x \in \mathbb{Z}, y \in \mathbb{Z}\}$. We seek to define a random walk on $\mathbb{Z}^2$, which we will denote by $(W_n^{(2)} : n \geq 0)$. Here the superscript indicates the dimension of the process and we will use $W_n^{(1)}$ to denote a (one dimensional) simple symmetric random walk on $\mathbb{Z}$, formerly simply called W_n.

Each point $(x, y) \in \mathbb{Z}^2$ has four nearest neighbors, north, south, east, and west, that are a distance 1 from (x, y), namely: $(x, y+1)$, $(x, y-1)$, $(x+1, y)$, and $(x-1, y)$. We define a Markov chain $(W_n^{(2)} : n \geq 0)$ on state space $\mathbb{Z}^2$ as follows. Take $W_0^{(2)} = (0, 0)$, so the process starts at the origin. Then, given $W_n^{(2)} = (x, y)$, $W_{n+1}^{(2)}$ is one of $W_n^{(2)}$'s four nearest neighbors — each with probability $\frac{1}{4}$. Stated differently,

$$W_{n+1}^{(2)} = \begin{cases} W_n^{(2)} + (+1, 0) & \text{with probability } \frac{1}{4} \\ W_n^{(2)} + (-1, 0) & \text{with probability } \frac{1}{4} \\ W_n^{(2)} + (0, +1) & \text{with probability } \frac{1}{4} \\ W_n^{(2)} + (0, -1) & \text{with probability } \frac{1}{4}, \end{cases}$$

where the increment is independent of the process up to time n. We will let X_n and Y_n denote the first and second coordinates of $W_n^{(2)}$, respectively, so $W_n^{(2)} = (X_n, Y_n)$. Figure 25.1 shows a realization of the first 10 steps of such a walk. Here $W_{10} = (-1, -1)$, so $X_{10} = Y_{10} = -1$ (the '∘' in the figure).

Figure 25.1. An illustrative realization of $W_n^{(2)}$ for $0 \leq n \leq 10$.

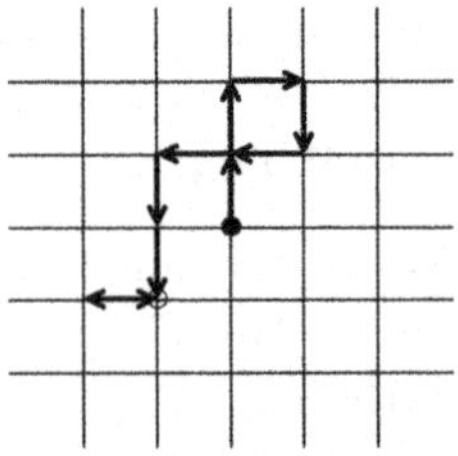

This defines a *simple symmetric random walk on* $\mathbb{Z}^2$. The state space $\mathbb{Z}^2$ is clearly irreducible under this process and, like the simple symmetric random walk on $\mathbb{Z}$, it has period 2. To verify this latter statement, note that $X_n + Y_n$ changes parity with each step. Like the simple symmetric random walk on $\mathbb{Z}$, we can only have $W_n^{(2)} = (0, 0)$ if n is even. Our goal here is to show that the simple symmetric random walk on $\mathbb{Z}^2$ is recurrent. We have

Theorem 25.2. $P[W_{2n}^{(2)} = (0, 0)] = (P[W_{2n}^{(1)} = 0])^2$, *so* $W_n^{(2)}$ *is recurrent.*

Proof. Let $\mathbf{u} = (\frac{1}{2}, \frac{1}{2})$ and $\mathbf{v} = (\frac{1}{2}, -\frac{1}{2})$. Then $\mathbf{u} + \mathbf{v} = (1, 0)$, $-\mathbf{u} - \mathbf{v} = (-1, 0)$, $\mathbf{u} - \mathbf{v} = (0, 1)$, and $-\mathbf{u} + \mathbf{v} = (0, -1)$. So if A and B are independent with

$P[A = \pm 1] = P[B = \pm 1] = \frac{1}{2}$, then $A\mathbf{u} + B\mathbf{v}$ is either $(1,0)$, $(-1,0)$, $(0,1)$, or $(0,-1)$ — each with probability $\frac{1}{4}$. Let $(A_i : i \geq 1)$ and $(B_i : i \geq 1)$ be independent iid sequences with $P[A_i = \pm 1] = P[B_i = \pm 1] = \frac{1}{2}$. By the above discussion,

$$W_n^{(2)} = (0,0) + (A_1\mathbf{u} + B_1\mathbf{v}) + \cdots + (A_n\mathbf{u} + B_n\mathbf{v})$$

is a SSRW on $\mathbb{Z}^2$ starting at the origin. But

$$\begin{aligned} W_{2n}^{(2)} = (0,0) &\iff (A_1 + \cdots + A_{2n})\mathbf{u} + (B_1 + \cdots + B_{2n})\mathbf{v} = (0,0) \\ &\iff A_1 + \cdots + A_{2n} = 0 \text{ and } B_1 + \cdots + B_{2n} = 0. \end{aligned}$$

Where the second ‘ $\iff$ ’ holds because $\mathbf{u}$ and $\mathbf{v}$ are linearly independent vectors. Now $W_n = A_1 + \cdots + A_n$ and $W_n^* = B_1 + \cdots + B_n$ define independent SSRWs on $\mathbb{Z}$ starting at 0 so

$$P[W_{2n}^{(2)} = (0,0)] = P[W_{2n} = 0 \text{ and } W_{2n}^* = 0] = (P[W_{2n}^{(1)} = 0])^2.$$

As for the assertion of recurrence, from (24.13), we see that

$$P[W_{2n}^{(2)} = (0,0)] = \frac{b_n^2}{n}$$

where $b_n^2 \to b^2 > 0$ as $n \to \infty$. Since $\sum_n \frac{1}{n}$ diverges, we get that $W_n^{(2)}$ is recurrent. □

This beautiful argument is surprisingly simple. Presently we offer a second proof that is more computational in nature. It too is nice because at the heart is an appealing combinatorial identity.

ALTERNATIVE PROOF. Fix some number n. Recall that X_i and Y_i denote the first and second coordinates of $W_i^{(2)}$, respectively, so $W_i^{(2)} = (X_i, Y_i)$. With each step in the process, either X_i or Y_i (but not both) change by ± 1. Let K (which is random) denote the number of steps i, with $0 \leq i < 2n$ where it is the first coordinate that changes, i.e. where $X_{i+1} = X_i \pm 1$ and $Y_{i+1} = Y_i$. Then $0 \leq K \leq 2n$, specifically $K \sim \text{Binomial}\left(2n, \frac{1}{2}\right)$. If K is odd, then X_{2n} is odd and cannot be 0 — therefore $W_{2n}^{(2)}$ cannot be (0,0). That is, $P[W_{2n}^{(2)} = (0,0) \cap K = k] = 0$ for odd k. Hence

$$\begin{aligned} P[W_{2n}^{(2)} = (0,0)] &= \sum_{k=0}^{2n} P[W_{2n}^{(2)} = (0,0) \cap K = k] \\ &= \sum_{k=0}^{n} P[W_{2n}^{(2)} = (0,0) \cap K = 2k]. \end{aligned}$$

If $2k$ of the $2n$ steps involve the X coordinate then k of those steps must be to the right ($X_{i+1} = X_i + 1$) and k to the left to have $X_{2n} = 0$. Similarly, $n - k$ of the $2n - 2k$ changes of the Y coordinate must be up ($Y_{i+1} = Y_i + 1$), and $n - k$ down, to have $Y_{2n} = 0$. To specify a walk of length $2n$ that begins and ends at $(0, 0)$ for which the X coordinate changes $2k$ times we must therefore: (i) specify which $2k$ of the $2n$ steps involve a change of the X coordinate; (ii) specify which k of those $2k$ steps involve a step to the right; and (iii) specify which $n-k$ of the remaining $2n-2k$ steps involve a step up. It follows that there are $\binom{2n}{2k}\binom{2k}{k}\binom{2n-2k}{n-k}$ such walks, each with probability $\left(\frac{1}{4}\right)^{2n} = \left(\frac{1}{2}\right)^{4n}$, yielding

$$P[W_{2n}^{(2)} = (0,0) \cap K = 2k] = \binom{2n}{2k}\binom{2k}{k}\binom{2n-2k}{n-k}\left(\frac{1}{2}\right)^{4n},$$

and thus

$$P[W_{2n}^{(2)} = (0,0)] = \left(\frac{1}{2}\right)^{4n}\sum_{k=0}^{n}\binom{2k}{k}\binom{2n-2k}{n-k}\binom{2n}{2k}.$$

Now $\binom{2k}{k}\binom{2n-2k}{n-k}\binom{2n}{2k} = \binom{2n}{n}\binom{n}{k}^2$ (work out both sides, they're both $\frac{(2n)!}{[k!(n-k)!]^2}$), so

$$\begin{aligned}
P[W_{2n}^{(2)} = (0,0)] &= \left(\frac{1}{2}\right)^{4n}\binom{2n}{n}\sum_{k=0}^{n}\binom{n}{k}^2 \\
&= \left(\frac{1}{2}\right)^{4n}\binom{2n}{n}^2 \\
&= \left[\left(\frac{1}{2}\right)^{2n}\binom{2n}{n}\right]^2 = P[W_{2n}^{(1)} = 0]^2.
\end{aligned}$$

The second equality rests on the combinatoric identity $\binom{2n}{n} = \sum_{k=0}^{n}\binom{n}{k}^2$. To see this, suppose we wish to select n bottles of wine from among $2n$ bottles. There are $\binom{2n}{n}$ ways to do this. Now suppose n of the $2n$ bottles are red wine and n are white wine, and let's organize the count a different way. Let k denote how many bottles of red we select, so we also select $n-k$ bottles of white. There are $\binom{n}{k}$ ways to select the k reds and $\binom{n}{n-k}$ ways to select the remaining $n - k$ whites. Hence, once k is selected, there are $\binom{n}{k}\cdot\binom{n}{n-k} = \binom{n}{k}^2$ ways to select the n wines in such a way that exactly k are red. Clearly we must have $0 \le k \le n$, so the number of ways to select the n wines is also given by $\sum_{k=0}^{n}\binom{n}{k}^2$. We have partitioned the count according to how many bottles of red are selected. □

EXERCISES

1. In the alternative proof of Theorem 25.2 show that, given $K = 2k$, X_{2n} and Y_{2n} are conditionally independent. That is, show that

$$P[X_{2n} = x, Y_{2n} = y \mid K = 2k] = P[X_{2n} = x \mid K = 2k] \cdot P[Y_{2n} = y \mid K = 2k].$$

2. For the two-dimensional random walk $W_n^{(2)} = (X_n, Y_n)$, show that the expected squared distance from the origin after n steps is $E[X_n^2 + Y_n^2] = n$.

3. Use that $W_n^{(2)} = (0,0) \implies X_n = 0$ and that the SSRW on $\mathbb{Z}$ is not positive recurrent to argue that $W_n^{(2)}$ is not positive recurrent.

4. Define a random walk, call it $\widetilde{W}$, on $\mathbb{Z}^2$ as follows. $\widetilde{W}_0 = (0,0)$ and

$$\widetilde{W}_{n+1} = \begin{cases} \widetilde{W}_n + (1,1) & \text{with probability } \frac{1}{4} \\ \widetilde{W}_n + (1,-1) & \text{with probability } \frac{1}{4} \\ \widetilde{W}_n + (-1,1) & \text{with probability } \frac{1}{4} \\ \widetilde{W}_n + (-1,-1) & \text{with probability } \frac{1}{4}, \end{cases}$$

where the increment is independent of the process up to time n. Show that $P[\widetilde{W}_n = 0] = P[W_n^{(2)} = 0]$.

• *Discrete Harmonic Functions.* Fix positive integer k and let

$$C = \{(m,n) : m \in \mathbb{Z}, n \in \mathbb{Z}, |m| \le k, |n| \le k\}.$$

Then C is a $(2k+1) \times (2k+1)$ square portion of $\mathbb{Z}^2$ with the origin at its center. Let ∂C denote the *boundary* of C (this is standard notation): $\partial C = \{(m,n) \in C : |m| = k \text{ or } |n| = k\}$. The *interior* of C, denoted C°, is everything else in C, so $C^\circ = C \setminus \partial C$. We show this here for $k = 3$:

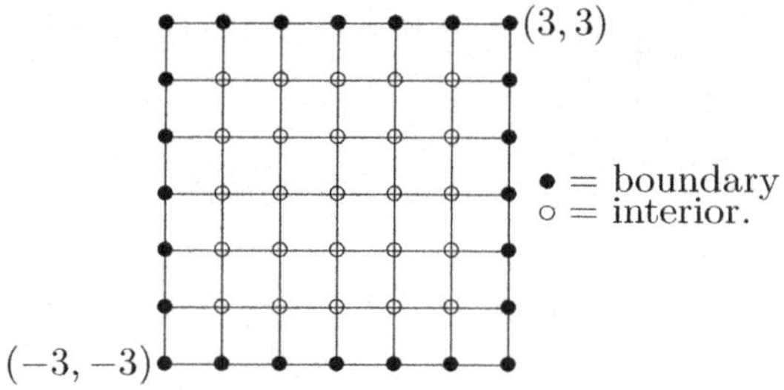

A *discrete harmonic* function on C is a function $f : C \to \mathbb{R}$ such that

$$\text{for } (m,n) \in C^\circ,$$

$$f(m,n) = \frac{f(m,n+1) + f(m,n-1) + f(m+1,n) + f(m-1,n)}{4}. \qquad (*)$$

In words, the value of $f(m,n)$ at an interior point is the average of the values at that point's four nearest $\mathbb{Z}^2$ neighbors.

5. Suppose $g : \partial C \to \mathbb{R}$ is any function, so g assigns numbers to the boundary points of C. Show that there is a discrete harmonic function $f : C \to \mathbb{R}$ with $f(m,n) = g(m,n)$ for all points $(m,n) \in \partial C$. Hint: Let $(W_i^{(2)}(m,n) : i \geq 0)$ be a SSRW on $\mathbb{Z}^2$ starting at (m,n) and let G denote the value of $g()$ at the location where this walk first hits ∂C.

6. Show that the harmonic function $f(m,n)$ satisfying this boundary condition is unique. Hint: follow a procedure similar to problem 1, §24.

7. For any function $f(x,y)$, let $\Delta_x f(x,y) = f(x+1,y) - f(x,y)$, and put $\Delta_x^2 f(x,y) = \Delta_x f(x,y) - \Delta_x f(x-1,y)$. Similarly define $\Delta_y^2 f(x,y)$, keeping x fixed and varying y. Show that the discrete harmonic condition $(*)$ is equivalent to $\Delta_x^2 f(m,n) + \Delta_y^2 f(m,n) = 0$. (A function $f(x,y)$ defined on an open region $R \subset \mathbb{R}^2$ is harmonic if $\frac{\partial^2 f}{\partial x^2} + \frac{\partial^2 f}{\partial y^2} = 0$ at each point in the region.)

§26. Random Walks in Three Dimensions

Here we study the recurrence/transience of the simple symmetric random walk in three dimensions. The state space for this walk is $\mathbb{Z}^3 = \{(x,y,z) : x \in \mathbb{Z},\ y \in \mathbb{Z},\ z \in \mathbb{Z}\}$. We denote this walk by $W_n^{(3)} = (X_n, Y_n, Z_n)$, $n \geq 0$, where $W_0^{(3)} = (0,0,0)$ and, for $n \geq 0$,

$$W_{n+1}^{(3)} = \begin{cases} W_n^{(3)} + (+1,0,0) & \text{with probability } \frac{1}{6} \\ W_n^{(3)} + (-1,0,0) & \text{with probability } \frac{1}{6} \\ W_n^{(3)} + (0,+1,0) & \text{with probability } \frac{1}{6} \\ W_n^{(3)} + (0,-1,0) & \text{with probability } \frac{1}{6} \\ W_n^{(3)} + (0,0,+1) & \text{with probability } \frac{1}{6} \\ W_n^{(3)} + (0,0,-1) & \text{with probability } \frac{1}{6}, \end{cases}$$

where the increment is independent of the process up to time n. We seek to establish that:

THEOREM 26.1. *For some number D, $P[W_{2n}^{(3)} = (0,0,0)] \leq \frac{D}{n^{3/2}}$ so $W_n^{(3)}$ is transient.*

REMARKS. Like the one and two dimensional cases, if n is odd, $W_n^{(3)}$ cannot be $(0,0,0)$ because the parity of $X_i + Y_i + Z_i$ changes with each step. The transience conclusion follows from the fact that $\sum_n \frac{1}{n^{3/2}} < \infty$ together with Theorem 20.5 or, if you wish, the Borel-Cantelli Lemma. One might expect, based on the two dimensional case, that $P[W_{2n}^{(3)} = (0,0,0)] = (P[W_{2n}^{(1)} = 0])^3$. This is not true, although a similar result holds. Specifically, $P[W_{2n}^{(3)} = (0,0,0)] = \frac{d_n}{n^{3/2}}$, where $d_n \to d$ as $n \to \infty$. Here $0 < d < \infty$. We leave a proof of this to the exercises — here only the weaker upper bound is needed to establish transience.

Informal Argument. The idea behind Theorem 26.1 is relatively simple; the details are a little more complicated. The basic idea is that with each step of the walk exactly one coordinate changes by ± 1. Roughly $\frac{1}{3}$ of the time it is the Z coordinate, and $\frac{2}{3}$ of the time it is the X or Y coordinate. Suppose after $2n$ steps of the walk, the X or Y coordinate has changed $2r \approx \frac{2}{3} \cdot 2n$ times and the Z coordinate has changed $2s \approx \frac{1}{3} \cdot 2n$ times. In order for $W_{2n}^{(3)}$ to be $(0,0,0)$, we must have $(X_{2n}, Y_{2n}) = (0,0)$ and $Z_{2n} = 0$. But (X_{2n}, Y_{2n}) is distributed like a two dimensional random walk after $2r$ steps and Z_{2n} is distributed like a one dimensional random walk after $2s$ steps. Hence for appropriate constants we should have

$$\begin{aligned} P[W_{2n}^{(3)} = (0,0,0)] &\approx P[W_{2r}^{(2)} = (0,0)] \cdot P[W_{2s}^{(1)} = 0] \\ &\approx \textit{constant} \cdot \frac{1}{r} \cdot \frac{1}{\sqrt{s}} \\ &\approx \textit{new constant} \cdot \frac{1}{n} \cdot \frac{1}{\sqrt{n}}. \end{aligned}$$

A Large Deviation Detour. First we make precise the sense in which "roughly $\frac{1}{3}$ of the time it is the Z coordinate". We begin with:

LEMMA 26.2. *Suppose $X_1, X_2, \ldots$ are iid with finite exponential moment. For $n \geq 1$, let $S_n = X_1 + \cdots + X_n$. Let $\mu = EX_i$, so $ES_n = n\mu$, and suppose $\epsilon > 0$. Then there is some number $\rho < 1$ such that $P[|S_n - n\mu| \geq n\epsilon] \leq 2\rho^n$ for each n.*

REMARK. Since the X_i's are iid, they have common variance σ^2 and the variance of S_n is $n\sigma^2$. Hence typical or "standard" deviations of S_n away from its mean $n\mu$ are of order $\sqrt{n}\sigma$. No matter how small ϵ is, deviations of order $n\epsilon$ are much larger than $\sqrt{n}\sigma$ for large values of n, and such deviations are called *large deviations.* We should expect them to occur with small probability.

PROOF OF LEMMA 26.2. Since the X_i have finite exponential moment, their moment generating function, call it $M_X(\theta)$, is defined for $|\theta| \leq c$ for some number $c > 0$. Then $M_{S_n}(\theta)$ is also defined for $|\theta| \leq c$ and $M_{S_n}(\theta) = M_X(\theta)^n$. By (11.5) — replacing X with S_n and a with $(\mu + \epsilon)n$ — we get that

$$P[S_n \geq (\mu+\epsilon)n] \leq M_{S_n}(\theta)e^{-\theta(\mu+\epsilon)n} = \left[M_X(\theta)e^{-(\mu+\epsilon)\theta}\right]^n \tag{26.3}$$

Letting $f(\theta) = M_X(\theta)e^{-(\mu+\epsilon)\theta}$, which is also defined for $|\theta| \leq c$, we see that $f(0) = 1$ and

$$f'(\theta) = -(\mu+\epsilon)M_X(\theta)e^{-(\mu+\epsilon)\theta} + e^{-(\mu+\epsilon)\theta}M_X'(\theta),$$

so $f'(0) = -(\mu + \epsilon) + \mu < 0$. (Here we have used that $M'_X(0) = EX = \mu$ — see Corollary 11.9). But $f(0) = 1$ and $f'(0) < 0$ imply that $f(\theta^*) < 1$ for small enough $\theta^* > 0$. With $\rho_1 = f(\theta^*)$ we get that

$$P[S_n \geq n\mu + n\epsilon] \leq \rho_1^n. \tag{26.4}$$

The same analysis applied to the random variables $-X_i$ (which are iid with finite exponential moment and mean $-\mu$) gives that $P[-S_n \geq -n\mu + n\epsilon] \leq \rho_2^n$ for some $\rho_2 < 1$, i.e.,

$$P[S_n \leq n\mu - n\epsilon] \leq \rho_2^n. \tag{26.5}$$

Then (26.4) and (26.5) give that

$$\begin{aligned} P[|S_n - n\mu| \geq n\epsilon] &= P[S_n \leq n\mu - n\epsilon] + P[S_n \geq n\mu + n\epsilon] \\ &\leq \rho_1^n + \rho_2^n \leq 2\rho^n, \end{aligned}$$

where $\rho = \max(\rho_1, \rho_2) < 1$. □

To apply this to our setting, fix n and let K_n denote the number of steps i, with $0 \leq i < n$, for which the Z coordinate of the random walk changes as we step from $W_i^{(3)}$ to $W_{i+1}^{(3)}$, i.e., for which $Z_{i+1} = Z_i \pm 1$. Since $K_n \sim \text{Binomial}\,(n, \frac{1}{3})$ we may think of K_n as the sum of n iid Bernoulli $(\frac{1}{3})$ random variables — these are the X_i of Lemma 26.2. Now fix any $\epsilon > 0$; applying the lemma, we get a number $\rho < 1$ such that

$$P\left[\left|K_n - \frac{n}{3}\right| \geq n\epsilon\right] \leq 2\rho^n. \tag{26.6}$$

Finishing the Proof. With this large deviation estimate in hand, we may complete the...

Proof of Theorem 26.1. Convergent sequences of numbers are bounded, so for some number B we have all $b_n \leq B$, where the b_n are from (24.13). Hence

$$P[W_{2n}^{(1)} = 0] = b_n \cdot \frac{1}{\sqrt{n}} \leq \frac{B}{\sqrt{n}} = \frac{\sqrt{2}B}{\sqrt{2n}},$$

while

$$P[W_{2n+1}^{(1)} = 0] = 0 \leq \frac{\sqrt{2}B}{\sqrt{2n+1}}.$$

Hence, for odd or even n we have

$$P[W_n^{(1)} = 0] \leq \frac{\sqrt{2}B}{\sqrt{n}}.$$

Since $P[W_n^{(2)} = (0,0)] = (P[W_n^{(1)} = 0])^2$ — we proved this for even n but for odd n they are both 0 — we have also that

$$P[W_n^{(2)} = (0,0)] \leq \frac{2B^2}{n}.$$

Recalling that K_n is the number of steps i, with $0 \leq i < n$, for which the Z coordinate of the random walk changes, the partition theorem gives that

$$P[W_n^{(3)} = (0,0,0)] = \sum_{k=0}^{n} P[W_n^{(3)} = (0,0,0) \mid K_n = k] \cdot P[K_n = k]. \qquad (26.7)$$

Now we split the range of the index of summation k into two cases. With $R = \{0, 1, \ldots, n\}$, put

$$R_1 = \left\{k \in R : \left|k - \frac{n}{3}\right| < \frac{n}{6}\right\}, \quad \text{and}$$
$$R_2 = \left\{k \in R : \left|k - \frac{n}{3}\right| \geq \frac{n}{6}\right\},$$

so $R = R_1 \cup R_2$. If we choose ρ to correspond to $\epsilon = \frac{1}{6}$ in (26.6), we get that $P[K_n \in R_2] \leq 2\rho^n$. Hence

$$\begin{aligned}\sum_{k\in R_2} P[W_n^{(3)} = (0,0,0) \mid K_n = k] \cdot P[K_n = k] &\leq \sum_{k\in R_2} P[K_n = k] \\ &= P[K_n \in R_2] \leq 2\rho^n. \end{aligned} \qquad (26.8)$$

Now given $K_n = k$, Z_n is distributed like $W_k^{(1)}$ and (X_n, Y_n) is distributed like $W_{n-k}^{(2)}$. Furthermore Z_n and (X_n, Y_n) are conditionally independent. Hence

$$\begin{aligned}P[W_n^{(3)} = (0,0,0) \mid K_n = k] &= P[W_{n-k}^{(2)} = (0,0)] \cdot P[W_k^{(1)} = 0] \\ &< 2\sqrt{2}B^3 \cdot \frac{1}{n-k} \cdot \frac{1}{\sqrt{k}}.\end{aligned}$$

If $k \in R_1$, then $\frac{n}{6} < k < \frac{n}{2}$, so $n - k > n - \frac{n}{2} = \frac{n}{2}$. Hence

$$P[W_n^{(3)} = (0,0,0) \mid K_n = k] \leq 2\sqrt{2}B^3 \cdot \frac{2}{n} \cdot \frac{1}{\sqrt{n/6}} = \frac{C}{n^{3/2}},$$

where $C = 4\sqrt{12}B^3$. Therefore

$$\begin{aligned}\sum_{k\in R_1} P[W_n^{(3)} = (0,0,0) \mid K_n = k] \cdot P[K_n = k] &\leq \sum_{k\in R_1} \frac{C}{n^{3/2}} \cdot P[K_n = k] \\ &= \frac{C}{n^{3/2}} \sum_{k\in R_1} P[K_n = k] \\ &\leq \frac{C}{n^{3/2}}. \end{aligned} \qquad (26.9)$$

Combining (26.7), (26.8), and (26.9) gives that

$$P[W_n^{(3)} = (0,0,0)] \leq \frac{C}{n^{3/2}} + 2\rho^n.$$

Now $\rho < 1$, so $2\rho^n n^{3/2} \to 0$ as $n \to \infty$ and hence the numbers $2\rho^n n^{3/2}$ for $n \geq 1$ are bounded by some number $\widetilde{C}$. Then $2\rho^n \leq \frac{\widetilde{C}}{n^{3/2}}$ and

$$P[W_n^{(3)} = (0,0,0)] \leq \frac{D}{n^{3/2}},$$

where $D = C + \widetilde{C}$. □

EXERCISES

• Here we modify the proof of Theorem 26.1 to establish the stronger result that $P[W_{2n}^{(3)} = (0,0,0)] = \frac{d_n}{n^{3/2}}$, where $d_n \to d$ as $n \to \infty$ for some d with with $0 < d < \infty$. These exercises are challenging.

1. Suppose $R_1, R_2, \ldots$ are subsets of the integers. For $n \geq 1$ and $k \in R_n$ let $u_{n,k}$ and $v_{n,k}$ be positive numbers. Suppose further that

$$\text{for any } \epsilon > 0,\ 1 - \epsilon \leq \frac{v_{n,k}}{u_{n,k}} \leq 1 + \epsilon \qquad (*)$$
$$\text{for all } k \in R_n \text{ provided } n \text{ is sufficiently large.}$$

If $\lim_{n\to\infty} \sum_{k\in R_n} u_{n,k} = L$, show that also $\lim_{n\to\infty} \sum_{k\in R_n} v_{n,k} = L$.

NOTATION. If $(*)$ holds, we will write that "for $k \in R_n$, $v_{n,k} \overset{\text{u}}{\approx} u_{n,k}$". The "u" stands for "uniformly" — the same ϵ works for all $k \in R_n$.

2. Argue that

$$P[W_{2n}^{(3)} = (0,0,0)] = \sum_{2k\in R_{n1}\cup R_{n2}} P[W_{2n}^{(3)} = (0,0,0) \mid K_{2n} = 2k] \cdot P[K_{2n} = 2k], \qquad (**)$$

where $R_n = \{0, 2, 4, \ldots, 2n\}$, $\epsilon > 0$, and

$$R_{n1} = \left\{2k \in R : \left|2k - \frac{2n}{3}\right| < n\epsilon\right\} \quad \text{and} \quad R_{n2} = \left\{2k \in R : \left|2k - \frac{2n}{3}\right| \geq n\epsilon\right\}.$$

Recall that K_m is the number of steps i, with $0 \leq i < m$, for which the Z coordinate changes as we step from $W_i^{(3)}$ to $W_{i+1}^{(3)}$.

3. Show that the sum in $(**)$ for $2k \in R_{n2}$ goes to 0 exponentially fast in n.

4. Show that for $2k \in R_{n1}$,

$$P[W^{(2)}_{2n-2k} = (0,0)] \overset{\mathrm{u}}{\approx} b^2 \cdot \frac{3}{2} \cdot \frac{1}{n} \quad \text{and} \quad P[W^{(1)}_{2k} = 0] \overset{\mathrm{u}}{\approx} b \cdot \sqrt{3} \cdot \frac{1}{\sqrt{n}}.$$

(Recall b is defined following (24.13).)

5. Show that for $2k \in R_{n1}$,

$$P[W^{(3)}_{2n} = (0,0,0) \,|\, K_{2n} = 2k] \overset{\mathrm{u}}{\approx} b^3 \cdot \frac{3\sqrt{3}}{2} \cdot \frac{1}{n^{3/2}}.$$

6. Argue that the parity of K_m, $m = 0, 1, 2, \ldots$ is a Markov chain with state space $S = \{\text{odd}, \text{even}\}$ with PTM $\mathbf{P} = \begin{pmatrix} 2/3 & 1/3 \\ 1/3 & 2/3 \end{pmatrix}$. Use this to show that $P[K_m \text{ is even}]$ goes to $1/2$ exponentially fast in m.

7. Show that $\sum_{2k \in R_{n1}} P[K_{2n} = 2k]$ goes to $1/2$ exponentially fast in n.

8. Put these various approximations together to show that $P[W^{(3)}_{2n} = (0,0,0)] = \frac{d_n}{n^{3/2}}$, where $d_n \to d = \frac{b^3 3\sqrt{3}}{4}$ as $n \to \infty$. Try to be rigorous. Note: using that $b = \frac{1}{\sqrt{\pi}}$ we get that $d = (3/\pi)^{3/2}/4$.

§27. Stirling's Formula

Here we develop Stirling's formula, our estimate for $n!$. The estimate says that

$$n! \;=\; c_n \Big(\frac{n}{e}\Big)^n \sqrt{n}, \tag{27.1}$$

where $c_n \to \sqrt{2\pi}$ as $n \to \infty$. We will show something a little weaker, which is all we needed above. In particular, we will show that $c_n \to c$ as $n \to \infty$ where $0 < c < \infty$. Our proof will not identify the limiting value of c_n as $\sqrt{2\pi}$ (but see exercises).

If we *define* the quantity c_n so that (27.1) becomes an equality, then we must prove that c_n has a positive limit c as $n \to \infty$. Taking the log of both sides of (27.1) and putting $a_n = \log c_n$ gives that

$$\begin{aligned} \ln(n!) &= a_n + n\big(\ln(n) - 1\big) + \frac{1}{2}\ln(n) \\ &= a_n + \Big(n + \frac{1}{2}\Big)\ln(n) - n. \end{aligned}$$

We must show that

$$a_n \;=\; \ln(n!) - \Big(n + \frac{1}{2}\Big)\ln(n) + n$$

has a finite limit a as $n \to \infty$. For then we will have

$$c_n = e^{a_n} \to e^a,$$

as $n \to \infty$, so $c = e^a > 0$. A crucial point here is that a_n does not run off to either $-\infty$ or $+\infty$, for then we would have $c = 0$ or $c = \infty$, respectively. Now

$$\begin{aligned} a_{n+1} - a_n &= \ln\left((n+1)!\right) - \left(n+1+\frac{1}{2}\right)\ln(n+1) + n + 1 \\ &\quad - \left[\ln(n!) - \left(n+\frac{1}{2}\right)\ln(n) + n\right] \\ &= \ln(n+1) - \left(n+\frac{1}{2}\right)\left(\ln(n+1) - \ln(n)\right) - \ln(n+1) + 1 \\ &= -\left(n+\frac{1}{2}\right)\ln\left(1+\frac{1}{n}\right) + 1. \end{aligned}$$

Using that $\ln(1+x) = x - \frac{1}{2}x^2 + O(x^3)$ (see Appendix II, Example 45.2) gives that

$$a_{n+1} - a_n = -\left(n+\frac{1}{2}\right)\left(\frac{1}{n} - \frac{1}{2}\frac{1}{n^2} + \epsilon_n\right) + 1,$$

where $|\epsilon_n| \le C\frac{1}{n^3}$ for some constant C and for all sufficiently large n. Clearly we may assume that $C \ge \frac{1}{4}$ (if $C < \frac{1}{4}$, taking $C = \frac{1}{4}$ will work *a fortiori*). Multiplying out and cancelling appropriate terms gives that

$$a_{n+1} - a_n = -n\epsilon_n + \frac{1}{4}\frac{1}{n^2} - \frac{1}{2}\epsilon_n.$$

For sufficiently large n, each of the three summands on the right above is no bigger in absolute value than $C\frac{1}{n^2}$ yielding that $|a_{n+1} - a_n| \le 3C\frac{1}{n^2}$. Since $\sum_{n=1}^{\infty} \frac{1}{n^2} < \infty$, the comparison test shows that the infinite series

$$\overbrace{a_1 + (a_2 - a_1) + (a_3 - a_2) + (a_4 - a_3) + \cdots + (a_n - a_{n-1})}^{\text{everything cancels except } a_n} + \cdots$$

converges to a finite number. But this is a telescoping sum and the partial sum of the first n terms (shown above) is just a_n, so $\lim_{n\to\infty} a_n$ exists and is finite. □

EXERCISES

• Here we show that $\sqrt{2\pi}$ is the limiting value of the c_n as $n \to \infty$. It takes some work, but the argument is interesting because it provides a proof of the earliest version of the CLT, due to LaPlace. Here we let $W_1, W_2, \ldots$ denote the one dimensional SSRW starting at 0. Recall that $P[W_n = 0] = 0$ for odd values of n and that W_n is even for even values of n.

1. For $|k| < n$, show that

$$P[W_{2n} = 2k] = \binom{2n}{n+k}\left(\frac{1}{2}\right)^{2n} \\ = \frac{c_{2n}}{c_{n+k}c_{n-k}}\left(1+\frac{k}{n}\right)^{-k}\left(1-\frac{k}{n}\right)^{k}\left(1-\frac{k^2}{n^2}\right)^{-n-\frac{1}{2}}\sqrt{\frac{2}{n}}. \qquad (*)$$

2. Fix any $a > 0$ and let $R_n = \{2k \in \mathbb{Z} : |2k| < a\sqrt{2n}\}$. Use Chebyshev's inequality to show that

$$1 - \frac{1}{a^2} \leq P[W_{2n} \in R_n] \leq 1.$$

3. Show that for $2k \in R_n$, $\frac{c_{2n}}{c_{n+k}c_{n-k}} \overset{\mathrm{u}}{\approx} \frac{1}{c}$, where $c = \lim_{n\to\infty} c_n$.

4. Let $\Delta x = \sqrt{2/n}$ and put $x_k = k\Delta x$. (This notation suppresses the fact that x_k and Δx depend on n, but this should be kept in mind.) Show that for $2k \in R_n$, $|x_k| \leq a$ and that as $2k$ runs through R_n, x_k runs from $-a$ $(\pm\Delta x)$ to a $(\pm\Delta x)$ in steps of Δx.

5. Show that for $2k \in R_n$, both $\left(1+\frac{k}{n}\right)^{-k} \overset{\mathrm{u}}{\approx} e^{-x_k^2/2}$ and $\left(1-\frac{k}{n}\right)^{k} \overset{\mathrm{u}}{\approx} e^{-x_k^2/2}$. Hints: (i) take natural logs and use that $\ln(1+x) = x + O(x^2)$ as $x \to 0$; (ii) the first approximation implies the second as $2k \in R_n \iff -2k \in R_n$, and $x_{-k} = -x_k$.

6. Show that for all $2k \in R_n$, $\left(1-\frac{k^2}{n^2}\right)^{-n} \overset{\mathrm{u}}{\approx} e^{x_k^2/2}$ and $\left(1-\frac{k^2}{n^2}\right)^{-\frac{1}{2}} \overset{\mathrm{u}}{\approx} 1$.

7. Show that for all $2k \in R_n$, $P[W_{2n} = 2k] \overset{\mathrm{u}}{\approx} \frac{1}{c}e^{-x_k^2/2}\Delta x$.

8. Use problem 1 from §26 to show that

$$P[W_{2n} \in R_n] \to \int_{-a}^{a} \frac{1}{c}e^{-x^2/2}\,dx$$

as $n \to \infty$.

9. Use problem 2 above with to conclude that $\int_{-\infty}^{\infty} \frac{1}{c}e^{-x^2/2}\,dx = 1$ and therefore that $c = \sqrt{2\pi}$.

REMARK. As promised, this offers a version of the CLT. Since $EW_{2n} = 0$ and $\operatorname{Std} W_{2n} = \sqrt{2n}$, problems 8 and 9 show that for $a > 0$, that

$$P\left[\left|\frac{W_{2n} - EW_{2n}}{\operatorname{Std} W_{2n}}\right| \leq a\right] \to P[|\text{Normal}(0,1)| \leq a] \qquad (**)$$

as $n \to \infty$.

10. Show that $(**)$ implies that for any a (positive or not)

$$P\left[\frac{W_{2n} - EW_{2n}}{\operatorname{Std} W_{2n}} \leq a\right] \to P[\text{Normal}\,(0,1) \leq a]$$

as $n \to \infty$. Hint: you may wish to use that the distributions of W_{2n} and Normal $(0, 1)$ are both symmetric about 0.

11. Argue why the restriction in problem 10 to an even number of walk steps can be dropped. Hint: $W_{2n+1} - W_{2n} = \pm 1$, which is small compared to $\operatorname{Std} W_{2n}$.

CHAPTER V

ARRIVAL PROCESSES AND POISSON POINT PROCESSES

Arrival processes come up in many real-world settings, including (to mention only a few) insurance, queueing theory, inventory management, and finance (credit default swaps, e.g.). In each setting, something "arrives": insurance claims arrive at the insurer, customers arrive at a bank waiting to be serviced by the next available teller, consumers arrive to purchase a product off the shelf, corporate defaults arrive. A common feature of these different settings is the randomness of the time of the next arrival.

§28. A Motivating Example

Arrival processes work as follows. Let $T_1, T_2, \ldots$ be a sequence of positive random variables, called the *inter-arrival times.* When we accumulate them, as in

$$\tau_n = T_1 + \cdots + T_n,$$

we get the time of the nth arrival. Then $0 < \tau_1 < \tau_2 < \tau_3 < \cdots$ is the sequence of *arrival times* and, with $\tau_0 = 0$, $T_n = \tau_n - \tau_{n-1}$ is the nth inter-arrival time. We define the associated *counting process* by

$$\begin{aligned} N_t &= \max\{n : \tau_n \leq t\} \\ &= \textit{number of arrivals by time } t. \end{aligned}$$

The three related processes, $(T_1, T_2, \ldots)$, $(\tau_1, \tau_2, \ldots)$ and $(N_t : t \geq 0)$ may be viewed as different ways to report the same information. If handed a realization of any one of them, the other two can be deduced. In this section we consider a simple arrival process. The results we obtain will motivate the homogeneous Poisson process, a fundamental process widely used in applications.

Fix a parameter value $\lambda > 0$, to be interpreted as an *arrival rate* as explained presently, and fix any time period Δt small enough so that $0 < \lambda\Delta t < 1$; shortly we will let $\Delta t \to 0$. Let $t_n = n\Delta t$ so $0 < t_1 < t_2 < \cdots$ and $t_{n+1} - t_n = \Delta t$. In this simple model, we allow arrivals only at the times $\{t_n\}$, and at each time t_n there can be 0 or 1 arrival — multiple arrivals at the same time are prohibited.

At time t_n there is an arrival with probability $\lambda\Delta t$ independent of previous arrival history. In other words, if

$$X_n = \begin{cases} 1 & \text{if there is an arrival at time } t_n \\ 0 & \text{otherwise,} \end{cases}$$

then $X_1, X_2, \ldots$ are iid Bernoulli $(\lambda\Delta t)$. Our hope is that the distributional properties of this process will converge (in some sense) as the increment of time per period, Δt, becomes small.

Arrivals on a Fixed Time Interval. Let I denote the interval of time $(s, t]$, so I is fixed and will not change as we vary Δt. Let m denote the number of times $t_n = n\Delta t$ that fall in the interval I:

$$m = \#\{n : s < n\Delta t \leq t\},$$

so, for example, $m = 15$ in the Figure 28.1 below. Then $N_t - N_s$, the number of arrivals on the interval I (5 in the figure), is distributed like Binomial $(m, \lambda\Delta t)$. Letting $\ell = t - s$ denote the length of the interval, we see that $m\Delta t \to \ell$ as $\Delta t \to 0$ (prove this!). Hence $m \cdot \lambda\Delta t \to \lambda\ell$ as $\Delta t \to 0$, and it will follow from Theorem 28.2 below (which is of independent interest) that the distribution of $N_t - N_s$ converges to a Poisson distribution with mean $\lambda\ell$.

Figure 28.1. Arrivals are indicated with $\bullet$'s.

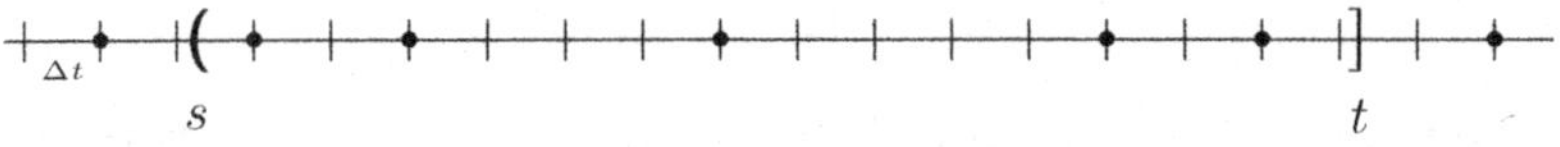

THEOREM 28.2. *If $p_1, p_2, \ldots$ are probabilities with $mp_m \to \theta$ as $m \to \infty$, then*

$$\text{Binomial}\,(m, p_m) \to \text{Poisson}\,(\theta) \quad \textit{as } m \to \infty$$

in the sense that, for any $k \geq 0$,

$$P[\text{Binomial}\,(m, p_m) = k] \to P[\text{Poisson}\,(\theta) = k] \quad \textit{as } m \to \infty.$$

PROOF. Suppose $p_1, p_2, \ldots$ are probabilities with $mp_m \to \theta$ as $m \to \infty$. Then, for any fixed k,

$$\begin{aligned} P[\text{Binomial}\,(m, p_m) = k] &= \binom{m}{k} p_m^k (1 - p_m)^{m-k} \\ &= \frac{m!}{k!(m-k)!m^k} \left(\frac{mp_m}{1 - p_m}\right)^k (1 - p_m)^m. \end{aligned}$$

Now

$$\begin{aligned}\frac{m!}{k!(m-k)!m^k} &= \frac{1}{k!}\frac{m\cdot(m-1)\cdots(m-(k-1))}{m^k}\\ &\to \frac{1}{k!}, \quad \text{as } m\to\infty,\end{aligned}$$

since k is fixed and, for $0 \le i < k$, $\frac{m-i}{m} = 1 - \frac{i}{m} \to 1$, as $m \to \infty$. By hypothesis $mp_m \to \theta$, so we must have $p_m \to 0$ and

$$\left(\frac{mp_m}{1-p_m}\right)^k \to \theta^k, \quad \text{as } m\to\infty.$$

For the last piece, let $\theta_m = mp_m$ so

$$(1-p_m)^m = \left(1+\frac{-\theta_m}{m}\right)^m \to e^{-\theta}, \quad \text{as } m\to\infty,$$

since $\theta_m \to \theta$ (see Appendix II, Theorem 45.4). Putting the three pieces together gives that

$$P[\text{Binomial}\,(m,p_m)=k] \to \frac{1}{k!}\theta^k e^{-\theta} = P[\text{Poisson}\,(\theta)=k]. \quad \square$$

With Theorem 28.2 we have shown that as we shorten the time interval Δt, the distribution of the number of arrivals on a time interval of fixed length ℓ converges to the Poisson $(\lambda\ell)$ distribution. Notice that this distribution does not depend on the starting time s of the interval, i.e., it does not depend on where the interval is situated in time. Notice also that, by construction, the number of arrivals on the interval I is independent of the history of the process up to time s.

The Inter-Arrival Times. Suppose, as we watch the arrival process unfold with time, that the $n-1^{\text{st}}$ arrival occurs at time t_k. Then we watch times t_{k+1}, t_{k+2}, and so forth waiting for the next arrival. The probability that the next arrival occurs at time t_{k+i} is $(1-\lambda\Delta t)^{i-1}\cdot\lambda\Delta t$, since we must have $i-1$ periods with no arrivals followed by an arrival. In other words, the number of *periods* until the next arrival is distributed like Geometric $(\lambda\Delta t)$. Since each period is of length Δt, the next inter-arrival *time*, T_n, satisfies $T_n \sim \Delta t \cdot \text{Geometric}\,(\lambda\Delta t)$, and

$$\begin{aligned}P[T_n > t] &= P[\text{Geometric}\,(\lambda\Delta t) > t/\Delta t]\\ &= P\big[\text{Geometric}\,(\lambda\Delta t) > \lfloor t/\Delta t\rfloor\big]\\ &= (1-\lambda\Delta t)^{\lfloor t/\Delta t\rfloor}.\end{aligned}$$

The second equality above holds since geometric random variables are integer valued. Letting $m = \lfloor t/\Delta t \rfloor$ we have: (i) $m \to \infty$ as $\Delta t \to 0$; and $m \le \frac{t}{\Delta t} < m + 1$, so (ii) $m\Delta t \to t$ as $\Delta t \to 0$. Therefore,

$$(1 - \lambda\Delta t)^{\lfloor t/\Delta t \rfloor} = \left(1 + \frac{-\lambda \cdot m\Delta t}{m}\right)^m \to e^{-\lambda t} \text{ as } \Delta t \to 0,$$

again via Theorem 45.4. It follows that the inter-arrival times, which are independent by virtue of the construction of the process, are asymptotically exponential with mean $1/\lambda$.

EXERCISE

1. Show that $m\Delta t \to \ell$ as $\Delta t \to 0$ as claimed in the paragraph just above Figure 28.1.

§29. The Homogeneous Poisson Process in One Dimension

With this example in mind, we proceed to characterize the homogeneous Poisson process. For any $0 \le s < t$, $N_t - N_s$ gives the number of arrivals on the interval $(s, t]$. We say that an arrival process has *stationary increments* if, for all intervals $(s, t]$, the distribution of $N_t - N_s$ depends on $\ell = t - s$ but does not depend on s. In other words, the distribution of the number of arrivals on a time interval depends only on the interval's length, and not on where the interval is situated in time. We say that the arrival process has *independent increments* if, for all intervals $(s, t]$, the random variable $N_t - N_s$ is independent of the history of the process up to time s. An equivalent, and more precisely stated, formulation of independent increments is the following. With

$$0 = t_0 < t_1 < t_2 < \cdots < t_n, \tag{29.1}$$

let $N_i = N_{t_i} - N_{t_{i-1}}$ denote the number of arrivals on the ith interval $(t_{i-1}, t_i]$. Note that by construction these intervals are disjoint. The process has *independent increments* if, under these circumstances, $N_1, \ldots, N_n$ are independent random variables. Each N_i is independent of the history $N_1, \ldots, N_{i-1}$.

Fix some number $\lambda > 0$ and put $h = 1/\lambda$. If the inter-arrival times $T_1, T_2, \ldots$ are iid exponentials with mean h, the resulting arrival process is called the *homogeneous Poisson process* with *arrival rate* λ, or, equivalently, with *mean inter-arrival time* h. Our main result concerning this process is:

THEOREM 29.2. *The following three characterizations of an arrival process are equivalent:*

(i) *the inter-arrival times are iid* Exponential(h);

(ii) *the increments are stationary and independent with* $N_t \sim$ Poisson(λt);

(iii) *the increments are stationary and independent with* $P[N_{\Delta t} = 1] = \lambda\Delta t + o(\Delta t)$ *and* $P[N_{\Delta t} > 1] = o(\Delta t)$ *as* $\Delta t \to 0$.

Below we will show that (i) $\Longrightarrow$ (ii) $\Longrightarrow$ (iii) $\Longrightarrow$ (i). The following two key lemmas do the heavy lifting.

LEMMA 29.3. *For the homogeneous Poisson process with arrival rate* λ, $N_t \sim$ Poisson (λt).

PROOF. The proof of this is purely computational. We will show that

$$P[\tau_n > t] \;=\; e^{-\lambda t}\left(1 + \lambda t + \frac{(\lambda t)^2}{2!} + \cdots + \frac{(\lambda t)^{n-1}}{(n-1)!}\right), \text{ for all } t \geq 0. \qquad (29.4)$$

Since $\{N_t = n\} = \{\tau_n \leq t < \tau_{n+1}\}$, (29.4) gives that

$$P[N_t = n] \;=\; P[\tau_{n+1} > t] - P[\tau_n > t] \;=\; e^{-\lambda t}\frac{(\lambda t)^n}{n!},$$

as required.

We will prove (29.4) by induction on n. When $n = 1$, (29.4) says that $P[\tau_1 > t] > e^{-\lambda t}$. Since $\tau_1 = T_1 \sim$ Exponential $(1/\lambda)$, (29.4) holds for $n = 1$. Now assume (29.4) holds for some positive integer n. We have

$$\begin{aligned} P[\tau_{n+1} > t] \;&=\; \int_0^\infty P[\tau_{n+1} > t \,|\, T_{n+1} = s]\,\lambda e^{-\lambda s}\,ds \\ &=\; \int_0^\infty P[\tau_n > t - s]\,\lambda e^{-\lambda s}\,ds \\ &=\; \int_0^t P[\tau_n > t - s]\,\lambda e^{-\lambda s}\,ds \;+\; \int_t^\infty \lambda e^{-\lambda s}\,ds \\ &=\; \int_0^t P[\tau_n > s]\,\lambda e^{-\lambda(t-s)}\,ds \;+\; e^{-\lambda t} \end{aligned}$$

The second equality above holds because $\tau_{n+1} = \tau_n + T_{n+1}$ and τ_n is independent of T_{n+1}. The third holds because $P[\tau_n > t - s] = 1$ if $s > t$. The last equality is a simple substitution of s for $t - s$. Using the induction hypothesis (29.4) as applied to $P[\tau_n > s]$ gives, after simplification and integration (see exercises), that

$$P[\tau_{n+1} > t] \;=\; e^{-\lambda t}\left(1 + \lambda t + \frac{(\lambda t)^2}{2!} + \cdots + \frac{(\lambda t)^n}{n!}\right),$$

as required to complete the induction. $\square$

This second key lemma plays a role in many settings.

LEMMA 29.5 (EXPONENTIALS ARE MEMORYLESS). *Let* $X \sim$ Exponential (h) *and fix* $x > 0$. *Then, conditioned on* $\{X > x\}$, $X - x \sim$ Exponential (h).

PROOF. The computations establishing this lemma could not be more straightforward. With $\lambda = 1/h$,

$$\begin{aligned} P[X - x > t \,|\, X > x] \;&=\; \frac{P[\{X - x > t\} \cap \{X > x\}]}{P[X > x]} \\ &=\; \frac{P[X > x + t]}{P[X > x]} \\ &=\; \frac{e^{-\lambda\cdot(x+t)}}{e^{-\lambda x}} \;=\; e^{-\lambda t}. \quad \square \end{aligned}$$

If inter-arrival times are independent exponentials with mean 10 minutes and it's been, say, 30 minutes since the last arrival, you can still expect to wait (on average) 10 more minutes until the next arrival!

PROOF OF THEOREM 29.2. (i) $\Longrightarrow$ (ii). Lemma 29.3 shows that $N_t \sim$ Poisson (λt). Now fix $0 < s < t$ and let $\ell = t - s$ denote the length of the interval $(s, t]$. Suppose the history (call it H) of the process up to time s is that there are n arrivals occurring at times $t_1 < t_2 < \cdots < t_n < s$ (see Figure 29.6). Let $\widetilde{T}_1 = T_{n+1} - (s - t_n)$ and, for $i > 1$, put $\widetilde{T}_i = T_{n+i}$, as shown in Figure 29.6. By Lemma 29.5, conditioned on H, $\widetilde{T}_1$ is exponential with mean h. Furthermore, $\widetilde{T}_2, \widetilde{T}_3, \ldots$ are additional independent Exponential (h)'s. In summary, conditioned on H,

$$N_t - N_s = \max\{n : \widetilde{T}_1 + \cdots + \widetilde{T}_n \leq \ell\},$$

where $\widetilde{T}_1, \widetilde{T}_2, \ldots$ are iid Exponential (h)'s. It follows, as in Lemma 29.3, that conditioned on H, $N_t - N_s \sim$ Poisson $(\lambda\ell)$. Since the conditional distribution of $N_t - N_s$ is Poisson $(\lambda\ell)$ regardless of the history H, we conclude that the unconditional distribution of $N_t - N_s$ is Poisson $(\lambda\ell)$ and that $N_t - N_s$ is independent of H. The increments are therefore stationary and independent.

Figure 29.6.

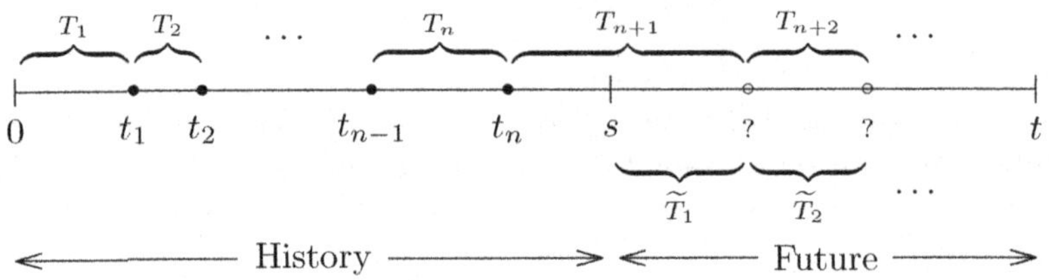

(ii) $\Longrightarrow$ (iii). That the increments are stationary and independent is inherited from (ii); we must only show that $P[N_{\Delta t} = 1] = \lambda\Delta t + o(\Delta t)$ and $P[N_{\Delta t} > 1] = o(\Delta t)$, as $\Delta t \to 0$. But, by (ii),

$$\begin{aligned} P[N_{\Delta t} = 0] &= e^{-\lambda\Delta t} = 1 - \lambda\Delta t + O(\Delta t^2), \quad \text{and} \\ P[N_{\Delta t} = 1] &= e^{-\lambda\Delta t}\lambda\Delta t = (1 + O(\Delta t))\lambda\Delta t. \end{aligned}$$

Now $O(\Delta t) \cdot \lambda\Delta t \Longrightarrow O(\Delta t^2) \Longrightarrow o(\Delta t)$, so $P[N_{\Delta t} = 1] = \lambda\Delta t + o(\Delta t)$ and $P[N_{\Delta t} = 0] = 1 - \lambda\Delta t + o(\Delta t)$. Hence

$$P[N_{\Delta t} > 1] = 1 - (P[N_{\Delta t} = 0] + P[N_{\Delta t} = 1]) = o(\Delta t),$$

as required.

(iii) $\Longrightarrow$ (i). Suppose that as time elapses and the arrival process unfolds we observe that $T_1 = t_1, T_2 = t_2, \ldots, T_n = t_n$ — let us call this history H. We will show that, conditioned on H, $T_{n+1} \sim$ Exponential(h). Since this holds regardless of the history, i.e., regardless of the values of $t_1, \ldots, t_n$, this will establish that T_{n+1} is independent of $T_1, \ldots, T_n$ and that the unconditional distribution of T_{n+1} is Exponential(h).

Put $s = t_1 + \cdots + t_n$ and fix $t > 0$. Divide the interval $(s, s+t]$ into m disjoint subintervals of length $\Delta t = t/m$. Let $A_1, A_2, \ldots, A_m$ denote the number of arrivals in each of the respective subintervals (see Figure 29.7).

Figure 29.7.

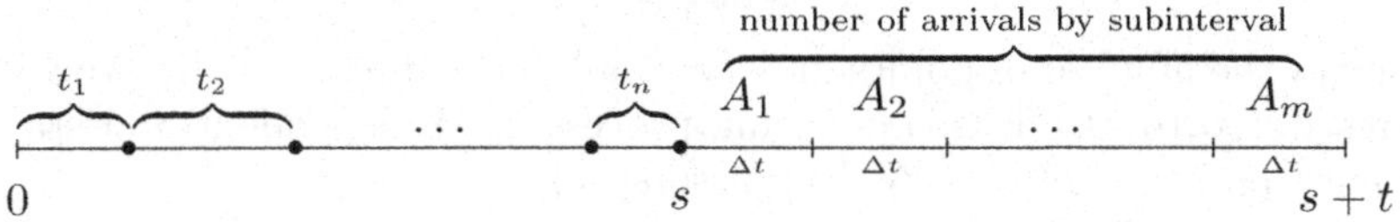

Since the increments are independent and stationary, the random variables A_i are independent and $P[A_i = 0] = P[N_{\Delta t} = 0] = 1 - \lambda\Delta t + o(\Delta t)$. Thus, for any $m > 0$,

$$\begin{aligned} P[T_{n+1} > t \,|\, H] &= P[\text{every } A_i = 0 \,|\, H] \\ &= P[\text{every } A_i = 0] \\ &= (1 - \lambda\Delta t + o(\Delta t))^m \\ &= \left(1 + \frac{-\lambda t + o(\Delta t)m}{m}\right)^m. \end{aligned}$$

In the second equality we use that the random variables A_i are independent of H; in the third that they are independent of each other. As $m \to \infty$, $\Delta t \to 0$ and $o(\Delta t)m = t\frac{o(\Delta t)}{\Delta t} \to 0$, so $-\lambda t + o(\Delta t)m \to -\lambda t$ and, again via Theorem 45.4,

$$\left(1 + \frac{-\lambda t + o(\Delta t)m}{m}\right)^m \to e^{-\lambda t}.$$

Hence $P[T_{n+1} > t \,|\, H] = e^{-\lambda t}$ and we have shown that, conditioned on H, $T_{n+1} \sim$ Exponential(h). □

Theorem 29.2 gives us two ways to construct a homogeneous Poisson process. The most straightforward follows (i). Use the iid sequence $U_1, U_2, \ldots$ on Ω to generate, via inverse transform, iid exponentials $T_n = -h\ln(U_n)$. Then use these inter-arrival times to construct the arrival times τ_n and the counting process N_t.

The Point Process Construction. The second approach is less straightforward, but generalizes nicely to higher dimensions. It takes advantage of (ii). Under this approach, the Poisson process is constructed restricted to a finite time interval $[0, t^*]$. Let $N = N_{t^*}$ denote the (random) number of arrivals on the time interval $[0, t^*]$. Here we view the arrival times $\{\tau_1, \tau_2, \ldots, \tau_N\}$ as a *random subset* of the interval $[0, t^*]$. In our second approach, rather than directly constructing the inter-arrival times, we construct a randomly generated subset of points in $[0, t^*]$. Some refer to this as a *Poisson point process.*

This works as follows. First, since $N = N_{t^*} \sim \text{Poisson}(\lambda t^*)$, we construct N from a Uniform$(0,1)$ via the inverse transform method. We now must choose N points from the interval $[0, t^*]$. We draw them independently and uniformly, i.e., for independent uniforms $U_1, \ldots, U_N$, we put $X_i = t^* \cdot U_i$. This gives us a random subset $\mathcal{P} = \{X_1, X_2, \ldots, X_N\} \subset [0, t^*]$. For $0 \le t \le t^*$, we then define

$$N_t = \#\{x \in \mathcal{P} : x \le t\},$$

so N_t gives the number of points in $\mathcal{P}$ that fall in the interval $[0, t]$. We will show that, restricted to $[0, t^*]$, this counting process has independent and stationary increments and that $N_{s+\ell} - N_s \sim \text{Poisson}(\lambda \ell)$.

We need a preliminary fact that is interesting in its own right. Suppose a bank has N customers any particular day where $N \sim \text{Poisson}(\theta)$. Further suppose that each customer either makes a deposit or a withdrawal. The customers make deposits with probability α and withdrawals with probability $\beta = 1 - \alpha$, independently and independent of N. Let A be the number of customers making deposits and B the number making withdrawals. We will show that $A \sim \text{Poisson}(\alpha\theta)$, $B \sim \text{Poisson}(\beta\theta)$, and that A and B are independent. We state this somewhat more formally as

THEOREM 29.8. *Suppose $N, I_1, I_2, \ldots$ are independent with $N \sim$* Poisson(θ) *and each $I_i \sim$* Bernoulli(α). *Put $J_i = 1 - I_i$, $A = I_1 + I_2 + \cdots + I_N$, and $B = J_1 + J_2 + \cdots + J_N$. Then $A \sim$* Poisson$(\alpha\theta)$, *$B \sim$* Poisson$(\beta\theta)$, *and A and B are independent.*

REMARK. Here you should think of I_i as 1 if the ith customer makes a deposit and 0 if not; J_i is 1 if the ith customer makes a withdrawal and 0 if not. Then A is the total number of customers making deposits and B the total number making withdrawals.

PROOF. Conditioned on $\{N = n\}$, A is the sum of n iid Bernoulli(α)'s and is distributed like Binomial(n, α). In particular, $P[A = a \mid N = n] = 0$ if $n < a$. With this in mind, we compute

$$\begin{aligned} P[A = a] &= \sum_{n=0}^{\infty} P[A = a \mid N = n]\, P[N = n] \\ &= \sum_{n=a}^{\infty} P[A = a \mid N = n]\, P[N = n] \\ &= \sum_{n=a}^{\infty} \binom{n}{a} \alpha^a \beta^{n-a} e^{-\theta} \frac{\theta^n}{n!}. \end{aligned}$$

Continuing, we have

$$\begin{aligned} P[A=a] &= \sum_{n=a}^{\infty} \binom{n}{a} \alpha^a \beta^{n-a} e^{-\theta} \frac{\theta^n}{n!} \\ &= e^{-\alpha\theta} \frac{(\alpha\theta)^a}{a!} e^{-\beta\theta} \sum_{n=a}^{\infty} \frac{(\beta\theta)^{n-a}}{(n-a)!} \\ &= e^{-\alpha\theta} \frac{(\alpha\theta)^a}{a!} \cdot \boxed{e^{-\beta\theta} \sum_{n=0}^{\infty} \frac{(\beta\theta)^n}{n!}} \\ &= e^{-\alpha\theta} \frac{(\alpha\theta)^a}{a!}. \end{aligned}$$

The final equality holds because the quantity in the box is one. Hence $A \sim$ Poisson $(\alpha\theta)$. An entirely similar calculation shows that $B \sim$ Poisson $(\beta\theta)$.

To show that A and B are independent (which is somewhat surprising), we note that $A + B = N$ and we compute

$$\begin{aligned} P[A=a, B=b] &= P[A=a, N=a+b] \\ &= P[A=a \mid N=a+b] \cdot P[N=a+b] \\ &= \binom{a+b}{a} \alpha^a \beta^b \cdot e^{-\theta} \frac{\theta^{a+b}}{(a+b)!} \\ &= e^{-\alpha\theta} \frac{(\alpha\theta)^a}{a!} \cdot e^{-\beta\theta} \frac{(\beta\theta)^b}{b!} \\ &= P[A=a] \cdot P[B=b]. \quad \square \end{aligned}$$

This result generalizes. Suppose the bank has m disjoint types of customers, call them type 1, type 2, and so forth. Again let $N \sim$ Poisson (θ) denote the total number of customers on any particular day. Suppose each customer is independently type i with probability α_i, where $\alpha_1 + \cdots + \alpha_m = 1$. Let A_i denote the number of customers among the N that are type i, so $A_1 + \cdots + A_m = N$. Then each $A_i \sim$ Poisson $(\alpha_i\theta)$ and $A_1, A_2, \ldots, A_m$ are independent. A proof uses Theorem 29.8 (where $m = 2$) and induction on m (see exercises).

We now return to our Poisson point process.

The Distribution of $N_{s+\ell} - N_s$. Suppose $0 \le s < s + \ell \le t^*$, and let $M = N_{s+\ell} - N_s$ denote the number of points from $\mathcal{P}$ that fall in $(s, s+\ell]$. Since each X_i is uniformly distributed on $[0, t^*]$, the probability that it lies in the subinterval $(s, s+\ell]$ is $\alpha = \ell/t^*$. Since the X_i's are independent and there are $N \sim$ Poisson (λt^*) of them, an application of our lemma gives that $M \sim$ Poisson $(\lambda\ell)$. The counting process is stationary as this distribution depends on ℓ but not s.

Independent Increments. Suppose (29.1) holds with $t_n = t^*$ and let $N_i = N_{t_i} - N_{t_{i-1}}$ denote the number of points in $\mathcal{P}$ that fall in $(t_{i-1}, t_i]$. For each of the $X_k \in \mathcal{P}$, we will say that X_k is "of type i" if X_k lies in the ith subinterval $(t_{i-1}, t_i]$. For $1 \le i \le n$, let $\alpha_i = (t_i - t_{i-1})/t^*$. Since $t_0 = 0$ and $t_n = t^*$ we have $\sum_{i=1}^{n} \alpha_i = 1$. Also, since the X_k's are uniformly distributed on $[0, t^*]$ we have $P[X_k \text{ is of type } i] = \alpha_i$. In the context of our (generalized) Theorem 29.8, N_i is the number of X_k's that are of type i. Since the X_k's are independent, the theorem insures that the N_i are independent.

The Compound Poisson Process. Consider a bagel shop where customers arrive according to a homogeneous Poisson process. Customers buy varying numbers of bagels — say the number of bagels purchased by a given customer is distributed like Binomial $(5, \frac{1}{3})$, for example. What is the probability distribution of the number of bagels sold in any particular hour of business? Or consider an insurance company where claims arrive according to a Poisson process and the size of each claim follows some underlying probability distribution. What is the probability distribution of the aggregate claims over a year?

These are two examples of a *compound Poisson process.* More formally, suppose $N, X_1, X_2, \ldots$ are independent with $N \sim \text{Poisson}\,(\theta)$ and each $X_i \sim F$, for some underlying distribution F. Now put

$$S = X_1 + \cdots + X_N.$$

Then S is a compound Poisson process. The precise distribution of S will depend on F. Here we simply compute the mean and variance of S. We do this by conditioning on the value of N. Let $S_n = X_1 + \cdots + X_n$, so $ES_n = nEX$ and $\text{Var}\, S_n = n\text{Var}\, X$. Regarding the mean,

$$\begin{aligned} ES &= \sum_{n=0}^{\infty} E[S \mid N = n]\, P[N = n] \\ &= \sum_{n=0}^{\infty} E[S_n \mid N = n]\, P[N = n] \\ &= \sum_{n=0}^{\infty} E[S_n]\, P[N = n] \\ &= \sum_{n=0}^{\infty} nEX\, P[N = n] \;=\; EX \sum_{n=0}^{\infty} nP[N = n] \;=\; EX \cdot EN. \end{aligned} \tag{29.9}$$

The second equality holds because, given $\{N = n\}$, we have $S = S_n$; the third holds because the X_i's, and hence their sum S_n, are independent of N.

Regarding the variance, we compute the second moment of S:

$$\begin{aligned} ES^2 &= \sum_{n=0}^{\infty} E[S^2 \mid N = n]\, P[N = n] \\ &= \sum_{n=0}^{\infty} ES_n^2\, P[N = n] \\ &= \sum_{n=0}^{\infty} \left(\operatorname{Var} S_n + (ES_n)^2\right) P[N = n] \\ &= \sum_{n=0}^{\infty} n \operatorname{Var} X\, P[N = n] \;+\; \sum_{n=0}^{\infty} n^2 (EX)^2\, P[N = n] \\ &= \operatorname{Var} X \cdot EN \;+\; (EX)^2 EN^2, \end{aligned}$$

which in turn gives

$$\begin{aligned} \operatorname{Var} S &= ES^2 - (ES)^2 \\ &= \operatorname{Var} X \cdot EN \;+\; (EX)^2 EN^2 \;-\; (EX)^2 (EN)^2 \\ &= \operatorname{Var} X \cdot EN \;+\; (EX)^2 \operatorname{Var} N. \end{aligned} \tag{29.10}$$

The computations of (29.9) and (29.10) work for any non-negative integer-valued random variable N, not just $N \sim$ Poisson. Let us consider (29.10) for two extreme cases. First, suppose $N = n$ is constant, so $\operatorname{Var} N = 0$ and $EN = n$. Then $S = S_n$ and (29.10) gives that $\operatorname{Var} S = n \operatorname{Var} X$, as required. Next, suppose that each $X_i = x$ is constant but N is not, so $\operatorname{Var} X = 0$ and $EX = x$. In this extreme,

$$S \;=\; \overbrace{x + \cdots + x}^{N \text{ times}} \;=\; xN,$$

so (29.10) should produce $\operatorname{Var} S = x^2 \operatorname{Var} N$, which it does. In general, the variability of S comes from two sources — the variability of N and the variability of the X_i's, as reflected in the two terms in (29.10).

Specializing now to the case where $N \sim$ Poisson (θ), we have $EN = \operatorname{Var} N = \theta$. Also let $\mu = EX$ and $\sigma^2 = \operatorname{Var} X$. Then (29.9) gives $ES = \mu\theta$, while (29.10) gives

$$\operatorname{Var} S \;=\; \sigma^2\theta + \mu^2\theta \;=\; \theta(\sigma^2 + \mu^2) \;=\; \theta E X^2.$$

EXERCISES

1. Do the "simplification and integration" alluded to at the end of the proof of Lemma 29.3.

2. If X and Y are independent exponentials with means 2 and 3, respectively, what is the distribution of $Z = \min(X, Y)$?

3. I've made appointments to see two students, one at noon and the other at 12:15. The duration of the appointments are independent exponentials with mean 15 minutes. I cannot meet with the second student until I'm done with the first. What is the expected amount of time the 12:15 student spends at my office? Assume the students arrive at their appointed times.

4. You have two dogs, Mutt and Jeff, whose remaining lifespans are independent exponentials with means 4 years and 7 years. When the first dies, what is the remaining lifespan of the other? Answer this same question if 5 years have elapsed without the death of either dog.

5. A turtle takes 45 minutes to cross the street. If cars arrive according to a Poisson process with mean arrival rate 3 per hour, what is the probability that the turtle crosses safely?

6. Consider the following game. I have two bells that ring independently at random times. Bell A's rings follow a Poisson process with mean rate 2 per minute. Bell B's rings follow a Poisson process with mean rate 1 per minute. Every time bell A rings, I pay you \$1. You can shout "stop" at any time ending the game — you keep your winnings. But if Bell B rings before you shout stop, you must give the money back to me and the game ends. For example, if Bell B rings before Bell A, you get paid nothing and the game is over.

(a) Suppose your strategy of play is to stop when you get the first dollar. What are your expected winnings?

(b) Suppose your strategy is to stop when you get \$2. What are your expected winnings?

(c) Is there a better strategy?

7. Generalize the proof of Theorem 29.8 to the general case $m \geq 2$.

§30. The Inhomogeneous Poisson Process

Here we consider a continuous-time arrival process where the rate of arrivals, $\lambda(t)$, is not constant in time. Let $\lambda(t)$ denote a non-negative function of time $t \geq 0$. A continuous-time arrival process is an *inhomogeneous Poisson process with arrival rate* $\lambda(t)$ if the counting process N_t satisfies:

$$\begin{aligned} &\text{the process } N_t \text{ has independent increments;} \\ P[N_{t+\Delta t} - N_t = 1] &= \lambda(t)\Delta t + o(\Delta t); \text{ and} \\ P[N_{t+\Delta t} - N_t > 1] &= o(\Delta t). \end{aligned} \tag{30.1}$$

The latter two conditions above hold as $\Delta t \to 0$. If $\lambda(t) = \lambda$ is constant then (30.1) describes the homogeneous Poisson process with arrival rate λ.

Formal Construction of the Process. We construct such a process on $\boldsymbol{\Omega}$ from our standard iid Uniform $(0,1)$ random variables. We will need two independent iid sequences of Uniform $(0,1)$'s — $(U_n : n \geq 1)$ and $(V_n : n \geq 1)$. We will also need to place some restrictions on $\lambda(t)$ other than non-negativity. Let us not worry about determining the least restrictive conditions that work. For now, simply assume that $\lambda(t)$ is continuous and is bounded by some number $\lambda_{\max}$. Use the sequence $(U_n : n \geq 1)$ to generate *candidate* inter-arrival times $(\widetilde{T}_n : n \geq 1)$ that are iid Exponential $(h_{\min})$, where $h_{\min} = 1/\lambda_{\max}$, and let $\widetilde{\tau}_n$ denote the time of the nth *candidate arrival*:

$$\widetilde{\tau}_n = \widetilde{T}_1 + \widetilde{T}_2 + \cdots + \widetilde{T}_n. \qquad (30.2)$$

The candidate times $\widetilde{\tau}_1, \widetilde{\tau}_2, \ldots$ arrive at rate $\lambda_{\max}$ — a rate that always exceeds (or at most equals) the variable arrival rate $\lambda(t)$. They must therefore be *thinned.* We use the independent iid sequence $(V_n : n \geq 1)$ to accomplish this. Let $\alpha(t) = \frac{\lambda(t)}{\lambda_{\max}}$, so $0 \leq \alpha(t) \leq 1$. We accept the candidate arrival $\widetilde{\tau}_n$ as an actual arrival if $V_n \leq \alpha(\widetilde{\tau}_n)$ — that is, we accept it with probability $\alpha(\widetilde{\tau}_n)$. The algorithm proceeds as follows: (i) Let $n = \widetilde{\tau} = 0$; (ii) generate $\widetilde{T} \sim$ Exponential $(h_{\min})$ and put $\widetilde{\tau} = \widetilde{\tau} + \widetilde{T}$; (iii) generate independent $V \sim$ Uniform $(0,1)$; (iv) if $V \leq \alpha(\widetilde{\tau})$ put $n = n+1$ and then $\tau_n = \widetilde{\tau}$; (v) go back to (ii).

Alternatively, we may construct the candidate arrival times, i.e. the $\widetilde{\tau}$'s, restricted to an interval $[0, t^*]$ using the Poisson point process described above. Specifically, generate $\widetilde{N} \sim$ Poisson $(\lambda_{\max} t^*)$ and then construct the point subset $\widetilde{\mathcal{P}} = \{\widetilde{X}_1, \ldots, \widetilde{X}_{\widetilde{N}}\} \subset [0, t^*]$ where each $\widetilde{X}_i$ is drawn uniformly and independently from $[0, t^*]$. From our previous discussion, we know that the point set $\widetilde{\mathcal{P}}$ has the same distribution as the point set $\{\text{the } \widetilde{\tau}_n\text{'s}\} \cap [0, t^*]$. We then thin $\widetilde{\mathcal{P}}$ as follows. For each i, we take $\widetilde{X}_i \in \mathcal{P}$ if and only if $V_i \leq \alpha(\widetilde{X}_i)$, where the V_i are independent uniform random variables.

This approach is summarized as: (i) Generate $\widetilde{N} \sim$ Poisson $(\lambda_{\max} t^*)$ and set $n = 1$ and $k = 0$; (ii) if $n > \widetilde{N}$, stop; (iii) generate $\widetilde{X} \sim$ Uniform $(0, t^*)$ and independent $V \sim$ Uniform $(0,1)$; (iv) if $V \leq \alpha(\widetilde{X})$ increment k by 1 and put $X_k = \widetilde{X}$; (v) increment n by 1 and go back to (ii). When this algorithm stops we have $\mathcal{P} = \{X_1, \ldots, X_k\}$. Of course, these numbers will not be in numerical order.

A Specific Example. Suppose RichMan's Auto Dealership sells Bentley automobiles. These are expensive cars and RichMan has, on average, 10 customers per year for Bentleys (I hope he sells other cars too). Business is seasonal, with customers arriving at a rate of 5 per year in the depth of winter and at a rate of 15 per year during the summer. The dealership models the arrival rate as $\lambda(t) = 10 - 5\cos(2\pi t)$, where $t = 0$ corresponds to January 1, and t measures elapsed time in years (so $t = 1$ corresponds to January 1 of the following year). See Figure 30.3 for one year of this seasonal cycle.

Figure 30.3. $\lambda(t) = 10 - 5\cos(2\pi t)$.

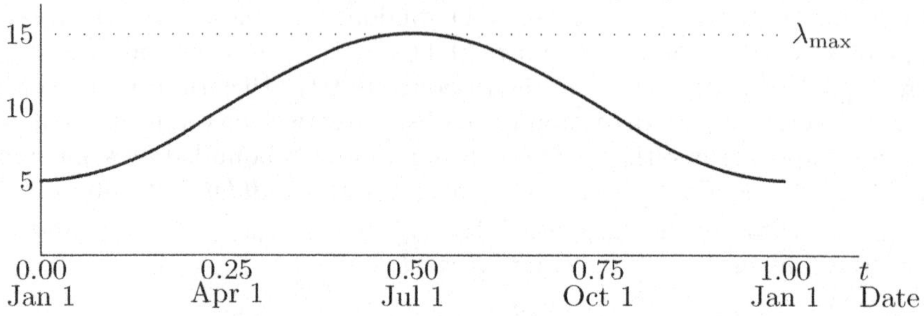

Figure 30.4 illustrates the construction of the process through the first year. The columns labeled U_n and V_n show a realization of the independent sequences of iid Uniform $(0,1)$ random variables. Here, $\lambda_{max} = 15$ and the candidate inter-arrival times $\widetilde{T}_n$ are computed from the U_n's via inverse transform: $\widetilde{T}_n = \frac{-1}{15}\ln(U_n)$. The $\widetilde{T}_n$'s are then iid Exponential $(1/15)$ random variables. The $\widetilde{\tau}_n$'s accumulate the candidate inter-arrival times. For example, $\widetilde{\tau}_3 = .0236 + .0827 + .1091 = \widetilde{\tau}_2 + .1091$. They give the candidate arrival times. Since the candidate arrivals arrive at a rate of 15 per year, they generally arrive too fast (except at July 1st) and the candidate arrivals must be thinned. The probability that the nth arrival is accepted as an actual arrival is given by $\alpha(\widetilde{\tau}_n)$, where $\alpha(t) = \lambda(t)/\lambda_{max} = (10 - 5\cos(2\pi t))/15$. These numbers are presented in the column labeled $\alpha(\widetilde{\tau}_n)$. The V_n, which are iid Uniform $(0,1)$'s, are used to make the accept/reject decision. If $V_n \leq \alpha(\widetilde{\tau}_n)$ (which happens with probability $\alpha(\widetilde{\tau}_n)$, then the nth candidate arrival is accepted as an arrival — otherwise not. Here, the actual arrivals (limited to the first year) begin with .1063, .2472, .4694, .4797, .6245, .7042 — as highlighted in Figure 30.4.

Figure 30.4. One year of the Bentley process.

n	U_n	$\widetilde{T}_n$	$\widetilde{\tau}_n$	$\alpha(\widetilde{\tau}_n)$	V_n	Accepted?
1	.7016	.0236	.0236	.3370	.7254	No
2	.2894	.0827	.1063	.4050	.3872	Yes
3	.1948	.1091	.2154	.5942	.9603	No
4	.6208	.0318	.2472	.6608	.1115	Yes
5	.0357	.2222	.4694	.9939	.1075	Yes
6	.8571	.0103	.4797	.9973	.9781	Yes
7	.1232	.1396	.6193	.9107	.9203	No
8	.9253	.0052	.6245	.9031	.7658	Yes
9	.4225	.0574	.6819	.8050	.9758	No
10	.7158	.0223	.7042	.7613	.0733	Yes
11	.0135	.2870	.9912	.3338	.9225	No
12	.3126	.0775	1.0687	.3639	.8514	No

Why this Construction Works. Here we argue informally why the construction of the previous section produces an arrival process satisfying (30.1).

That $N_t - N_s$, for $t > s$, is independent of the history of the process up to time s follows from two observations. First, the candidate arrivals are a homogeneous Poisson process so the realization of candidate arrivals after time s is independent of the process up to time s. Second, the accept/reject decisions for these candidate arrivals are made using independent numbers from the V_n sequence.

To see that $P[N_{t+\Delta t} - N_t > 1] = o(\Delta t)$, let $\widetilde{N}_t$ denote the counting process associated with the candidate arrivals. Then, as noted above, $\widetilde{N}_t$ is a homogeneous Poisson process so we have $P[\widetilde{N}_{t+\Delta t} - \widetilde{N}_t > 1] = o(\Delta t)$. But necessarily $N_{t+\Delta t} - N_t \le \widetilde{N}_{t+\Delta t} - \widetilde{N}_t$, as the first process is obtained by thinning the second.

Finally we argue that $P[N_{t+\Delta t} - N_t = 1] = \lambda(t)\Delta t + o(\Delta t)$. Let E_1 denote the event that there is exactly one candidate arrival in the time interval $(t, t+\Delta t]$, E_2 the event that there are two or more candidate arrivals in $(t, t+\Delta t]$, and A the event that the next candidate arrival subsequent to time t is accepted as an actual arrival. Then

$$E_1 \cap A \subset \{N_{t+\Delta t} - N_t = 1\} \subset (E_1 \cap A) \cup E_2,$$

and, since $PE_2 = o(\Delta t)$ we have $P[N_{t+\Delta t} - N_t = 1] = PE_1 \cdot P[A|E_1] + o(\Delta t)$. Again using that the candidate arrival process is Poisson with rate $\lambda_{\max}$, we have that $PE_1 = \lambda_{\max}\Delta t + o(\Delta t)$. If there is exactly one candidate arrival on $(t, t+\Delta t]$, it is accepted with some probability $\alpha(s) = \frac{\lambda(s)}{\lambda_{\max}}$ where $t < s \le t+\Delta t$ and s is the time of that one arrival. Since $\lambda(t)$ is continuous, we have $\alpha(s) \to \alpha(t)$ as $\Delta t \to 0$ so $P[A|E_1] = \alpha(t) + o(1)$ as $\Delta t \to 0$. Putting this together and using that $\lambda_{\max}\alpha(t) = \lambda(t)$, we get

$$\begin{aligned} P[N_{t+\Delta t} - N_t = 1] &= (\lambda_{\max}\Delta t + o(\Delta t)) \cdot (\alpha(t) + o(1)) + o(\Delta t) \\ &= \lambda(t)\Delta t + o(\Delta t), \end{aligned}$$

as $\Delta t \to 0$.

The Distribution of $N_t - N_s$. For the homogeneous case with arrival rate λ, $N_t - N_s$, the number of arrivals on the interval $(s, t]$, is distributed like Poisson $(\lambda \cdot (t-s))$. For the inhomogeneous process we have:

THEOREM 30.5. *If the process constructed above has bounded and continuously varying arrival rate $\lambda(t)$ then $N_t - N_s \sim$ Poisson (θ), where $\theta = \int_s^t \lambda(x)\,dx$.*

REMARK. This agrees when applied to the homogeneous case as $\int_s^t \lambda\,dx = \lambda \cdot (t-s)$. This theorem holds for piecewise continuous functions $\lambda(t)$, not just continuous functions, although (30.1) may not hold at λ's discontinuities (see exercises). In Step 1 below we establish the theorem for step functions, which are decidedly not continuous.

PROOF OF THEOREM 30.5. *Step 1.* First we establish this when $\lambda(x)$ is a step function, as depicted in Figure 30.6 below. In this case there are numbers $s = x_0 < x_1 < \cdots < x_m = t$ and corresponding function values $\lambda_1, \ldots, \lambda_m$ with $\lambda(x) = \lambda_k$ for $x_{k-1} < x \leq x_k$. Here $1 \leq k \leq m$ — in the Figure $m = 4$.

Figure 30.6. The case where $\lambda(x)$ is a step function.

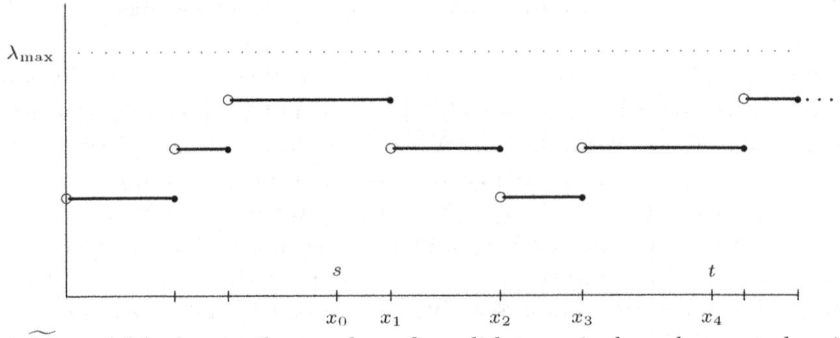

Let $\widetilde{M_k}$ and M_k denote the number of candidate arrivals and accepted arrivals, respectively, on the time interval $(x_{k-1}, x_k]$. Then

$$\begin{aligned} N_t - N_s &= \text{the number of accepted arrivals on } (s, t] \\ &= M_1 + \cdots + M_m. \end{aligned}$$

Previously we established that this arrival process has independent increments, so $M_1, \ldots, M_m$ are independent. Furthermore, each $\widetilde{M_k} \sim$ Poisson $(\lambda_{\max} \cdot (x_k - x_{k-1}))$ and the candidate arrivals on $(x_{k-1}, x_k]$ are independently accepted with probability $\frac{\lambda_k}{\lambda_{\max}}$. From Theorem 29.8 we see that the number of accepted arrivals, M_k, is distributed like Poisson $(\lambda_k \cdot (x_k - x_{k-1}))$. Presently we need:

LEMMA 30.7. *If $M_1, \ldots, M_m$ are independent with each $M_k \sim$* Poisson (θ_k) *then $M_1 + \cdots + M_m \sim$* Poisson (θ), *where $\theta = \theta_1 + \cdots + \theta_m$.*

PROOF. See exercises. □

From this we get that $N_t - N_s \sim$ Poisson (θ), where

$$\theta = \sum_{k=1}^{m} \lambda_k \cdot (x_{k+1} - x_k) = \int_s^t \lambda(x)\, dx.$$

This concludes step 1. We will need the following lemma for step 2:

LEMMA 30.8. *Suppose N and A_i, B_i, $i \geq 1$, are non-negative integer valued random variables satisfying:* (i) $A_i \leq N \leq B_i$; *and* (ii), *for each $n \geq 0$ we have $P[A_i = n] \to p_n$ and $P[B_i = n] \to p_n$, both as $i \to \infty$. Then $P[N = n] = p_n$.*

PROOF. By (i), $\{B_i \leq n\} \subset \{N \leq n\} \subset \{A_i \leq n\}$, so

$$P[B_i \leq n] \;\leq\; P[N \leq n] \;\leq\; P[A_i \leq n].$$

Now, by (ii),

$$P[B_i \leq n] \;=\; \sum_{m=0}^{n} P[B_i = m] \;\rightarrow\; \sum_{m=0}^{n} p_m \quad \text{as} \quad i \rightarrow \infty.$$

Similarly, $P[A_i \leq n] \rightarrow \sum_{m=0}^{n} p_m$ as $i \rightarrow \infty$. It follows that $P[N \leq n] = \sum_{m=0}^{n} p_m$ for all $n \geq 0$, so

$$P[N = n] \;=\; P[N \leq n] - P[N \leq n-1] \;=\; \sum_{m=0}^{n} p_m - \sum_{m=0}^{n-1} p_m \;=\; p_n. \quad \square$$

Step 2 — General continuous $\lambda(x)$. By hypothesis, $\lambda(x)$ is continuous on $[s,t]$ and hence uniformly continuous on $[s,t]$ (see Theorem 44.5). Fix any $\epsilon > 0$ and choose δ so that $|\lambda(x) - \lambda(x')| < \epsilon$ whenever $|x - x'| < \delta$ with $x, x' \in [s,t]$. Let $P = \{x_0 < x_1 < \cdots < x_m\}$ denote a partition of $[s,t]$ with $|x_{k-1} - x_k| < \delta$ for each k. Now put

$$\begin{aligned} \lambda_k^- &= \min\{\lambda(x) : x_{k-1} \leq x \leq x_k\} \quad \text{and} \\ \lambda_k^+ &= \max\{\lambda(x) : x_{k-1} \leq x \leq x_k\}, \end{aligned}$$

so, for $x_{k-1} \leq x \leq x_k$, we have $\lambda_k^- \leq \lambda(x) \leq \lambda_k^+$. By our choice of δ we have $\lambda_k^+ \leq \lambda_k^- + \epsilon$.

Defining step functions $\lambda^-(x)$ and $\lambda^+(x)$ on $(s,t]$ where $\lambda^-(x) = \lambda_k^-$ and $\lambda^+(x) = \lambda_k^+$ for $x_{k-1} < x \leq x_k$, we have that

$$\lambda^-(x) \;\leq\; \lambda(x) \;\leq\; \lambda^+(x) \;\leq\; \lambda^-(x) + \epsilon, \tag{30.9}$$

for all $x \in (s,t]$. Then

$$\begin{aligned} \int_s^t \lambda^-(x)\,dx \;\leq\; \int_t^s \lambda(x)\,dx \;&\leq\; \int_s^t \lambda^+(x)\,dx \\ &\leq\; \int_s^t \lambda^-(x)\,dx + \epsilon\cdot(t-s). \end{aligned} \tag{30.10}$$

This construction is based on an arbitrary $\epsilon > 0$. We do this construction for $\epsilon = \frac{1}{i}$, for $i = 1, 2, 3, \ldots$ and let $\lambda_i^-(x)$ and $\lambda_i^+(x)$ denote the corresponding step functions as constructed above. Now let A_i denote the number of arrivals accepted from the candidate arrivals on the time interval $(s,t]$ if a candidate arrival at time x is accepted with probability $\lambda_i^-(x)/\lambda_{\max}$ and let B_i denote the number of candidate arrivals accepted if an arrival at time x is accepted with probability $\lambda_i^+(x)/\lambda_{\max}$. That is, make the acceptance/rejection decision for candidate arrivals with $\lambda(x)$ replaced by $\lambda_i^-(x)$ for A_i, and by $\lambda_i^+(x)$ for B_i. By (30.9), we have that $A_i \leq N_t - N_s \leq B_i$. By Step 1, we see that

$$A_i \sim \text{Poisson}\left(\int_s^t \lambda_i^-(x)\,dx\right), \quad B_i \sim \text{Poisson}\left(\int_s^t \lambda_i^+(x)\,dx\right).$$

Since $\frac{1}{i}\cdot(t-s) \to 0$, (30.10) gives that the two Poisson parameters above converge to $\theta = \int_s^t \lambda(x)\,dx$ as $i \to \infty$ and Lemma 30.8 yields $N_t - N_s \sim \text{Poisson}(\theta)$. □

Higher Dimensional Versions. The one dimensional Poisson point process has higher dimensional analogs. On a line, the distance between consecutive points may be viewed as inter-arrival times — the T_n's. On a plane or in space there is no such interpretation and we must view the process as a point process, i.e, a random subset of a plane or space.

Suppose, for example, we wish to construct such a point process on the unit square $\mathcal{S} = [0,1] \times [0,1] \subset \mathbb{R}^2$. That is, we wish to construct a random subset of points $\mathcal{P} \subset \mathcal{S}$. Suppose we have a non-negative function $\lambda : \mathcal{S} \to \mathbb{R}^+$, where $\lambda(x,y)$ represents the *density* of the point process at location $(x,y) \in \mathcal{S}$. Specifically, if $\mathcal{R}$ is a small square patch of area ΔA sitting inside of $\mathcal{S}$ with $(x,y) \in \mathcal{R}$ and N is the number of points from $\mathcal{P}$ that fall inside of $\mathcal{R}$, we wish to have $P[N=1] = \lambda(x,y)\Delta A + o(\Delta A)$ and $P[N>1] = o(\Delta A)$. Additionally, if $\mathcal{R}_1, \ldots, \mathcal{R}_n$ are disjoint regions of $\mathcal{S}$ and $N_1, \ldots, N_n$ are the respective number of points from $\mathcal{P}$ falling inside these regions, we wish for $N_1, \ldots, N_n$ to be independent.

The construction, and the reason it works, is identical to the one dimensional point process. We first construct a homogeneous point process on $\mathcal{S}$ and then thin it to reflect the density function $\lambda(x,y)$. Suppose $\lambda_{\max}$ is a bound for $\lambda(x,y)$ on $\mathcal{S}$: all $\lambda(x,y) \leq \lambda_{\max}$. Construct $\widetilde{N} \sim \text{Poisson}(\lambda_{\max} \cdot \text{Area}(\mathcal{S}))$ — this is the number of candidate points that fall in $\mathcal{S}$. (Here, of course, $\text{Area}(\mathcal{S}) = 1$.) Then situate $\widetilde{N}$ points uniformly on $\mathcal{S}$. This may be accomplished by repeatedly ($\widetilde{N}$ times) constructing independent Uniform$(0,1)$'s X and Y and adding the point (X,Y) to $\widetilde{\mathcal{P}}$. Next we thin $\widetilde{\mathcal{P}}$. For each point $(X,Y) \in \widetilde{\mathcal{P}}$ compute $\alpha = \frac{\lambda(X,Y)}{\lambda_{\max}}$. Generate independent $V \sim \text{Uniform}(0,1)$ and accept (X,Y) as an actual point (put (X,Y) into $\mathcal{P}$) if $V \leq \alpha$.

An example of this is shown in Figure 30.11. Here, the two squares are the unit square $\mathcal{S}$ and, for $(x,y) \in \mathcal{S}$ we take $\lambda(x,y) = 10000e^{-10d(x,y)}$, where $d(x,y)$ is the distance from the point (x,y) to the boundary of the square. For example, $d(\frac{1}{2},\frac{1}{2}) = \frac{1}{2}$ so $\lambda(\frac{1}{2},\frac{1}{2}) = 10000e^{-5} \approx 67$, Along the boundary of the square we have $d(x,y) = 0$ and $\lambda(x,y) = 10000$, so $\lambda_{\max} = 10000$. Near the center of the square, where $\lambda(x,y)$ is minimized, we have $\alpha \approx \frac{67}{10000}$ and only one out of each (roughly) 149 candidate points is accepted as an actual point in the thinned process. The homogeneous Poisson process with density 10000 is shown to the left; to the right is the thinned process.

Figure 30.11. Candidate and thinned point processes.

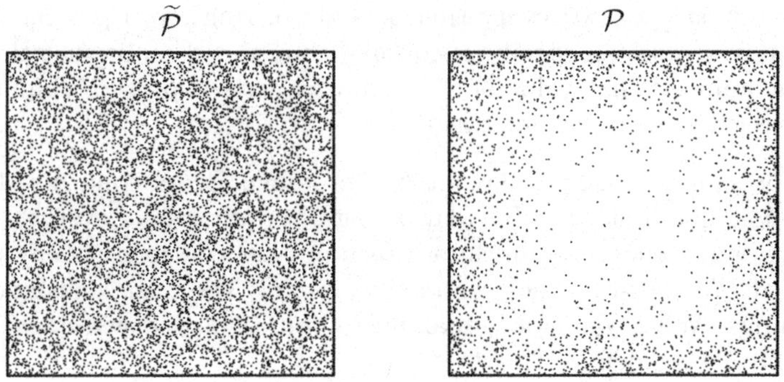

EXERCISES

1. Prove Lemma 30.7 for $m = 2$ by direct computation.

2. Prove the lemma for general m by induction.

3. Show that (30.1) will not hold at time $t = s$ if $\lim_{t\to s^+} \lambda(t) \neq \lambda(s)$.

4. Suppose $T \geq 0$ is a non-negative random variable with continuous density function supported on the interval $(0, t^*)$, where possibly $t^* = \infty$. Then $f(t)$ satisfies $P[t < T \leq t + \Delta t] = f(t)\Delta t + o(\Delta t)$. Let $F(t)$ denote its distribution function. The *hazard rate function* $\lambda(t)$ for T is given by $\lambda(t) = \frac{f(t)}{1-F(t)}$ for $0 \leq t < t^*$.

(a) Show that the hazard rate function satisfies $P[t < T \leq t + \Delta t \,|\, T > t] = \lambda(t)\Delta t + o(\Delta t)$, for $0 \leq t < T$.

(b) Compute the hazard rate function if $T \sim$ Exponential (h). (Here $t^* = \infty$.)

(c) Compute the hazard rate function if $T \sim$ Uniform $(0, t^*)$, where $t^* < \infty$.

(d) Show that in general $P[T > t] = \exp\left(-\int_0^t \lambda(s)\, ds\right)$.

(e) For $f(t)$ to be a bona-fide density function we must (in addition to non-negativity) have $\int_0^{t^*} f(t)\, dt = 1$. What analogous condition must $\lambda(t)$ satisfy to be a bona-fide hazard rate function?

5. Suppose T_1 is the first inter-arrival time of an inhomogeneous Poisson process with continuous arrival rate function $\lambda(t)$. Show that $\lambda(t)$ is T_1's hazard rate function.

6. Clock A's alarm sounds according to an inhomogeneous Poisson process with arrival rate function $a(t) = 10(1-t)$ for $0 \le t \le 1$ (in minutes) and $a(t) = 0$ thereafter. The rate function for clock B's alarm is $b(t) = 10t$ for $0 \le t \le 1$, $b(t) = 10$ thereafter. Suppose we play the following game. Every time clock A's alarm rings, I pay you \$1; every time B's alarm rings, you pay me \$1. Before the game starts, you can specify a stopping time $t^* \le 1$ — at that time we stop playing the game. What value of t^* maximizes your expected winnings? What are your expected winnings for this t^*?

7. With the same clocks, we play a different game. Every time A's alarm rings, I pay you \$1. The first time B's alarm rings, you must give everything back to me and the game ends. Again, in advance of play you may specify a stopping time t^* — if B's alarm has not sounded by time t^* the game ends and you keep your winnings. What value of t^* maximizes your expected winnings? What are your expected winnings for this t^*?

8. Here we play with the same clocks and the same rules as problem 7. The only difference is that you do not have to specify t^* in advance of play, but rather may decide dynamically when to stop. A stopping strategy is a function $s(n,t) \to \{\text{stop}, \text{continue}\}$ where $0 \le t \le 1$ is the elapsed play-time and $n \ge 0$ is how many dollars I've paid you by time t. For example, if play has not stopped by time $t = 0.5$ (30 seconds) and A's alarm has sounded 4 times, $s(4, 0.5)$ tells you whether to stop at that moment or continue play. Determine a stopping strategy that is optimal in the sense that it maximizes your expected winnings. (Actually computing the expected winnings is intractable, but should it be less than or greater than your answer to 7?)

9. Suppose each customer arriving at RichMan's Auto (refer to 30.3 and 30.4) independently buys 1, 2, or 3 Bentleys with probability 0.5, 0.25, and 0.25, respectively. (His clientele is *very* rich.) Let S denote the number of Bentleys sold during the first quarter of the year. Compute ES and $\operatorname{Var} S$.

10. Use Markov's inequality to put an upper bound on $P[S \ge 6]$. Use Corollary 9.3 with $n = 2$ to do the same. Which bound is better?

§31. When Inter-Arrival Times are IID but not Exponential

In this final section we return to the one dimensional setting and view the process as an arrival, rather than point, process. We relax the assumption that the inter-arrival times are exponential (but keep them iid) and study the long-term behavior of the *waiting time process.*

The Waiting Time Process. For any continuous-time arrival process we may compute the waiting time process $(W_t : t \ge 0)$ defined as

$$W_t = \min\{\tau_n : \tau_n > t\} - t.$$

For any time t, W_t measures the amount of additional time, starting at time t, that one must wait for the next arrival. Figure 31.2 illustrates this process for a particular realization of the underlying arrival process up to some deterministic time s. Recall that $N(s) = N_s$ denotes the number of arrivals by time s and $T_1, T_2, \ldots$ are the successive inter-arrival times. (We will use the notation $N(s)$ and N_s interchangeably, with usage determined by readability considerations.)

Here we study the average waiting time over the time interval $[0, s]$, defined as

$$A(s) = \frac{1}{s}\int_0^s W_t\,dt, \tag{31.1}$$

in the case where the inter-arrival times are iid but not necessarily exponential. We seek to show that $A(s)$ converges to a limiting value as $s \to \infty$ and to identify that limit.

Figure 31.2. Illustrative realization of the waiting time process.

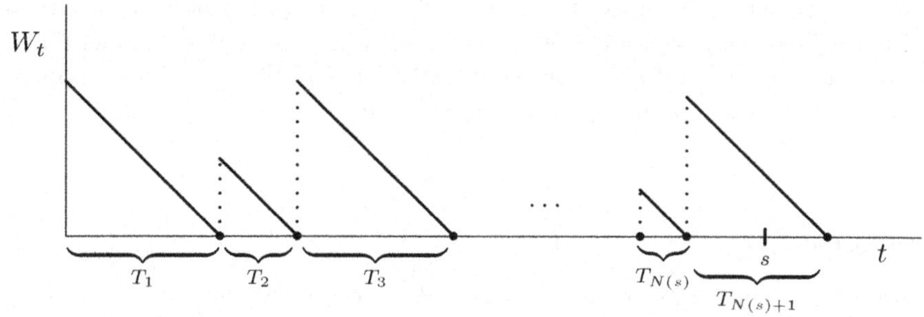

Note that the kth triangular piece in the figure has both base and height T_k. Therefore, by adding triangular areas, we have:

$$\sum_{k=1}^{N(s)} \frac{1}{2} T_k^2 < \int_0^s W_t\,dt < \sum_{k=1}^{N(s)+1} \frac{1}{2} T_k^2.$$

Dividing through by s yields that:

$$\frac{1}{s}\sum_{k=1}^{N(s)} \frac{1}{2} T_k^2 \le A(s) \le \frac{1}{s}\sum_{k=1}^{N(s)+1} \frac{1}{2} T_k^2. \tag{31.3}$$

The left and right sides above differ at most by the area of the last triangle divided by s — an amount that should get small as $s \to \infty$. The left hand side may be re-written as

$$\frac{1}{s}\sum_{k=1}^{N(s)} \frac{1}{2} T_k^2 = \frac{N(s)}{s} \cdot \frac{1}{N(s)}\sum_{k=1}^{N(s)} \frac{1}{2} T_k^2. \tag{31.4}$$

For generic inter-arrival time T, let $\mu = ET$ and $\sigma^2 = \text{Var}(T)$, so $E[T^2] = \sigma^2 + \mu^2$. Now $N(s)$ is the number of arrivals on a time interval of length s, so we expect that $\frac{s}{N(s)} \to \mu$ as $s \to \infty$, or $\frac{N(s)}{s} \to \frac{1}{\mu}$ as $s \to \infty$. As for the second factor on the right side of (31.4), the Strong Law of Large Numbers gives that

$$\frac{1}{n}\sum_{k=1}^{n} \frac{1}{2} T_k^2 \;\to\; \frac{1}{2}E[T^2] \;=\; \frac{\mu^2 + \sigma^2}{2},$$

as $n \to \infty$. This all (correctly) suggests that we might have:

$$A(s) \;\to\; \frac{1}{\mu}\frac{\mu^2 + \sigma^2}{2} \;=\; \frac{\mu}{2} + \frac{\sigma^2}{2\mu} \tag{31.5}$$

as $s \to \infty$.

Before rigorously proving this, we offer the following example and remark.

EXAMPLE 31.6. Suppose the inter-arrival times are exponential with mean μ. In this situation, we know from the memoryless property of exponentials that, for any time t, $W_t \sim \text{Exponential}(\mu)$ and hence $EW_t = \mu$. For exponentials, the variance is the square of the mean: $\sigma^2 = \mu^2$, so (31.5) gives that

$$A(s) \;\to\; \frac{\mu}{2} + \frac{\mu}{2} \;=\; \mu,$$

as expected.

REMARK. One might expect in general for the mean waiting time to be just $\frac{\mu}{2}$ — half of a typical inter-arrival time. Equation (31.5) shows that if there is *any* randomness to the inter-arrival times then the mean waiting time exceeds this value. Why is this true? Imagine a time t substantially in the future. Let $\tau^-(t)$ be time of the last arrival prior to time t and $\tau^+(t)$ be time of the first arrival after time t. Now put $T(t) = \tau^+(t) - \tau^-(t)$, so $T(t)$ is the inter-arrival time corresponding to the two arrivals bracketing time t. Under quite general conditions it *is* true that EW_t and $\frac{1}{2}ET(t)$ have the same limit as $t \to \infty$. However, $T(t)$ and a generic T_k (for any deterministic k) are not drawn from the same distribution even though $T(t)$ is one of the T_k. The inter-arrival time bracketing t is *biased* to be larger than a typical inter-arrival time.

Presently we state and prove:

THEOREM 31.7. *Suppose the inter-arrival times T_k are iid and have finite exponential moment. Then $A(s) \to \frac{\mu}{2} + \frac{\sigma^2}{2\mu}$ (almost surely) as $s \to \infty$.*

PROOF. We follow the outline of above.

Step 1. Here we show that $\frac{N(s)}{s} \to \frac{1}{\mu}$ as $s \to \infty$. Under the theorem's hypotheses, we know from the SLLN that

$$\frac{T_1 + \cdots + T_n}{n} \to \mu, \qquad \text{as } n \to \infty.$$

So, with probability one, if $0 < \epsilon < \mu$ we will have

$$\left| \frac{T_1 + \cdots + T_n}{n} - \mu \right| < \epsilon \quad \text{for all sufficiently large } n,$$

which after re-arranging implies that

$$n \cdot (\mu - \epsilon) < T_1 + \cdots + T_n < n \cdot (\mu + \epsilon) \qquad (31.8)$$
$$\text{for all sufficiently large } n.$$

The first inequality in (31.8) implies that $N(n \cdot (\mu - \epsilon)) < n$ or, since $N(t)$ is always an integer, that $N(n \cdot (\mu - \epsilon)) \leq n - 1$ and by substituting $n + 1$ for n we see that

$$N((n+1) \cdot (\mu - \epsilon)) \leq n \qquad (31.9)$$

— all for sufficiently large n.

Now, for any time s, let n_s be the integer satisfying $n_s \cdot (\mu - \epsilon) \leq s < (n_s + 1) \cdot (\mu - \epsilon)$ so, in particular, $n_s \to \infty$ as $s \to \infty$. Then by (31.9), for large s,

$$N(s) \leq N((n_s + 1) \cdot (\mu - \epsilon)) \leq n_s,$$

so

$$\frac{N(s)}{s} \leq \frac{n_s}{s} \leq \frac{n_s}{n_s \cdot (\mu - \epsilon)} = \frac{1}{\mu - \epsilon}. \qquad (31.10)$$

Using the second inequality in (31.8), with similar reasoning (see exercises), yields that

$$\frac{1}{\mu + \epsilon} \leq \frac{N(s)}{s} \leq \frac{1}{\mu - \epsilon}$$

for all sufficiently large s. Since ϵ may be taken arbitrarily small, we have established step 1.

Step 2. Here we show that

$$\frac{1}{N(s)} \sum_{k=1}^{N(s)} \frac{1}{2} T_k^2 \to \frac{1}{2} E[T^2] = \frac{\sigma^2 + \mu^2}{2} \quad \text{as } s \to \infty.$$

The issue here is that $N(s)$ is *random*, so the SLLN does not immediately apply. However, we have:

LEMMA 31.11. *Suppose X_n is a sequence of random variables that converges almost surely to the limit ℓ as $n \to \infty$. Suppose further that $N_s = N(s)$ is a positive integer-valued stochastic process with $N(s) \to \infty$ with probability one as $s \to \infty$. Then $X_{N(s)} \to \ell$ with probability one as $s \to \infty$.*

PROOF. This lemma is essentially the same as Lemma 19.7. The proof is virtually identical. □

Step 2, continued. This applies to our setting immediately by letting

$$X_n = \frac{1}{n}\sum_{k=1}^{n}\frac{1}{2}T_k^2. \tag{31.12}$$

Then, by the SLLN, $X_n \to \frac{1}{2}ET^2$ as $n \to \infty$. Also, by step 1 we know that $N(s) \to \infty$ as $s \to \infty$. Hence $X_{N(s)} \to \frac{1}{2}ET^2 = \frac{\mu^2+\sigma^2}{2}$ as $n \to \infty$. But, with X_n as in (31.12), we have

$$X_{N(s)} = \frac{1}{N(s)}\sum_{k=1}^{N(s)}\frac{1}{2}T_k^2.$$

This establishes step 2.

Step 3. Here we use (31.3) to complete the proof. Clearly steps 1 and 2 show that the left side of (31.3) converges to $\frac{1}{\mu}\cdot\frac{\mu^2+\sigma^2}{2} = \frac{\mu}{2}+\frac{\sigma^2}{2\mu}$ as $s \to \infty$. On the other hand, the right side is

$$\frac{1}{s}\sum_{k=1}^{N(s)+1}\frac{1}{2}T_k^2 = \frac{N(s)}{s}\cdot\frac{N(s)+1}{N(s)}\cdot\frac{1}{N(s)+1}\sum_{k=1}^{N(s)+1}\frac{1}{2}T_k^2.$$

The first factor on the right above converges to $\frac{1}{\mu}$, the second converges to 1, and the third converges to $\frac{\mu^s+\sigma^2}{2}$. Hence the right side of (31.3) also converges to $\frac{\mu}{2}+\frac{\sigma^2}{2\mu}$ as $s \to \infty$. It follows that $A(s)$, caught in between, converges to $\frac{\mu}{2}+\frac{\sigma^2}{2\mu}$ as $s \to \infty$. □

EXERCISES

1. Fill in the details in step 1 of the proof of Theorem 31.7 by using the second inequality of (31.8) to show that $\frac{1}{\mu+\epsilon} \le \frac{N(s)}{s}$ holds for all sufficiently large s.

2. Suppose buses arrive at a bus stop with inter-arrival times that are uniformly distributed from 10 minutes to 20 minutes. Over the course of a day, what is the mean waiting time for a bus?

3. Suppose a continuous-time arrival process has inter-arrival times that are uniformly distributed on $\{1, 2, 3, 4\}$.

(a) Find the long-term average waiting time for this process:

$$A = \lim_{s\to\infty}\frac{1}{s}\int_0^s W_t\,dt.$$

(b) If we restrict our attention to *integer* times, then $W_0, W_1, W_2, \ldots$ is a Markov chain with $S = \{1, 2, 3, 4\}$. Compute the PTM $\mathbf{P}$.

(c) Show that $\mathbf{P}$ is ergodic and find the invariant distribution μ for $\mathbf{P}$.

(d) Use the invariant distribution to compute the long-run average waiting time *at integer times*:

$$\widetilde{A} = \lim_{n\to\infty} \frac{1}{n} \sum_{k=0}^{n-1} W_k.$$

(e) Your answers to (d) and (a) should differ by exactly 1/2, i.e. $\widetilde{A} - A = \frac{1}{2}$. Explain this discrepancy. Hint: draw a picture showing the areas corresponding to $\int_0^n W_t\,dt$ and $\sum_{k=0}^{n-1} W_k$.

4. Suppose a continuous-time arrival process has inter-arrival times $T_1, T_2, \ldots$ that are iid with each T_k distributed like the square of a mean-one exponential. Find the mean inter-arrival time for this process. Find the long-term average waiting time for this process using Theorem 31.7. What do you notice? (This distribution does not have finite exponential moment but Theorem 31.7 still holds — see the next problem.)

5. Show that our version of the SLLN (Theorem 10.7) may be used in the proof of Theorem 31.7 above if the hypothesis *finite exponential moment* in the statement of Theorem 31.7 is replaced with *finite eighth moment.* Using problem 5 from §11, justify the parenthetical in problem 4.

CHAPTER VI

BROWNIAN MOTION

§32. Introduction

Brownian motion is a stochastic process indexed by the non-negative real numbers $\mathbb{R}^+ = \{t \in \mathbb{R} : t \geq 0\}$. We will use the notation $(B(t) : t \geq 0)$ to denote this process. In applications the index variable t is often thought of as time (we shall often refer to "*time t*"). Since t is a continuous variable, in addition to $B(0), B(1), B(2), \ldots$, it also makes sense to refer to $B(\frac{1}{2})$, $B(\sqrt{2})$, for example. For each fixed $t \geq 0$, $B(t)$ is a random variable whose value at $\omega \in \mathbf{\Omega}$ is $B_\omega(t)$. For each fixed $\omega \in \mathbf{\Omega}$, $B_\omega(t)$ is a function of the variable t. The notation $B_t(\omega)$ is more consistent with the usual $X(\omega)$ notation for random variables, but in this chapter we wish to emphasize that for each outcome ω, B_ω is a function of a continuous variable t. In the final section of this chapter we show how to generate a realization $B_\omega(t)$ corresponding to each number $\omega \in \mathbf{\Omega}$.

The stochastic process $B(t)$ is a *standard Brownian motion* if it satisfies the following properties:

$$B_\omega(0) = 0; \tag{32.1}$$

$$\text{for each } \omega \in \mathbf{\Omega},\ B_\omega(t) \text{ is a continuous function of } t; \tag{32.2}$$

$$B(t) - B(s) \sim \text{Normal}\,(0, t - s) \text{ for } 0 \leq s < t; \text{ and} \tag{32.3}$$

$$\begin{gathered} B(t_1) - B(t_0),\ B(t_2) - B(t_1),\ \ldots,\ B(t_m) - B(t_{m-1}) \\ \text{are independent whenever } 0 = t_0 < t_1 < \cdots < t_m. \end{gathered} \tag{32.4}$$

Statements (32.1) and (32.2) say that each B_ω is a continuous path starting at 0. Statement (32.3) says that the Brownian increments are normally distributed with mean 0 and variance equal to the elapsed time. Statement (32.4) says that Brownian increments $\Delta B_{t_i} = B(t_{i+1}) - B(t_i)$ over disjoint time periods are independent. It follows from (32.1) and (32.3) that $B(t)$ is a normal random variable with mean 0 and variance t.

An equivalent way to state (32.1), (32.3) and (32.4) is that the joint density function of the m-dimensional random vector $(B(t_1), B(t_2), \ldots, B(t_m))$ is

$$f(b_1, b_2, \ldots, b_m) = \prod_{i=0}^{m-1} \psi(b_{i+1} - b_i; 0, \Delta t_i), \tag{32.5}$$

where $b_0 = 0$, $\Delta t_i = t_{i+1} - t_i$ and $\psi(x;\mu,\sigma^2)$ is the Normal(μ,σ^2) density function evaluated at x:

$$\psi(x;\mu,\sigma^2) = \frac{1}{\sqrt{2\pi}\sigma} e^{-(x-\mu)^2/2\sigma^2}. \tag{32.6}$$

(We will use a semicolon to distinguish the function argument (x) from function parameters (μ and σ^2).) This is true because:

$$\text{(i)} \quad \boxed{\begin{aligned} B(t_1) &= b_1, \\ B(t_2) &= b_2, \\ &\cdots \\ B(t_m) &= b_m \end{aligned}} \quad \Longleftrightarrow \quad \boxed{\begin{aligned} \Delta B_{t_0} &= b_1 - b_0, \\ \Delta B_{t_1} &= b_2 - b_1, \\ &\cdots \\ \Delta B_{t_{m-1}} &= b_m - b_{m-1} \end{aligned}} \; ;$$

(ii) each $\Delta B_{t_i} \sim$ Normal$(0, \Delta t_i)$; and (iii) the increments are independent.

If we are handed a sequence $N_0, N_1, N_2, \ldots$ of independent standard normals, we may use these to generate paths whose distributional properties approximate as closely as desired the properties of Brownian motion. Specifically, let us concentrate on generating paths over the time interval $0 \le t \le 1$. (The same idea works on the general interval $0 \le t \le T$.) Fix some positive integer n and divide the interval $[0,1]$ into n equal subintervals: put $\Delta t = \frac{1}{n}$, and $t_i = i\Delta t$ for $0 \le i \le n$. Let ΔB_{t_i} denote the Brownian increment over the time interval $[t_i, t_{i+1}]$. Then these increments are independent and distributed like Normal$(0,\Delta t)$ or, equivalently, like $\sqrt{\Delta t}\cdot$Normal$(0,1)$. Setting $\Delta B_{t_i} = \sqrt{\Delta t}\cdot N_i$ and then

$$B(t_i) = \Delta B_{t_0} + \Delta B_{t_1} + \cdots + \Delta B_{t_{i-1}} = B(t_{i-1}) + \Delta B_{t_{i-1}}, \quad 1 \le i \le n,$$

will therefore generate values at the times t_i with the correct joint distribution. (Of course we take $B(0) = 0$.) This scheme is executed in Figure 32.7 for a particular realization of $N_0, N_1, \ldots$ as simulated on a computer. The values $B(t_i)$ are computed as above and the path $B(t)$ is linearly interpolated between these times. This exercise is carried out for $n = 10, 50, 100$, and 200. We have used the same sequence $(N_i : i \ge 0)$ to generate each of the four paths. This is apparent upon close examination: for example, the "$n = 10$" path appears (rescaled and highlighted) in the first fifth of the "$n = 50$" path (which, in turn, appears in the first half of the "$n = 100$" path, etc.).

This process produces paths whose properties, in a sense that can be made rigorous, approximate arbitrarily closely the defining properties (32.1 – 32.4). However, it should be apparent that this methodology does not produce paths that converge to a *particular* path. It is natural to ask if a stochastic process with properties (32.1 – 32.4) actually exists. We will answer this (in the affirmative) in a subsequent section. There, a different methodology (still using a sequence of iid standard normals) is presented. This methodology (the Lévy construction) produces a sequence of paths that converge to a particular path having the desired properties. We hold off on this and first investigate certain properties of Brownian motion assuming (for the moment) that it does exist.

Figure 32.7. Brownian path approximations for $0 \le t \le 1$.

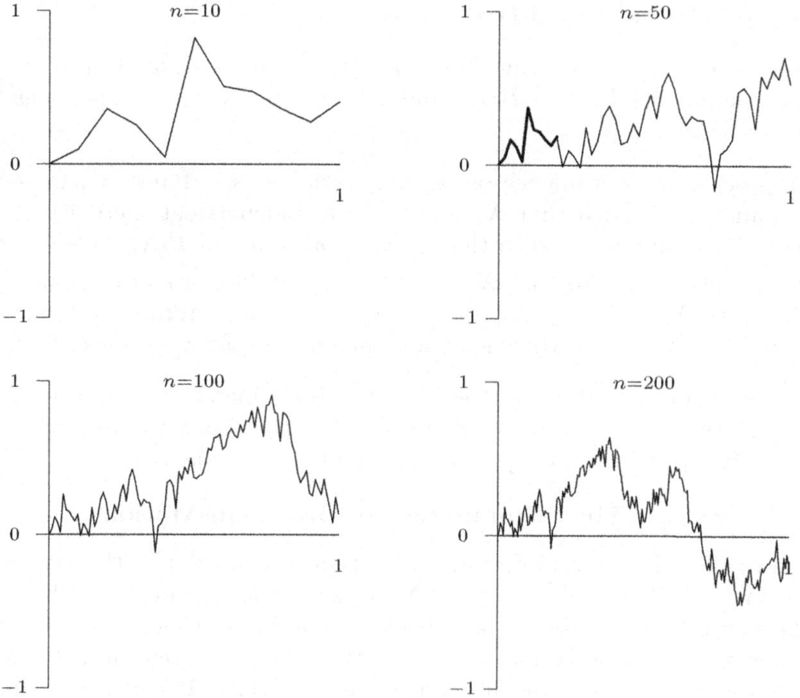

EXERCISES

1. Write down an iterated integral whose value is $P[B(1) \le -1,\ B(3) > 1]$.

• In problems 2 and 3 we show that the distributional properties of Brownian motion may be approximated by an appropriately scaled random walk. Let $W_0, W_1, \ldots$ be a SSRW on $\mathbb{Z}$ as defined in Chapter IV. Fractions of the form $i/2^n$ are called *dyadic rationals*. For $n \ge 0$, let $\mathcal{D}_n = \{\frac{0}{2^n}, \frac{1}{2^n}, \frac{2}{2^n}, \frac{3}{2^n}, \ldots\}$ and put $\mathcal{D} = \bigcup_n \mathcal{D}_n$. For $n \ge 0$, we define the nth Brownian approximant by

$$B_n\left(\frac{i}{2^n}\right) = \frac{W_i}{2^{n/2}}, \quad i = 0, 1, 2, \ldots,$$

with $B_n(t)$ linearly interpolated for values of t not in $\mathcal{D}_n$. Clearly (32.1) and (32.2) hold for the paths $B_n(t)$.

2. Show that (32.4) holds for $B_n(t)$ provided the times $t_1, \ldots, t_m$ are in $\mathcal{D}_n$. What about if they are not?

3. Show that (32.3) holds "asymptotically" in some sense as $n \to \infty$ for times s and t in $\mathcal{D}$. What if s and t are not in $\mathcal{D}$?

• Exercises 4 – 6 illustrate that the sequence of functions $B_n(t)$ cannot converge to any particular function $B(t)$ as $n \to \infty$.

4. Suppose $0 < s < t$ are times in $\mathcal{D}$ and n is a fixed positive integer. Show that the increments $B_n(t) - B_n(s)$ and $B_m(t) - B_m(s)$ are independent for all sufficiently large m.

5. Suppose X_n is a sequence of random variables satisfying: (i) for every n there is an $m > n$ such that X_n and X_m are independent; and (ii) there are numbers $a < b$ and $p > 0$ such that $P[X_n \le a] \ge p$ and $P[X_n \ge b] \ge p$ for all sufficiently large n. Show that $X_n \xrightarrow{\text{P}} X$ is not possible for any random variable X. Hints: (i) $|X_n - X_m| \le |X_n - X| + |X_m - X|$; (ii) determine a lower bound for $P[|X_n - X_m| \ge b - a]$ that holds for some $m > n$, for all sufficiently large n.

6. Use problems 4 and 5 and the fact that both increments in 4 are asymptotically Normal$(0, t - s)$ to argue that $B_n(t) - B_n(s)$ cannot converge (almost surely or even just in probability) to any random variable as $n \to \infty$.

§33. The "Bumpiness" of Brownian Motion

As is readily visible from Figure 32.7, a typical Brownian path looks nothing like a smooth (differentiable) curve. (At the scale that the eye sees, the $n = 200$ path looks pretty much like a typical Brownian path.) Rather it exhibits extreme "bumpiness". In this section we explore this property rigorously. The first result, easy enough to prove, illustrates the situation. Recall that a function f defined on an interval $[a, b]$ is *increasing* on $[a, b]$ if $f(t_1) \le f(t_2)$ whenever $a \le t_1 < t_2 \le b$.

THEOREM 33.1. *With probability one, a Brownian path $B_\omega(t)$ possesses no interval on which it is increasing and no interval on which it is decreasing.*

REMARK. This is in contrast to a polynomial $p(t)$, for example, for which one can partition the line into intervals on which p is increasing (where $p'(t) > 0$) and intervals on which p is decreasing (where $p'(t) < 0$). This is a standard calculus exercise.

PROOF. We prove the "increasing" part of the theorem; the "decreasing" part follows from a similar argument. Let

$$E = \{\omega : \text{there is an interval } [a, b] \text{ on which } B_\omega(t) \text{ is increasing}\}.$$

We must show that $PE = 0$. Fix any interval $[a, b]$; we will show that

$$P\{\omega : B_\omega(t) \text{ is increasing on } [a, b]\} = 0. \tag{33.2}$$

Suppose (33.2) holds and let $\mathcal{R}$ denote the set of all intervals of $\mathbb{R}^+$ with rational endpoints: an interval $I = [r, s]$ makes it into $\mathcal{R}$ if $0 \le r < s$ and both r and s are rational. If $B(t)$ is increasing on some interval $[a, b]$ then, by selecting rational r between a and b and then rational s between r and b, we may produce a smaller interval $[r, s]$ that belongs to $\mathcal{R}$ on which $B(t)$ is also increasing. Hence

$$E \subset \bigcup_{I \in \mathcal{R}} \{\omega : B_\omega(t) \text{ is increasing on the interval } I\}. \tag{33.3}$$

The crucial point is that $\mathcal{R}$ is a *countable* collection of intervals so the union in (33.3) is a countable union. Invoking countable subadditivity of probability gives that

$$PE \le \sum_{I \in \mathcal{R}} P\{\omega : B_\omega(t) \text{ is increasing on the interval } I\}. \tag{33.4}$$

Since each of the summands in (33.4) is zero by (33.2), we get that $PE = 0$.

It remains to prove (33.2). Partition $[a, b]$ into n subintervals $[t_i, t_{i+1}]$ by taking, say, $t_i = a + \frac{i}{n}(b - a)$, for $0 \le i \le n$, and for $i > 0$ let $A_i = \{\omega : B_\omega(t_{i-1}) \le B_\omega(t_i)\}$. Then by (32.3) and (32.4), $A_1, A_2, \ldots, A_n$ are independent events with $P[A_i] = \frac{1}{2}$. Also,

$$\{\omega : B_\omega(t) \text{ is increasing on } [a, b]\} \subset A_1 \cap A_2 \cap \cdots \cap A_n,$$

from which it follows by taking probabilities that

$$P\{\omega : B_\omega(t) \text{ is increasing on } [a, b]\} \le \left(\frac{1}{2}\right)^n.$$

Since this holds for any n, (33.2) follows. □

Brownian Paths are not Rectifiable. You may recall from calculus that if a continuous function $f(t)$ has a continuous derivative $f'(t)$ on an interval $[a, b]$ we may compute its *arc length* as

$$\textit{arc length} = s^* = \int_a^b \sqrt{1 + f'(t)^2}\, dt. \tag{33.5}$$

Even if $f(t)$ is not differentiable we may make sense of the notion of arc length. Specifically, let $\mathcal{T} = \{t_0, t_1, \ldots, t_n\}$ be any partition of the interval $[a, b]$. (So $a = t_0$, $b = t_n$, and $t_i < t_{i+1}$.) Then clearly the arc length from the point $(a, f(a))$ to the point $(b, f(b))$ along the graph $y = f(t)$ must exceed the length of the polygonal path whose straight-line segments proceed from (t_0, y_0) to (t_1, y_1), then from (t_1, y_1) to (t_2, y_2), and so forth, where $y_i = f(t_i)$ — see Figure 33.7. Letting $\Delta t_i = t_{i+1} - t_i$, $\Delta y_i = y_{i+1} - y_i$, and $\Delta s_i = \sqrt{\Delta t_i^2 + \Delta y_i^2}$, we get that the length of this polygonal path is $s(\mathcal{T}) = \sum_{i=0}^{n-1} \Delta s_i$. The arc length of $f(t)$ from $(a, f(a))$ to $(b, f(b))$ is *defined* to be the least upper bound of these numbers:

$$\textit{arc length} \;=\; s \;=\; \sup\{s(\mathcal{T}) : \mathcal{T} \text{ is a partition of } [a,b]\}. \tag{33.6}$$

Figure 33.7: A polygonal path approximating $y = 1 + \sin(t)$.
$[a,b] = [0, 2\pi]$, $\mathcal{T} = \{\frac{2\pi i}{5} : 0 \le i \le 5\}$

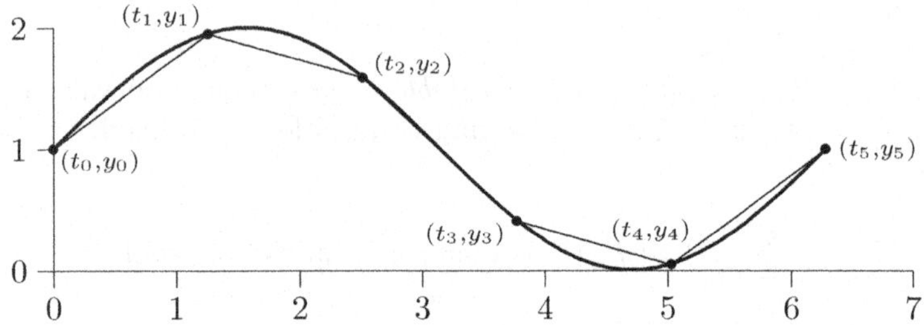

If $f(t)$ is continuously differentiable, the two quantities defined in (33.5) and (33.6) agree, i.e., $s^* = s$. To see this, first note that $\Delta y_i = f'(c_i)\Delta t_i$, for some number c_i on the interval (t_i, t_{i+1}). This is assured by the Mean Value Theorem. A little algebra then gives that

$$s(\mathcal{T}) \;=\; \sum_{i=0}^{n-1} \sqrt{1 + f'(c_i)^2}\Delta t_i,$$

which is a Riemann sum for the integral in (33.5). Hence, $s(\mathcal{T}) \to s^*$ as $||\mathcal{T}|| \to 0$. (Here $||\mathcal{T}||$ denotes the *mesh* or *norm* of the partition: $||\mathcal{T}|| = \max\{\Delta t_i : 0 \le i < n\}$.) By definition, $s(\mathcal{T}) \le s$ for any partition $\mathcal{T}$ and hence $s^* \le s$. To establish that $s \le s^*$, let r be any number less than s and choose $\epsilon > 0$ so that also $r + \epsilon < s$. Then pick $\delta > 0$ such $|s(\mathcal{T}) - s^*| < \epsilon$ whenever $||\mathcal{T}|| < \delta$ (see Figure 33.8 below). Now $r+\epsilon$ is not an upper bound for the numbers $s(\mathcal{T})$ so we may choose a particular partition $\mathcal{T}^*$ with $s(\mathcal{T}^*) > r + \epsilon$. Now add additional points to $\mathcal{T}^*$ as needed to form a finer partition $\mathcal{T}^{**}$ with $||\mathcal{T}^{**}|| < \delta$. Note that $s(\mathcal{T}^*) \le s(\mathcal{T}^{**})$ since each additional point added increases the length of the polygonal path due to the triangle inequality. Then

$$r + \epsilon \;<\; s(\mathcal{T}^*) \;\le\; s(\mathcal{T}^{**}) \quad \text{and} \quad |s(\mathcal{T}^{**}) - s^*| \;<\; \epsilon,$$

yielding that $r < s^*$. Since r was arbitrary (with $r < s$), we have $s \le s^*$ and, therefore, $s = s^*$.

Figure 33.8.

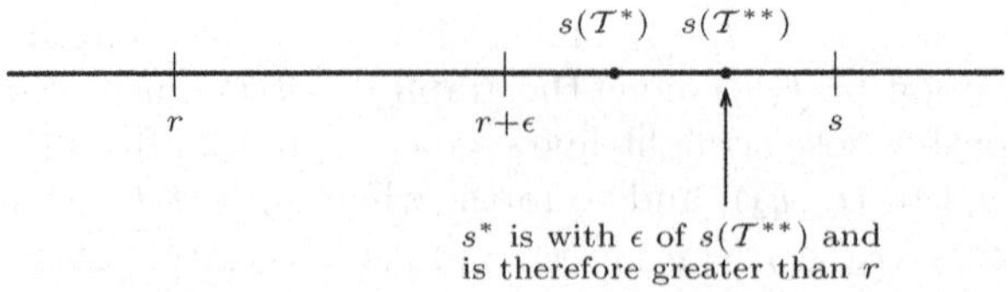

Total Variation. A quantity closely related to arc length is the *total variation* of $f(t)$ on an interval $[a, b]$. For any partition $\mathcal{T}$ of $[a, b]$, put $v(\mathcal{T}) = \sum_{i=0}^{n-1} |\Delta y_i|$. The total variation is then defined to be

$$\textit{total variation} \;=\; v \;=\; \sup\bigl\{v(\mathcal{T}) : \mathcal{T} \text{ is a partition of } [a, b]\bigr\}. \tag{33.9}$$

For each i, we have

$$|\Delta y_i| \;\leq\; \sqrt{\Delta t_i^2 + \Delta y_i^2} \;\leq\; \Delta t_i + |\Delta y_i|.$$

(Note that $\Delta t_i > 0$; the second inequality above follows from the triangle inequality.) Summing this over i ranging from 0 to $n - 1$ gives that

$$v(\mathcal{T}) \;\leq\; s(\mathcal{T}) \;\leq\; (b - a) + v(\mathcal{T}).$$

It follows from this that v is finite if and only if s is finite. If $v < \infty$, the function $f(t)$ is said to be *rectifiable* over the interval $[a, b]$. In particular, functions with continuous derivatives are rectifiable. The virtue of total variation is that it is easier to work with analytically than arc length.

Returning to Brownian motion, we will show that, with probability one, a Brownian path $B(t)$, is *not* rectifiable over any interval $[a, b]$, $a \geq 0$. In other words Brownian paths over finite intervals of time are infinitely long! We state this below with $a = 0$ and $b = 1$, but the same argument works over any interval $[a, b]$.

THEOREM 33.10. *Let v denote the total variation of $B(t)$ over the interval $[0, 1]$. Then $P[v = \infty] = 1$.*

PROOF. We will construct a deterministic sequence of partitions $\mathcal{T}_n$ of the interval $[0, 1]$ such that $P[v(\mathcal{T}_n) \to \infty] = 1$. Fix a positive integer n and let the numbers $t_i = \frac{i}{2^n}$, for $0 \leq i \leq 2^n$, generate the partition $\mathcal{T}_n$. Then

$$v(\mathcal{T}_n) \;=\; \sum_{i=0}^{2^n - 1} |\Delta B_{t_i}|, \tag{33.11}$$

where $\Delta B_{t_i} = B(t_{i+1}) - B(t_i)$. Now $\Delta B_{t_i} \sim \text{Normal}\,(0, 2^{-n})$, since $t_{i+1} - t_i = 2^{-n}$, giving also that $\Delta B_{t_i} \sim 2^{-n/2}N$ where $N \sim \text{Normal}\,(0, 1)$. Therefore $|\Delta B_{t_i}| \sim 2^{-n/2}|N|$ and hence $E\bigl[|\Delta B_{t_i}|\bigr] = 2^{-n/2}E|N| = \alpha 2^{-n/2}$. (The reader may verify that $\alpha = \sqrt{2/\pi}$. We use only that $0 < \alpha < \infty$.) From this and (33.11), which has 2^n summands, we obtain

$$E[v(\mathcal{T}_n)] \;=\; 2^n \cdot \alpha 2^{-n/2} \;=\; \alpha 2^{n/2}.$$

Clearly, then, $E[v(\mathcal{T}_n)] \to \infty$ as $n \to \infty$, but this does not quite establish that $P[v(\mathcal{T}_n) \to \infty] = 1$ (see exercises.)

Next we use Chebyshev's inequality to control the distribution of $v(\mathcal{T}_n)$ about its mean. Observe that the ΔB_{t_i} are independent (property (32.4)), so (33.11) gives that

$$\operatorname{Var}[v(\mathcal{T}_n)] = \sum_{i=0}^{2^n-1} \operatorname{Var}\left[|\Delta B_{t_i}|\right].$$

In general, $\operatorname{Var}|X| \le \operatorname{Var} X$ (see exercises), so

$$\operatorname{Var}\left[|\Delta B_{t_i}|\right] \le \operatorname{Var}\left[\Delta B_{t_i}\right] = 2^{-n},$$

and $\operatorname{Var}[v(\mathcal{T}_n)] \le 2^n \cdot 2^{-n} = 1$. Now

$$v(\mathcal{T}_n) \le \frac{\alpha}{2} 2^{n/2} \implies \left|v(\mathcal{T}_n) - \alpha 2^{n/2}\right| \ge \frac{\alpha}{2} 2^{n/2},$$

so Chebyshev's inequality gives that

$$\begin{aligned} P\left[v(\mathcal{T}_n) \le \frac{\alpha}{2} 2^{n/2}\right] &\le P\left[\left|v(\mathcal{T}_n) - \alpha 2^{n/2}\right| \ge \frac{\alpha}{2} 2^{n/2}\right] \\ &\le \frac{\operatorname{Var}[v(\mathcal{T}_n)]}{(\alpha/2)^2 \cdot 2^n} \le \frac{4}{\alpha^2} \cdot \frac{1}{2^n}. \end{aligned}$$

Since $\sum_n \frac{1}{2^n} < \infty$, the Borel-Cantelli Lemma gives that

$$P\left[v(\mathcal{T}_n) \le \frac{\alpha}{2} 2^{n/2} \text{ for infinitely many } n\right] = 0.$$

But if $v(\mathcal{T}_n) > \frac{\alpha}{2} 2^{n/2}$ for all but finitely many n, necessarily $v(\mathcal{T}_n) \to \infty$. Hence $P[v(\mathcal{T}_n) \to \infty] = 1$. Finally, since

$$v(\mathcal{T}_n) \to \infty \implies v = \infty,$$

we have that $P[v = \infty] = 1$. □

Brownian Paths are Nowhere Differentiable. In 1872 Karl Weierstrass (1815-1897) published the first example of a continuous function that is nowhere differentiable. Prior to that, such functions were suspected not to exist. We have seen that Brownian motion is sufficiently bumpy that it has infinite arc length. Here we prove the following, equally bizarre, consequence of Brownian motion's bumpiness.

THEOREM 33.12. *With probability one, Brownian motion is nowhere differentiable. Specifically,*

$$P\left[\textit{there exists an } s \in [0,1] \textit{ such that } B'_\omega(s) \textit{ exists}\right] = 0.$$

REMARK. Here we restrict our attention to the interval $[0, 1]$, but the argument presented below works on any interval: with probability one, no interval contains a point at which $B_\omega(t)$ is differentiable. In a probabilistic sense, nowhere differentiable continuous functions are the rule, not the exception.

PROOF. If $B'_\omega(s)$ exists for some $s \in [0, 1]$, i.e.,

$$\lim_{t \to s} \frac{B_\omega(t) - B_\omega(s)}{t - s} = L \quad \text{exists,}$$

then, for some $\delta > 0$ (the value of δ will depend on the path B_ω, i.e., it will depend on ω), we have

$$\left| \frac{B_\omega(t) - B_\omega(s)}{t - s} \right| < |L| + 1, \quad \text{provided } 0 < |t - s| < \delta.$$

Then

$$|B_\omega(t) - B_\omega(s)| \leq (|L| + 1) \cdot |t - s|, \quad \text{provided } |t - s| < \delta.$$

If n is large enough so that both $n > |L| + 1$ and $2^{-n} < \delta$, we will therefore have

$$|B_\omega(t) - B_\omega(s)| \leq n \cdot |t - s|, \quad \text{provided } |t - s| \leq 2^{-n}. \tag{33.13}$$

Let

$$D = \{\omega : \text{for some } s \in [0, 1],\ B'_\omega(s) \text{ exists}\},$$

and

$$D_n = \{\omega : \text{for some } s \in [0, 1],\ (33.13) \text{ holds}\}.$$

The above reasoning implies that $D \subset D_n$ for all sufficiently large n. Exactly how large n must be will depend on the particular path B_ω, i.e., n is outcome dependent.

Suppose $\omega \in D_n$ and let s satisfy (33.13). If I is an interval of length 2^{-n} that contains s, and both t and $\tilde{t}$ also lie in I, then

$$\begin{aligned} |B_\omega(t) - B_\omega(\tilde{t})| &\leq |B_\omega(t) - B_\omega(s)| + |B_\omega(s) - B_\omega(\tilde{t})| \\ &\leq n \cdot |t - s| + n \cdot |s - \tilde{t}| \\ &\leq 2n \cdot 2^{-n} = a_n, \end{aligned} \tag{33.14}$$

where the equality defines a_n.

For each positive integer n, we seek to define an event A_n. Partition $[0, 1]$ into 2^n subintervals of length 2^{-n} and let I_i, $0 \leq i < 2^n$, denote the ith subinterval: $I_i = [i2^{-n}, (i + 1)2^{-n}]$. Now partition I_i further into three equal subintervals with endpoints t_0, t_1, t_2, and t_3, where

$$t_k = i2^{-n} + \frac{k}{3} 2^{-n}, \quad k = 0, 1, 2, 3.$$

(See Figure 33.17.) Let C_i be the event defined by

$$C_i = \left\{\omega : \max\begin{pmatrix} |B_\omega(t_1) - B_\omega(t_0)|, \\ |B_\omega(t_2) - B_\omega(t_1)|, \\ |B_\omega(t_3) - B_\omega(t_2)| \end{pmatrix} \le a_n\right\}, \tag{33.15}$$

and finally put

$$A_n = C_0 \cup C_1 \cup \cdots \cup C_{2^n-1}. \tag{33.16}$$

Figure 33.17.

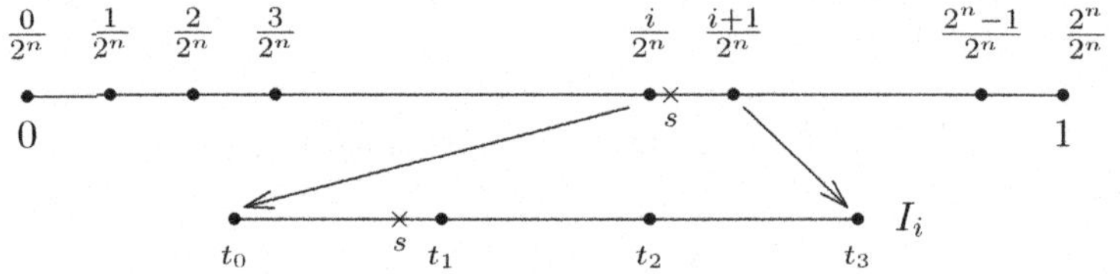

We claim that $D_n \subset A_n$. To see this, let $\omega \in D_n$ and suppose s satisfies (33.13). Since $0 \le s \le 1$, $s \in I_i$ for some i. For that i, each of the numbers $|B_\omega(t_1) - B_\omega(t_0)|$, $|B_\omega(t_2) - B_\omega(t_1)|$, and $|B_\omega(t_3) - B_\omega(t_2)|$ are all no bigger than a_n by (33.14). This means that $\omega \in C_i$ and therefore $\omega \in A_n$.

Since $D \subset D_n$ for all large n, it follows that $D \subset A_n$ for all large n, and therefore

$$D \subset \{A_n \text{ happens for infinitely many } n\}.$$

We will show that $\sum_n P[A_n] < \infty$, so the Borel-Cantelli lemma will give that $P[A_n \text{ infinitely often}] = 0$ and therefore $PD = 0$, completing the proof. First we estimate the $P[C_i]$. The three absolute Brownian increments appearing in (33.15) are independent (as $t_0 < t_1 < t_2 < t_3$) and all drawn from the same distribution:

$$\begin{aligned} |B_\omega(t_1) - B_\omega(t_0)| &\sim |B_\omega(t_2) - B_\omega(t_1)| \\ &\sim |B_\omega(t_3) - B_\omega(t_2)| \\ &\sim |\text{Normal}\,(0, 2^{-n}/3)| \sim \frac{2^{-n/2}}{\sqrt{3}} |\text{Normal}\,(0,1)|, \end{aligned}$$

as each $t_{k+1} - t_k = 2^{-n}/3$. Using independence and the fact that $P[|\text{Normal}\,(0,1)| \le \delta] \le \delta$ (see exercises), we see that

$$\begin{aligned} P[C_i] &= \left(P\left[\frac{2^{-n/2}}{\sqrt{3}}|\text{Normal}\,(0,1)| \le a_n\right]\right)^3 \\ &= \left(P\left[|\text{Normal}\,(0,1)| \le \sqrt{3}\cdot 2^{n/2}\cdot a_n\right]\right)^3 \\ &\le \left(\sqrt{3}\cdot 2^{n/2}\cdot a_n\right)^3 = (2\sqrt{3})^3\, n^3\, 2^{-3n/2}. \end{aligned}$$

(Refer to (33.14) for the definition of a_n.) Applying the subadditivity of probability to (33.16) then gives

$$P[A_n] \le \sum_{i=0}^{2^n-1} P[C_i] \le 2^n \cdot (2\sqrt{3})^3 \, n^3 \, 2^{-3n/2} = (2\sqrt{3})^3 \, n^3 \, 2^{-n/2}.$$

Now $(2\sqrt{3})^3 \, n^3 < 2^{n/4}$ for all large n (exponential growth eventually exceeds polynomial growth), so $(2\sqrt{3})^3 \, n^3 \, 2^{-n/2} < 2^{-n/4}$ for all large n. Since $\sum_{n=1}^{\infty} 2^{-n/4} < \infty$, we see that $\sum_{n=1}^{\infty} P[A_n] < \infty$. □

EXERCISES

1. Show that $\operatorname{Var}|X| \le \operatorname{Var} X$. Under what conditions are they equal?

2. To show that the total variation of $B_\omega(t)$ on the interval $[0,1]$ is infinite, we constructed a sequence of partitions $\mathcal{T}_n$ with $v(\mathcal{T}_n) \to \infty$. We used that $E[v(\mathcal{T}_n)] = \alpha 2^{n/2} \to \infty$ (*very quickly*) and $\operatorname{Var}[v(\mathcal{T}_n)] \le 1$ to establish this. In the following example we have:

$$\begin{gathered} EX_n \to \infty, \quad \operatorname{Var} X_n = 1, \quad \text{but} \\ P[X_n \to \infty] = 0. \end{gathered} \tag{$*$}$$

The key is that in this counterexample $EX_n \to \infty$ quite slowly. Divide the positive integers into blocks of consecutive integers so that block number k, call it B_k, contains k^2+1 numbers:

$$\overbrace{1\ 2}^{B_1}\ \overbrace{3\ 4\ 5\ 6\ 7}^{B_2}\ \overbrace{8\ 9\ 10\ 11\ 12\ 13\ 14\ 15\ 16\ 17}^{B_3} \cdots .$$

Now randomly select a distinguished number from each block, with all numbers in a block equally likely to be selected. For the numbers $n \in B_k$ define the random variable X_n as follows:

$$X_n = \begin{cases} 0 & \text{if } n \text{ is the distinguished number in } B_k \\ \frac{k^2+1}{k} & \text{otherwise.} \end{cases}$$

Verify that the three statements in $(*)$ are true.

3. Show that $P[|\text{Normal}\,(0,1)| \le \delta] \le \delta$, where $\delta > 0$.

4. Fix any number h where $0 < h < 1$. Show that, given $B(1) = b$, $B(h)$ is normally distributed with mean hb and variance $h \cdot (1-h)$.

5. If $0 < s < t$, show that $\operatorname{Cov}[B(s), B(t)] = s$.

6. Since $B_\omega(t)$ is a continuous function of t, we may compute the integral

$$Y(\omega) = \int_0^1 B_\omega(t)\, dt.$$

Show that $EY = 0$ and $\operatorname{Var} Y = 1/3$. You will need to exchange the order of expectation and integration — you need not justify this step.

§34. Recurrence

In this section we explore some recurrence properties of Brownian motion. Our Brownian path $B(t)$, now defined for all $t \geq 0$, starts at $B(0) = 0$. Here we show that $B(t)$ is recurrent in the sense of:

THEOREM 34.1 (RECURRENCE OF $B(t)$). *Let R denote the event that there is a (random) sequence of times $0 < \tau_1 < \tau_2 < \cdots$, with $\tau_n \to \infty$ and $B(\tau_n) = 0$. Then $P[R] = 1$.*

The basic idea is quite simple. Fix $t > 0$ and suppose we've watched the Brownian path evolve to time t and that $B_t = b$. We then let the path evolve an additional time s. If s is very large: (i) the Brownian increment $\Delta B = B_{t+s} - B_t$ is likely to be large compared to b; and (ii) with probability $1/2$, ΔB will have the opposite sign of b. If (i) and (ii) occur, the Brownian path (which is continuous) must pass through 0 between time t and time $t + s$. Now replace t with $t + s$ and repeat the exercise indefinitely.

In the ensuing paragraphs we make this rigorous. We state and prove two lemmas first, then pull everything together.

LEMMA 34.2. *Suppose $N_1 \sim \text{Normal}(0, \sigma_1^2)$. For any $\epsilon > 0$ there is a number v such that if $N_2 \sim \text{Normal}(0, \sigma_2^2)$ with $\sigma_2^2 \geq v$, then $P[|N_1| \geq |N_2|] < \epsilon$.*

PROOF. Fix $\epsilon > 0$. For any $L > 0$ we have

$$\{|N_1| \geq |N_2|\} \subset \{|N_1| \geq L\} \cup \{|N_2| \leq L\}.$$

Since $P[|N_1| \geq L] \leq \frac{E|N_1|}{L} \to 0$ as $L \to \infty$, we may choose L so that $P[|N_1| \geq L] < \frac{\epsilon}{2}$. If $N_2 \sim \text{Normal}(0, \sigma_2^2)$, then its density function never exceeds $1/\sqrt{2\pi\sigma_2^2}$, so $P[|N_2| \leq L] \leq 2L/\sqrt{2\pi\sigma_2^2}$. Hence for large enough σ_2^2 (specifically, for $\sigma_2^2 \geq v = \frac{8L^2}{\pi\epsilon^2}$) we have $P[|N_2| \leq L] \leq \frac{\epsilon}{2}$. But then $\sigma_2^2 \geq v \implies P[|N_1| \geq |N_2|] < \frac{\epsilon}{2} + \frac{\epsilon}{2} = \epsilon$. □

LEMMA 34.3. *There is a sequence of times $0 < t_1 < t_2 < \cdots$ such that*

$$P[|B(t_n)| \geq |B(t_{n+1}) - B(t_n)| \textit{ infinitely often}] = 0.$$

PROOF. Take $t_1 = 1$ and suppose $t_1 < t_2 < \cdots < t_n$ have been chosen. We pick t_{n+1} as follows. We have $B(t_n) \sim \text{Normal}(0, t_n)$ and, for $t_{n+1} > t_n$, $B(t_{n+1}) - B(t_n) \sim \text{Normal}(0, t_{n+1} - t_n)$. By Lemma 34.2 if $t_{n+1} - t_n$ is sufficiently large we will have $P[|B(t_n)| \geq |B(t_{n+1}) - B(t_n)|] < \frac{1}{2^n}$. The Borel-Cantelli Lemma then insures the result. □

PROOF OF THEOREM 34.1. For the times t_n of Lemma 34.3, let A_n be the event that the sign of $B(t_n)$ is different from the sign of $B(t_{n+1}) - B(t_n)$ and let B_n denote the event that $|B(t_n)| < |B(t_{n+1}) - B(t_n)|$. If A_n and B_n both happen, then $B(t)$ changes sign on the interval $[t_n, t_{n+1}]$, which by continuity of the Brownian path means that $B(\tau) = 0$ for some $\tau \in [t_n, t_{n+1}]$. Now $A_1, A_2, \ldots$ are independent since, for each n, $A_1, \ldots, A_{n-1}$ depend on the Brownian path up to time t_n and the sign of the increment $B(t_{n+1}) - B(t_n)$ is independent of the path to time t_n. Thus each A_n is independent of $A_1, \ldots, A_{n-1}$. Since $P[A_n] = \frac{1}{2}$ we have $P[A_n \text{ infinitely often}] = 1$. (It's like independently repeatedly tossing a fair coin — with probability one you will obtain infinitely many heads.) Summarizing,

$$\begin{aligned} P[A_n \text{ happens infinitely often}] &= 1, \text{ and} \\ P[B_n \text{ happens for all sufficiently large } n] &= 1, \end{aligned}$$

hence $P[A_n \cap B_n \text{ happens infinitely often}] = 1$. Thus, with probability one, the Brownian path $B(t)$ crosses the t-axis on infinitely many of the intervals $[t_n, t_{n+1}]$. □

REMARK. We have essentially used that $\mathbb{R}\setminus\{0\}$ is topologically disconnected, so it is impossible for a path starting in $\{x \in \mathbb{R} : x > 0\}$ to end up in $\{x < 0\}$ without passing through 0.

Higher Dimensions. One may study higher dimensional Brownian paths. Here we report, without proof, what is true about recurrence in this context. Suppose $B(t)$ and $\widetilde{B}(t)$ are two independent (one dimensional) Brownian processes constructed on $\boldsymbol{\Omega}$ via the Lévy procedure. Then

$$\Big((B(t), \widetilde{B}(t)) : t \geq 0\Big)$$

defines a two dimensional Brownian path. This path starts out at the origin $(0, 0)$ and it makes sense to ask questions like: Will it ever return to the origin? As it turns out, it is too much to ask that both $B(t) = 0$ and $\widetilde{B}(t) = 0$ occur simultaneously (for the same t). (The set $\mathbb{R}^2 \setminus \{(0,0)\}$ is topologically connected so we cannot use our one dimensional approach.) However, for $\epsilon > 0$ put

$$\mathcal{D} = \{(x, y) \in \mathbb{R}^2 : \sqrt{x^2 + y^2} < \epsilon\}.$$

Then, no matter how small ϵ is, the Brownian path $\big(B(t), \widetilde{B}(t) : t \geq 0\big)$ will return to $\mathcal{D}$ for arbitrarily large values of t with probability one. In this sense the Brownian path in two dimensions is recurrent.

In three dimensions the situation is different. (This should not be surprising in light of our experience with random walks.) If $B(t)$, $\widetilde{B}(t)$ and $\widehat{B}(t)$ are three independent one dimensional Brownian processes, then the path

$$\Big((B(t), \widetilde{B}(t), \widehat{B}(t)) : t \geq 0\Big)$$

in $\mathbb{R}^3$, which begins at $(0,0,0)$, will eventually leave the ball

$$\mathcal{D} = \{(x,y,z) \in \mathbb{R}^3 : \sqrt{x^2+y^2+z^2} < \epsilon\}$$

never to return, no matter how *large* ϵ is.

EXERCISES

• In exercises 1 – 7 we develop an alternative proof that $B(t)$ is recurrent modeled on our gambler's ruin proof that a random walk in one dimension is recurrent.

1. Fix $b > 0$ and $0 \leq a \leq b$ and consider the Brownian path $a + B(t)$ that starts at a. Show that $P[0 \leq a + B(t) \leq b] \to 0$ as $t \to \infty$. Use this to show that

$$P[a + B(t) \text{ remains between } 0 \text{ and } b \text{ forever}] = 0.$$

2. Let $s(a) = P[a + B(t) \text{ hits } 0 \text{ before it hits } b]$. If $a \pm \epsilon \in [0,b]$, show that $(a-\epsilon, s(a-\epsilon))$, $(a, s(a))$, and $(a+\epsilon, s(a+\epsilon))$ are collinear in the x-y plane.

3. Use problem 2 to show that $s(\lambda b) = 1 - \lambda$ for all *rational* $\lambda \in [0,1]$.

4. Argue that $s(a)$ is monotonic on $[0,b]$. Try to be rigorous.

5. Use problem 4 to show that $s(\lambda b) = 1 - \lambda$ for *all* $\lambda \in [0,1]$.

6. Thinking now of $a > 0$ as fixed and $b > a$ as variable, show that $P[a + B(t) \text{ hits } 0 \text{ before it hits } b] \to 1$ as $b \to \infty$.

7. Use this to establish that $B(t)$ is recurrent: show that regardless of the value a of the Brownian path $B(t)$ at a time $t = t^*$, with probability one $B(t) = 0$ at some time $t > t^*$.

• Here is yet another proof.

8. Define a sequence of random times $T_0 < T_1 < T_2 < \cdots$ as follows. Take $T_0 = 0$, and for $n \geq 0$ put $T_{n+1} = \min\{t > T_n : B(t) = B(T_n) \pm 1\}$. Show that this definition makes sense and that $W_n = B(T_n)$, $n \geq 0$, is a simple symmetric random walk on $\mathbb{Z}$.

9. Use the recurrence of W_n to deduce the recurrence of $B(t)$.

• Let $M(t) = \max\{B(s) : 0 \leq s \leq t\}$. Then for $a > 0$ we have $P[M(t) \geq a] = 2P[B(t) \geq a]$.

10. Following Lemma 24.21 and Theorem 24.22, formulate a reflection principle for Brownian motion and make an informal argument in support of this.

§35*. The Lévy Construction of Brownian Motion

In this section we present a construction of a Brownian path $B(t) = B_\omega(t)$ on the time interval $0 \le t \le 1$. We later extend this to all $t \ge 0$. We require as input a sequence of independent standard normal random variables $N_1(\omega), N_2(\omega), \ldots$, which may be defined on the space $\mathbf{\Omega} = (0,1)$ via inverse transform. The plan is to use them to construct a sequence of approximations $B_1(t), B_2(t), B_3(t), \ldots$ that converge to a limiting function $B(t)$ with the desired properties.

The Lévy construction will involve repeatedly selecting a value for the Brownian path at an additional time t^* after the path's values at other times have already been specified. We must draw the value $B(t^*)$ from the correct conditional distribution. Fix times $0 = t_0 < t_1 < \cdots < t_m = 1$ where, for some $0 < \ell < m$, t_ℓ is the midpoint of $[t_{\ell-1}, t_{\ell+1}]$. Suppose further that the joint distribution of $(B(t_i) : 1 \le i \le m)$ satisfies (32.5) with $B(t_0) = B(0) = b_0 = 0$. Here we are interested in the distribution of $B(t_\ell)$ conditioned on:

$$\{B(t_i) = b_i, \quad \text{for } i = 0, 1, \ldots, \ell-1, \ell+1, \ldots, m\}, \tag{35.1}$$

where the b_i are fixed ("pinned") values. To this end we have:

Lemma 35.2 (Pinned Brownian motion). *Suppose t_ℓ is the midpoint of $[t_{\ell-1}, t_{\ell+1}]$. Given* (35.1),

$$B(t_\ell) \sim \text{Normal}\left(\frac{b_{\ell-1} + b_{\ell+1}}{2}, \frac{t_{\ell+1} - t_{\ell-1}}{4}\right).$$

Proof. Let f denote the conditional distribution of $B(t_\ell)$ conditioned as hypothesized, h denote the joint distribution of

$$(B(t_1), \ldots, B(t_{\ell-1}), B(t_{\ell+1}), \ldots, B(t_m)),$$

and g denote the joint distribution of $(B(t_1), \ldots, B(t_m))$. Using (32.5),

$$\begin{aligned} f(b) &= \frac{g(b_1, \ldots, b_{\ell-1}, b, b_{\ell+1}, \ldots, b_m)}{h(b_1, \ldots, b_{\ell-1}, b_{\ell+1}, \ldots, b_m)} \\ &= \frac{\psi(b - b_{\ell-1}; 0, t_\ell - t_{\ell-1}) \cdot \psi(b_{\ell+1} - b; 0, t_{\ell+1} - t_\ell)}{\psi(b_{\ell+1} - b_{\ell-1}; 0, t_{\ell+1} - t_{\ell-1})} \\ &= \psi\left(b; \frac{b_{\ell-1} + b_{\ell+1}}{2}, \frac{t_{\ell+1} - t_{\ell-1}}{4}\right). \end{aligned}$$

In the second inequality we have cancelled from numerator and denominator terms of the form $\psi(b_{i+1} - b_i; 0, t_{i+1} - t_i)$ for $0 \le i \le \ell - 2$ and $\ell + 1 \le i \le m - 1$. The final equality results when $\psi()$ is replaced with its definition (see (32.6)) after substantial simplification and using that t_ℓ is the midpoint of $[t_{\ell-1}, t_{\ell+1}]$. The details are left to the intrepid reader (see exercises). □

We need a few additional preliminaries from analysis. First:

LEMMA 35.3. *Suppose $f_1, f_2, \ldots$ is a sequence of continuous functions defined on the interval $[0,1]$. Further, suppose that for all $\epsilon > 0$ there is a number n_0 such that $|f_n(t) - f_{n_0}(t)| < \epsilon$ whenever $n \geq n_0$ and $t \in [0,1]$. Then there is a continuous function $f(t)$ defined on $[0,1]$ such that for each $t \in [0,1]$, $f_n(t) \to f(t)$ as $n \to \infty$.*

PROOF. First we show that the sequence of numbers $(f_n(t) : n \geq 1)$ is Cauchy (see §44) for any fixed t. To this end, pick $\epsilon > 0$ and choose n_0 such that $|f_n(t) - f_{n_0}(t)| < \frac{\epsilon}{2}$ whenever $n \geq n_0$. If both $n, m \geq n_0$, we have

$$|f_n(t) - f_m(t)| \leq |f_n(t) - f_{n_0}(t)| + |f_{n_0}(t) - f_m(t)| < \frac{\epsilon}{2} + \frac{\epsilon}{2} = \epsilon,$$

and the Cauchy condition is satisfied. It follows that $(f_n(t) : n \geq 1)$ has a limiting value which we'll call, naturally, $f(t)$. That is, $f_n(t) \to f(t)$ as $n \to \infty$.

Next we show that the *function* $f(t)$ is continuous in t. Fix $\epsilon > 0$ and choose n_0 such that $|f_n(t) - f_{n_0}(t)| < \frac{\epsilon}{3}$ whenever $n \geq n_0$ and $t \in [0,1]$. By letting $n \to \infty$ we get that $|f(t) - f_{n_0}(t)| \leq \frac{\epsilon}{3}$ whenever $t \in [0,1]$. Fix $t_0 \in [0,1]$; we presently show the continuity of $f(t)$ at t_0. Now

$$\begin{aligned} |f(t) - f(t_0)| &\leq |f(t) - f_{n_0}(t)| + |f_{n_0}(t) - f_{n_0}(t_0)| + |f_{n_0}(t_0) - f(t_0)| \\ &\leq \frac{\epsilon}{3} + |f_{n_0}(t) - f_{n_0}(t_0)| + \frac{\epsilon}{3}, \end{aligned}$$

and since $f_{n_0}(t)$ is continuous we may choose $\delta > 0$ such that $|f_{n_0}(t) - f_{n_0}(t_0)| < \frac{\epsilon}{3}$ provided $|t - t_0| < \delta$. But then $|t - t_0| < \delta \implies |f(t) - f(t_0)| < \epsilon$, and $f(t)$ is continuous at t_0. □

REMARK. The conditions of this theorem are satisfied if the functions f_n are continuous and, for some number ρ with $0 < \rho < 1$ and some positive integer k_0, we have $|f_{k+1}(t) - f_k(t)| < \rho^k$ whenever $t \in [0,1]$ and $k \geq k_0$. Under these circumstances for $n \geq n_0 \geq k_0$ we have

$$\begin{aligned} |f_n(t) - f_{n_0}(t)| &\leq \sum_{k=n_0}^{n-1} |f_{k+1}(t) - f_k(t)| \quad \text{(triangle inequality)} \\ &\leq \sum_{k=n_0}^{\infty} \rho^k = \frac{\rho^{n_0}}{1-\rho}. \end{aligned}$$

For any $\epsilon > 0$ we may pick $n_0 \geq k_0$ such that $\rho^{n_0} < \epsilon \cdot (1-\rho)$; then $|f_n(t) - f_{n_0}(t)| < \epsilon$ whenever $n \geq n_0$.

Another analytical building block that we will need is the following:

LEMMA 35.4. *If $0 \leq x \leq 1$ and $m \geq 1$ then $1 - (1-x)^m \leq mx$.*

PROOF. Let $\ell(x) = 1 - (1-x)^m$ and $r(x) = mx$. Then $\ell(0) = r(0) = 0$ and

$$\ell'(x) = m(1-x)^{m-1} \leq m = r'(x),$$

where the inequality holds for $0 \le x \le 1$ and $m \ge 1$. Then, for $0 \le x \le 1$,

$$\ell(x) = \ell(x) - \ell(0) = \int_0^x \ell'(t)\,dt$$
$$\le \int_0^x r'(t)\,dt = r(x) - r(0) = r(x). \quad \square$$

The Lévy Construction. The *dyadic rationals* are fractions whose denominators are powers of 2: $\frac{1}{8}$, $\frac{13}{32}$, $\frac{35}{64}$, for example. Let $\mathcal{D}_n$ denote all fractions of the form $\frac{i}{2^n}$ for $0 \le i \le 2^n$. For example, $\mathcal{D}_0 = \{0, 1\}$ and $\mathcal{D}_1 = \{0, \frac{1}{2}, 1\}$. Now put $\mathcal{D} = \bigcup_n \mathcal{D}_n$, so $\mathcal{D}$ comprises all dyadic rationals on the interval $[0, 1]$. Clearly any $t \in [0, 1]$ can be approximated to within $1/2^{n+1}$ by the number in $\mathcal{D}_n$ that is closest to t. We'll refer to the elements of $\mathcal{D}_n$ as the *level n times*.

The construction works as follows. First, we specify (i.e., pin) the value of the Brownian path at the level 0 times: $t = 0$ and $t = 1$. Since $B(0) = 0$, really only $B(1)$ must be pinned. We know that $B(1) \sim \text{Normal}(0, 1)$, so we take $B(1) = N_1$ — the first standard normal in our list of independent standard normals. Next, at level 1, we pin the value of the path at time $t = \frac{1}{2}$, i.e., we specify $B(\frac{1}{2})$. At this point $B(0), B(\frac{1}{2})$, and $B(1)$ are pinned. Proceeding in this fashion, at level 2, we pin the value of the path at times $t = \frac{1}{4}$ and $t = \frac{3}{4}$ so now $B(0), B(\frac{1}{4}), B(\frac{1}{2}), B(\frac{3}{4})$, and $B(1)$ are all determined. In general, at level n, the values of the Brownian path have been pinned at all the level n times $t \in \mathcal{D}_n$. The level n approximation, which we'll call $B_n(t)$, to the Brownian path $B(t)$ is defined such that: (i) $B_n(t) = B(t)$ for the level n times; and (ii) $B_n(t)$ is interpolated linearly for times t that lie between two such consecutive level n times. The function $B_n(t)$ is therefore a piecewise linear path that agrees with the $B(t)$ at the level n times $t = \frac{i}{2^n}$. (See Figure 35.5, where the solid line corresponds to a path at level 3.)

Figure 35.5. A realization of levels 3 and 4 of the Lévy construction.

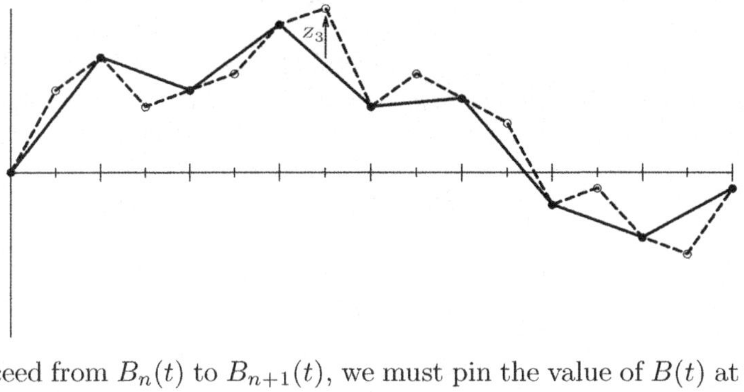

To proceed from $B_n(t)$ to $B_{n+1}(t)$, we must pin the value of $B(t)$ at the times

$$t \in \mathcal{D}_{n+1} \setminus \mathcal{D}_n = \left\{ \frac{i + 0.5}{m} : 0 \le i < m \right\}, \quad \text{where } m = 2^n.$$

Let

$$s = \frac{i}{m}, \quad s' = \frac{i+1}{m}, \quad \text{and} \quad t^* = \frac{s+s'}{2} = \frac{i+0.5}{m},$$

so t^* is the midpoint of $[s, s']$. Let b and b' denote the values of $B(s)$ and $B(s')$, respectively, as pinned by level n. Now, at this point in the construction, the value of $B(t)$ has been pinned at $t = s$ and $t = s'$ as well as at multiple times before s and after s'. By Lemma 35.2 (with $t_{\ell-1} = s$ and $t_{\ell+1} = s'$), the conditional distribution of $B(t^*)$ depends only on b and b' and is given by

$$\begin{aligned} B(t^*) &\sim \text{Normal}\left(\frac{b+b'}{2}, \frac{1}{4m}\right) \\ &\sim \frac{b+b'}{2} + \text{Normal}\left(0, \frac{1}{4m}\right) \sim B_n(t^*) + \text{Normal}\left(0, \frac{1}{4m}\right). \end{aligned}$$

Here we use that $B_n(s) = b$, $B_n(s') = b'$, and $B_n(t)$ is linearly interpolated for values of t between s and s' (t^* lies midway between s and s'). Letting $Z_0, Z_1, \ldots, Z_{m-1}$ denote the *perturbations* to B_n that take place at the midpoint of each subinterval going from level n to level $n+1$, we have

$$B_{n+1}(t^*) = B_n(t^*) + Z_i \quad \text{where } Z_i \sim \text{Normal}\left(0, \frac{1}{4m}\right).$$

To get Z_i, we use the next available independent standard normal, call it N_j, from our list and scale it by the appropriate factor: $Z_i = \sqrt{\frac{1}{4m}} N_j$. Figure 35.5 highlights the perturbation Z_3 in going from level 3 to level 4.

Convergence. Let Δ_n denote the largest absolute perturbation going from level n to $n+1$:

$$\Delta_n = \max\left\{|Z_0|, |Z_1|, \ldots, |Z_{m-1}|\right\}.$$

The reader will verify that, as t ranges over the ith subinterval $[\frac{i}{m}, \frac{i+1}{m}]$, the largest value of $|B_n(t) - B_{n+1}(t)|$ takes place at the midpoint of the interval, where the perturbation takes place, and this largest value is precisely $|Z_i|$. By considering each of the subintervals one at a time, it follows that the largest value of $|B_n(t) - B_{n+1}(t)|$ as t ranges over the full interval $[0, 1]$ is Δ_n. That is,

$$\max\left\{|B_n(t) - B_{n+1}(t)| : 0 \le t \le 1\right\} = \Delta_n. \tag{35.6}$$

Using that the Z_i's are independent and all distributed like Normal$(0, \frac{1}{4m})$, we see that

$$\begin{aligned} P[\Delta_n \ge a] &= 1 - P[\Delta_n < a] \\ &= 1 - \left(P[|Z_0| < a]\right)^m \\ &= 1 - \left(1 - P[|Z_0| \ge a]\right)^m \le mP[|Z_0| \ge a], \end{aligned} \tag{35.7}$$

where the inequality follows from Lemma 35.4. Now

$$\begin{aligned} P[|Z_0| \geq a] &= P\Big[\big|\text{Normal}\Big(0, \frac{1}{4m}\Big)\big| \geq a\Big] \\ &= P\big[|\text{Normal}(0,1)| \geq a \cdot 2\sqrt{m}\,\big] \;\leq\; 2e^{-2a^2 m}. \end{aligned} \tag{35.8}$$

Here we have used the estimate that $P[|\text{Normal}(0,1)| \geq b] \leq 2e^{-b^2/2}$ — see (11.6). Combining (35.7) with (35.8) and using that $m = 2^n$, we get, after some algebra, that

$$P[\Delta_n \geq a] \;\leq\; 2^{n+1}e^{-2^{n+1}a^2} \;\leq\; 2^{n+1-2^{n+1}a^2}. \tag{35.9}$$

(The second inequality holds because $e > 2$ and its exponent is negative.) This inequality holds for *any* $a > 0$. Let's take $a = (\frac{3}{4})^n$. Then

$$2^{n+1}a^2 \;=\; 2^{n+1}\Big(\frac{3}{4}\Big)^{2n} \;=\; 2\Big(\frac{9}{8}\Big)^n \;>\; 2n+1,$$

where the inequality holds for all large n because exponential growth exceeds linear growth. Combining this with (35.9), we get that, for all large n,

$$P\Big[\Delta_n \geq \Big(\frac{3}{4}\Big)^n\Big] \;\leq\; 2^{-n}.$$

Since $\sum_n 2^{-n} < \infty$, the Borel-Cantelli Lemma gives that

$$P\Big[\Delta_n \geq \Big(\frac{3}{4}\Big)^n \text{ for infinitely many } n\Big] \;=\; 0.$$

Referring back to (35.6), we see that, with probability one, for sufficiently large n we will have $|B_n(t) - B_{n+1}(t)| < (\frac{3}{4})^n$ for all $t \in [0,1]$. It then follows from Lemma 35.3 (see the remark following the lemma) that, with probability one, $B_n(t) \to B(t)$ as $n \to \infty$ for some continuous function $B(t)$. This establishes that properties (32.1) and (32.2) hold for our function $B(t)$.

REMARKS. It is easy to lose track of how $B(t) = B_\omega(t)$ depends on $\omega \in \mathbf{\Omega} = (0,1)$. Recall that $B(t)$ was constructed from the iid sequence $N_1, N_2, \ldots$ of standard normals. These, in turn, depend on ω: each $N_i = N_i(\omega)$. As a technical matter, what we have actually demonstrated above is that

$$P[\omega : \; B_\omega(t) \text{ is continuous in } t] \;=\; 1,$$

not that $B_\omega(t)$ is continuous for every ω. There are exceptions, that is, there is a set of outcomes E such that $B_\omega(t)$ is not continuous for $\omega \in E$. However, collectively these outcomes have probability 0: $PE = 0$. We can either work with the reduced outcome space $\tilde{\mathbf{\Omega}} = \mathbf{\Omega} \setminus E$ (which still has probability 1), or we may change the definition of $B(t)$ for those outcomes in E. For example, if $\omega \in E$, we may simply take $B_\omega(t) = 0$ for all t. Either way, (32.2) will hold for ALL outcomes.

The function $B(t)$ satisfies (32.3) and (32.4). If the times referred to in (32.3) and (32.4) are all dyadic rationals (they all belong to $\mathcal{D}$), then by construction (32.3) and (32.4) hold. This is because we construct $B(t)$ for each new $t \in \mathcal{D}$ with the correct conditional distribution given the values of B already constructed. But these times need not be in $\mathcal{D}$ and we must show by an approximation argument that (32.3) and (32.4) hold generally. We will need the following:

LEMMA 35.10. *Suppose* $X_n \sim \text{Normal}(0, \sigma_n^2)$ *with* $\sigma_n \to \sigma > 0$ *as* $n \to \infty$. *If* $X_n \xrightarrow{\text{P}} X$ *then* $X \sim \text{Normal}(0, \sigma^2)$.

PROOF. For any x and any $\epsilon > 0$ we have

$$\{X \le x\} \subset \{X_n \le x + \epsilon\} \cup \{|X - X_n| > \epsilon\},$$

so

$$P[X \le x] \le P[X_n \le x + \epsilon] + P[|X - X_n| > \epsilon]. \tag{35.11}$$

Now $P[|X - X_n| > \epsilon] \to 0$ as $n \to \infty$ and, for any number a,

$$\begin{aligned} P[X_n \le a] &= P\left[\text{Normal}(0, \sigma_n^2) \le a\right] \\ &= P\left[\text{Normal}(0,1) \le \frac{a}{\sigma_n}\right] \\ &\to P\left[\text{Normal}(0,1) \le \frac{a}{\sigma}\right] = P[\text{Normal}(0, \sigma^2) \le a], \end{aligned}$$

as $n \to \infty$. Here we have used that $\Psi(x)$ is continuous in x (in fact, differentiable). Hence $P[X_n \le x + \epsilon] \to \Psi(x + \epsilon; 0, \sigma^2)$ as $n \to \infty$. Combining these two facts with (35.11) gives $P[X \le x] \le \Psi(x + \epsilon; 0, \sigma^2)$. On the other hand,

$$\{X_n \le x - \epsilon\} \cap \{|X - X_n| < \epsilon\} \subset \{X \le x\},$$

and since $P[|X - X_n| < \epsilon] \to 1$ and $P[X_n \le x - \epsilon] \to \Psi(x - \epsilon; 0, \sigma^2)$, we get that $\Psi(x - \epsilon; 0, \sigma^2) \le P[X \le x]$.

Summarizing, we have

$$\Psi(x - \epsilon; 0, \sigma^2) \le P[X \le x] \le \Psi(x + \epsilon; 0, \sigma^2).$$

Letting $\epsilon \to 0$ and using that $\Psi(x; 0, \sigma^2)$ is continuous in x gives that $P[X \le x] = \Psi(x; 0, \sigma^2)$. □

Now fix $0 \le s < t$. For each level n, we approximate s from above and t from below by times in $\mathcal{D}_n$:

$$\begin{aligned} s_n &= \min\left\{\frac{k}{2^n} : s \le \frac{k}{2^n}\right\}, \text{ and} \\ t_n &= \max\left\{\frac{k}{2^n} : \frac{k}{2^n} \le t\right\}. \end{aligned}$$

For large n we will have $s \le s_n < t_n \le t$ (we will have $s_n = s$ or $t_n = t$ if $s \in \mathcal{D}_n$ or $t \in \mathcal{D}_n$, respectively). Also $|s_n - s| \le \frac{1}{2^n}$ so $s_n \to s$ and $B(s_n) \to B(s)$. Therefore we also have $B(s_n) \xrightarrow{\text{P}} B(s)$. Similarly $t_n \to t$ and $B(t_n) \xrightarrow{\text{P}} B(t)$, so $B(t_n) - B(s_n) \xrightarrow{\text{P}} B(t) - B(s)$. Now $B(t_n) - B(s_n) \sim \text{Normal}\,(0, t_n - s_n)$ as $s_n, t_n \in \mathcal{D}$. Since $t_n - s_n \to t - s > 0$, Lemma 35.10 gives that $B(t) - B(s) \sim \text{Normal}\,(0, t - s)$, establishing (32.3).

A similar argument shows that $B(t)$ satisfies (32.4) — here we sketch the argument for the case $m = 2$. Many of the steps below follow as in the argument establishing (32.3), above. Fix $0 = t_0 < t_1 < t_2 \le 1$. For each level n, we approximate each t_i from below and above by times in $\mathcal{D}_n$:

$$\begin{aligned} t_i^- &= \max\left\{\frac{k}{2^n} : \frac{k}{2^n} \le t_i\right\}, \text{ and} \\ t_i^+ &= \min\left\{\frac{k}{2^n} : t_i \le \frac{k}{2^n}\right\}. \end{aligned}$$

(The values of $t_i^\pm$ depend on n, although our notation suppresses that.) Now $t_i^\pm \to t_i$ as $n \to \infty$, and for large enough n we will have $t_i^+ < t_{i+1}^-$ for $i = 0$ and 1; in the discussion that follows we assume n is this large (see Figure 35.12 below).

Figure 35.12. Approximation by times in $\mathcal{D}_4$.

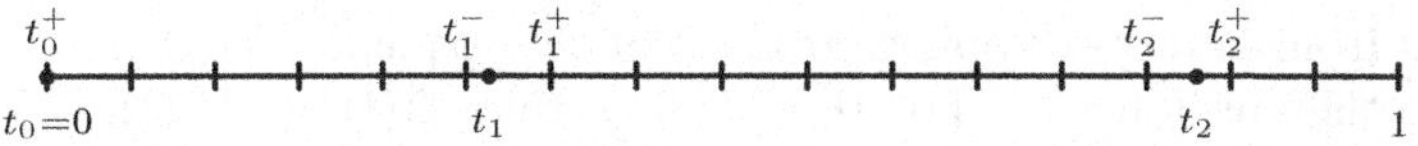

Let $X = B(t_1) - B(t_0)$ and $Y = B(t_2) - B(t_1)$ and also put $X_n = B(t_1^-) - B(t_0^+)$ and $Y_n = B(t_2^-) - B(t_1^+)$. Then, as the times $t_i^\pm$ belong to $\mathcal{D}$ and $t_0^+ < t_1^- \le t_1^+ < t_2^-$, we have $X_n \sim \text{Normal}\,(0, t_1^- - t_0^+)$, $Y_n \sim \text{Normal}\,(0, t_2^- - t_1^+)$, and X_n and Y_n are independent. Also, $X_n \xrightarrow{\text{P}} X$ and $Y_n \xrightarrow{\text{P}} Y$ as $n \to \infty$. Now, for any x and y and $\epsilon > 0$, we have

$$\begin{aligned} \{X \le x, Y \le y\} \subset\ & \{X_n \le x + \epsilon, Y_n \le y + \epsilon\} \\ & \cup \{|X_n - X| > \epsilon\} \cup \{|Y_n - Y| > \epsilon\}, \end{aligned}$$

from which we get (using the independence of X_n and Y_n and letting $n \to \infty$), that

$$P[X \le x, Y \le y] \le \Psi(x + \epsilon; 0, t_1 - t_0) \cdot \Psi(y + \epsilon; 0, t_2 - t_1).$$

A similar argument shows that

$$\Psi(x - \epsilon; 0, t_1 - t_0) \cdot \Psi(y - \epsilon; 0, t_2 - t_1) \le P[X \le x, Y \le y],$$

so, letting $\epsilon \to 0$

$$P[X \le x, Y \le y] = \Psi(x; 0, t_1 - t_0) \cdot \Psi(y; 0, t_2 - t_1) = P[X \le x] \cdot P[Y \le y],$$

where the second equality follows from (32.3). Thus X and Y are independent, establishing (32.4) for the case $m = 2$. This argument generalizes to $m > 2$.

Summarizing, we have:

THEOREM 35.13. *The Lévy construction produces a stochastic process $B_\omega(t)$ defined for $\omega \in \mathbf{\Omega}$ and $0 \le t \le 1$ satisfying conditions (32.1) – (32.4) for a Brownian motion.*

Extensions of the Construction. Via the Lévy construction we obtain a Brownian path $B(t)$ defined for $0 \le t \le 1$. This construction could easily have been adapted to any time interval $0 \le t \le T$. For some applications, will need to extend this to obtain a process satisfying (32.1)–(32.4) defined for all $t \ge 0$. The basic idea is to construct a sequence $B_1(t), B_2(t), \ldots$ of independent Brownian processes each defined for $0 \le t \le 1$ and then to "patch them together" to form $B(t)$, defined for all $t \ge 0$. Recall, to form a single Brownian process, call it $B_1(t)$, on $0 \le t \le 1$, we first constructed a sequence $N_1, N_2, \ldots$ of iid standard normals via inverse transform using the sequence $U_1, U_2, \ldots$ of iid Uniform $(0,1)$'s defined on $\mathbf{\Omega}$. We then used the Lévy construction with $N_1, N_2, \ldots$ as inputs to the procedure.

Here we will have to be more parsimonious with the iid normals. Specifically, lay them out in an array — see Figure 35.14 below (we've seen this trick before). Then use the first row of the array to generate $B_1(t)$ via Lévy, the second row to generate $B_2(t)$, and so forth. The $B_n(t)$'s are independent because they are generated from non-overlapping portions of the sequence $N_1, N_2, \ldots$. We patch them together as follows. For $0 \le t \le 1$ take $B(t) = B_1(t)$. Now on the interval $(1, 2]$, $B(t)$ must start where it left off, at $B(1)$. So for $1 < t \le 2$ take $B(t) = B(1) + B_2(t-1)$. In general, suppose we have constructed $B(t)$ in this manner for $0 \le t \le n$. Then for $n < t \le n+1$ take $B(t) = B(n) + B_{n+1}(t-n)$.

Clearly condition (32.1) for the standard Brownian process is satisfied. By virtue of the way we patched together the individually continuous $B_n(t)$'s, the overall path $B(t)$ is also continuous. It remains only to show that the distributional properties (32.3) and (32.4) are satisfied. We leave this to the exercises.

Figure 35.14. Parsimonious use of the sequence $N_1, N_2, \ldots$.

N_1	N_2	N_4	N_7	N_{11}	N_{16}	N_{22}	N_{29}	$\xrightarrow{\text{Lévy}}$	$B_1(t)$
N_3	N_5	N_8	N_{12}	N_{17}	N_{23}	$\iddots$		$\xrightarrow{\text{Lévy}}$	$B_2(t)$
N_6	N_9	N_{13}	N_{18}	N_{24}				$\xrightarrow{\text{Lévy}}$	$B_3(t)$
N_{10}	N_{14}	N_{19}	N_{25}					$\xrightarrow{\text{Lévy}}$	$B_4(t)$
N_{15}	N_{20}	N_{26}		$\ddots$				$\xrightarrow{\text{Lévy}}$	$B_5(t)$
N_{21}	N_{27}							$\xrightarrow{\text{Lévy}}$	$B_6(t)$
N_{28}								$\xrightarrow{\text{Lévy}}$	$B_7(t)$

EXERCISES

1. Establish the final equality in the proof of Lemma 35.2.

2. Fix $t_0 < t_2$ and let $t_1 = (t_0 + t_2)/2$. Let $f(t_0) = y_0$ and $f(t_2) = y_2$ with $f(t)$ linearly interpolated on $[t_0, t_2]$. Let $g(t_0) = y_0$, $g(t_1) = y_1$, and $g(t_2) = y_2$, with $g(t)$ linearly interpolated on the intervals $[t_0, t_1]$ and $[t_1, t_2]$. Show that

$$\max\{|f(t) - g(t)| : t_0 \le t \le t_2\} \;=\; |g(t_1) - f(t_1)|.$$

3. Verify that Brownian motion properties (32.3) and (32.4) are satisfied by the patched-together process $(B(t) : t \ge 0)$ described at the end of this section.

CHAPTER VII

A GLIMPSE OF STOCHASTIC CALCULUS

§36. Introduction

In the previous chapter we constructed a Brownian path $(B_\omega(t) : 0 \leq t)$ for each outcome $\omega \in \mathbf{\Omega}$. Here, to reduce notational clutter, we will typically use the more conventional notation B to denote our Brownian process and B_t to denote its value at time t. This, of course, suppresses the ω dependence of the path. In this chapter we illustrate how to use our Brownian process to construct other continuous-time stochastic processes. We give two applications, one from finance and another from macroeconomics.

The easiest way to construct new processes from B is to take $X_t = f(B_t)$ for some function $f(b)$. For example, we might take $f(b) = b^2$, yielding $X_t = B_t^2$. We will require slightly more generality than that, and in our first example we will take $X_t = f(t, B_t)$ where $f(t, b)$ is an ordinary continuous function of two variables. The continuity of the function f together with the continuity of B_t in the variable t insure the continuity of the process X_t in t. Notice also that the value of X_t can be computed with knowledge of the Brownian process at time t. If $t' > t$, the increment $B_{t'} - B_t$ and X_t are therefore independent random variables. In this setting we prove an elementary version of Itô's formula leading to a classic application in finance.

In our second example the formula defining the process is a little more complicated — we defer it to a later section. With this example we develop the basic properties of the Ornstein-Uhlenbeck (O-U) process — a mean reverting cousin of Brownian motion. We conclude the chapter with an economic application of the O-U process.

In both examples, the stochastic process X_t constructed from the Brownian process B satisfies the following conditions:

$$\begin{aligned}
&\text{(i) for each outcome } \omega\text{, the path } X_t = X_t(\omega) \text{ is continuous in } t;\\
&\text{(ii) the initial value } X_0 \text{ and the Brownian process } B \text{ are independent;}\\
&\text{(iii) for each } \omega\text{, the value of } X_t(\omega) \text{ can be computed from } X_0(\omega)\\
&\qquad\qquad \text{together with } (B_s(\omega) : 0 \leq s \leq t).
\end{aligned} \tag{36.1}$$

Henceforth all stochastic processes under consideration are assumed to satisfy these three conditions.

Loosely speaking the random variable X_0 determines the starting value of the X process and the Brownian path determines how it evolves from there as time elapses. In our first example, where $X_t = f(t, B_t)$, we have $X_0 = f(0,0)$ which is deterministic — hence (ii) is automatically satisfied. However, in our second example (the O-U process) we will consider random starting values for the process. Conditions (ii) and (iii) imply that for any $0 \le t < t'$ we have that X_t is independent of the Brownian increment $B_{t'} - B_t$. This is because X_t depends only on X_0 (which is independent of B) and the Brownian process up to time t (which is independent of subsequent Brownian increments).

EXERCISES

• In problems 1 – 6, determine which conditions in (36.1) are satisfied by the process.

1. $X_t = B_{t/2}$.

2. $X_t = \int_0^t B_s\, ds$. (This is an ordinary Riemann integral, whose value is random, since the integrand, B_s, is a continuous function of s.)

3. $X_t = \max\{B_s : 0 \le s \le t\}$.

4. $X_t = B_{t+1}$.

5. $X_t = \begin{cases} +1 & \text{if } B_t \ge 0, \\ -1 & \text{if } B_t < 0. \end{cases}$

6. $X_t = X + B_t$, where X is a random variable independent of B.

§37. Stochastic Differential and Integral Equations

Suppose that M_t, N_t and X_t are three stochastic processes satisfying the conditions of (36.1). For $t \ge 0$ and $\Delta t > 0$ put $\Delta B_t = B_{t+\Delta t} - B_t$ and $\Delta X_t = X_{t+\Delta t} - X_t$. If, for small Δt, we have

$$\Delta X_t \;\approx\; M_t \Delta t + N_t \Delta B_t \tag{37.1}$$

in a sense to be made precise momentarily, we write

$$dX_t \;=\; M_t dt + N_t dB_t. \tag{37.2}$$

Equation (37.2) is a *stochastic differential* equation that describes what are sometimes (especially in finance) referred to as the *dynamics* of the X process. To make rigorous sense of (37.2), let us define the random variable $\epsilon(t, \Delta t)$ by the relation

$$\overbrace{\Delta X_t}^{\text{exact increment}} \;=\; \overbrace{M_t \Delta t + N_t \Delta B_t}^{\text{approximate increment}} + \overbrace{\epsilon(t, \Delta t)}^{\text{error}}, \tag{37.3}$$

so $\epsilon(t, \Delta t)$, the error in the approximation, is a "plug". Fix any time $T > 0$ and suppose $0 \leq u < v \leq T$. Let $\mathcal{T} = \{t_0, t_1, \ldots, t_n\}$ partition the time interval $[u, v]$ into n subintervals $[t_i, t_{i+1}]$, so $u = t_0 < t_1 < \cdots < t_n = v$. Let $\Delta t_i = t_{i+1} - t_i$ and let $||\mathcal{T}|| = \max\{\Delta t_i : 0 \leq i < n\}$ denote the partition's mesh. Now let $\Delta X_{t_i} = X_{t_{i+1}} - X_{t_i}$ and $\Delta B_{t_i} = B_{t_{i+1}} - B_{t_i}$ so

$$X_v - X_u \;=\; X_{t_n} - X_{t_0} \;=\; \sum_{i=0}^{n-1} \Delta X_{t_i}$$

(write out a few terms, the right side is a telescoping sum). Finally, define the random variable $\epsilon(\mathcal{T})$ by the relation

$$\begin{aligned}
\epsilon(\mathcal{T}) &= (X_v - X_u) - \left(\sum_{i=0}^{n-1} [M_{t_i} \Delta t_i + N_{t_i} \Delta B_{t_i}] \right) \\
&= \sum_{i=0}^{n-1} \Delta X_{t_i} - \left(\sum_{i=0}^{n-1} [M_{t_i} \Delta t_i + N_{t_i} \Delta B_{t_i}] \right) \;=\; \sum_{i=0}^{n-1} \epsilon(t_i, \Delta t_i).
\end{aligned} \tag{37.4}$$

Here we have computed the error if we estimate the X increment over the time interval $[u, v]$ by summing the approximate increments, as given in (37.1), over the n time intervals $[t_i, t_{i+1}]$. We will say that (37.2) holds on $[0, T]$ if, for any deterministic time interval $[u, v] \subset [0, T]$, we have

$$\epsilon(\mathcal{T}) \xrightarrow{\text{P}} 0 \quad \text{as} \quad ||\mathcal{T}|| \to 0. \tag{37.5}$$

That is to say, if $\mathcal{T}_m$ is any deterministic sequence of partitions of $[u, v]$ with $||\mathcal{T}_m|| \to 0$, we have $\epsilon(\mathcal{T}_m) \xrightarrow{\text{P}} 0$ as $m \to \infty$. If (37.5) holds, then we also have

$$\sum_{i=0}^{n-1} [M_{t_i} \Delta t_i + N_{t_i} \Delta B_{t_i}] \xrightarrow{\text{P}} X_v - X_u \quad \text{as } ||\mathcal{T}|| \to 0, \tag{37.6}$$

and conversely. Now $\sum_{i=0}^{n-1} M_{t_i} \Delta t_i$ is a Riemann sum approximation of $\int_u^v M_t \, dt$, which is a perfectly legitimate Riemann integral as the integrand is a continuous (though random) function of t. Hence by ordinary calculus $\sum_{i=0}^{n-1} M_{t_i} \Delta t_i \to \int_u^v M_t \, dt$ as $||\mathcal{T}|| \to 0$. It follows from (37.6) that $\sum_{i=0}^{n-1} N_{t_i} \Delta B_{t_i}$ converges in probability to a (random) limit whose value, in fact, must be $X_v - X_u - \int_u^v M_t \, dt$. We use the notation $\int_u^v N_t \, dB_t$ for this limit-in-probability and, combining two integrals, we write

$$\int_u^v M_t \, dt + N_t \, dB_t \;=\; X_v - X_u, \tag{37.7}$$

which is called a *stochastic integral* equation.

EXAMPLE 37.8. Figure 37.9 shows a realization out to time $T = 2$ of a Brownian path B_t (left) and the resulting path of a stochastic process defined by $X_t = B_t^2 - t$. Put $u = 0.50$, $v = 1.00$ and consider the partition $\mathcal{T} = \{0.50, 0.75, 1.00\}$. The corresponding values of B_t at those three times are -1.44, -1.29, and -0.58, respectively. You may easily verify (with a calculator) that $X_u = 1.57$ and $X_v = -0.66$, so $X_v - X_u = -2.23$. This is the right-hand side of (37.6)). As we shall see shortly, the dynamics for this process are $M_t = 0$ (it is a *zero drift* process) and $N_t = 2B_t$ — so $dX_t = 2B_t dB_t$. With this data, partition, and dynamics, the left-hand side of (37.6) is

$$\sum_{i=0}^{1} 2B_{t_i}\Delta B_{t_i} =$$
$$\big[2 \times -1.44 \times \big(-1.29 - (-1.44)\big)\big] + \big[2 \times -1.29 \times \big(-0.58 - (-1.29)\big)\big]$$
$$= -2.26.$$

Referring to the top line in (37.4), we see that $\epsilon(\mathcal{T}) = -2.23 - (-2.26) = 0.03$. (That the error is this small for such a crude partition is purely coincidental.)

Figure 37.9. A realization of B_t and the resulting $X_t = B_t^2 - t$.

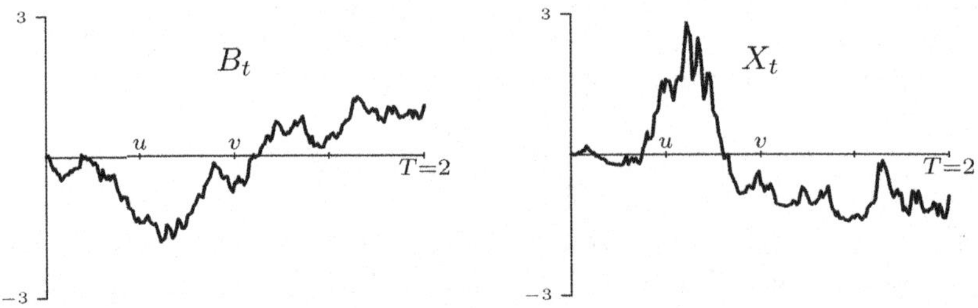

Stochastic calculus is a difficult and highly technical area of probability theory, requiring a healthy dose of measure theory for a full understanding. In the two examples of this chapter we consider relatively simple settings. There we can obtain fairly rigorous results without the machinery of measure theory.

EXERCISES

• Additional data points for the Brownian path from Example 37.8 are:

t	0.50	0.55	0.60	0.65	0.70	0.75	0.80	0.85	0.90	0.95	1.00
B_t	−1.44	−1.40	−1.71	−1.53	−1.36	−1.29	−0.94	−0.60	−0.24	−0.47	−0.58

1. Following Example 37.8, compute $\sum_{i=0}^{n-1} 2B_{t_i}\Delta B_{t_i}$ and $\epsilon(\mathcal{T})$ with $\mathcal{T} = \{0.50, 0.60, 0.70, 0.80, 0.90, 1.00\}$.

2. Same for $\mathcal{T} = \{0.50, 0.55, 0.60, 0.65, 0.70, 0.75, 0.80, 0.85, 0.90, 0.95, 1.00\}$.

§38. Itô's Formula for Brownian Motion

Here we develop an elementary version of Itô's formula in the context of Brownian motion. Suppose $f(t, b)$ is a continuous function of two variables. Then, as we have discussed, $X_t = f(t, B_t)$ describes a new stochastic process with continuous paths. For the discussion that follows to be rigorous, we impose at this juncture some technical restrictions. For convenience, we restrict attention to the X_t process on a finite time interval $[0, T]$. Additionally, we require that:

$$\begin{array}{c}\text{(i) the partial derivatives } f_{tt},\ f_{tb},\ f_{bbb} \text{ all exist and are continuous;}\\ \text{(ii) for some } h > 0,\quad E\left[(f_{bb}(t, B_t))^2\right] \le h \text{ for all } t \in [0, T].\end{array} \tag{38.1}$$

Condition (ii) is rather technical, but not terribly onerous. Presently we prove:

THEOREM 38.2 (ITÔ'S FORMULA). *If the function $f(t, b)$ satisfies (38.1), then on the interval $[0, T]$ the process $X_t = f(t, B_t)$ has dynamics given by*

$$dX_t = \left(f_t(t, B_t) + \frac{1}{2} f_{bb}(t, B_t)\right) dt + f_b(t, B_t) dB_t.$$

NOTATIONAL REMARK. There is some room for notational confusion here, as X_t denotes the value of the stochastic process X at time t while f_t denotes the partial derivative of f with respect to t. We will use capital letters to denote stochastic processes and lower case letters to denote functions and real variables. This should mitigate any confusion.

EXAMPLE 38.3. In Example 37.8 we have $f(t, b) = b^2 - t$, so $f_t(t, b) = -1$, $f_b(t, b) = 2b$, and $f_{bb}(t, b) = 2$. This gives

$$dX_t = \left(-1 + \frac{1}{2} \cdot 2\right) dt + 2B_t dB_t = 2B_t dB_t.$$

Here the drift term (M_t in (37.3)) is zero — this is a *zero drift* process.

PROOF OF THEOREM 38.2. For any Brownian path B let $L = \max\{|B_t| : 0 \le t \le T\}$. Then the rectangle $\mathcal{R}(T, L) = \{(t, b) : 0 \le t \le T \text{ and } |b| \le L\}$ contains the path. Since f's derivatives in (38.1) are continuous they are bounded in absolute value in this rectangle (see Figure 38.4). Letting C denote the least such upper bound, we observe that C is a random variable as it depends on ω through the path B that is associated with ω.

Figure 38.4. f's derivatives are bounded by C in $\mathcal{R}(T, L)$.

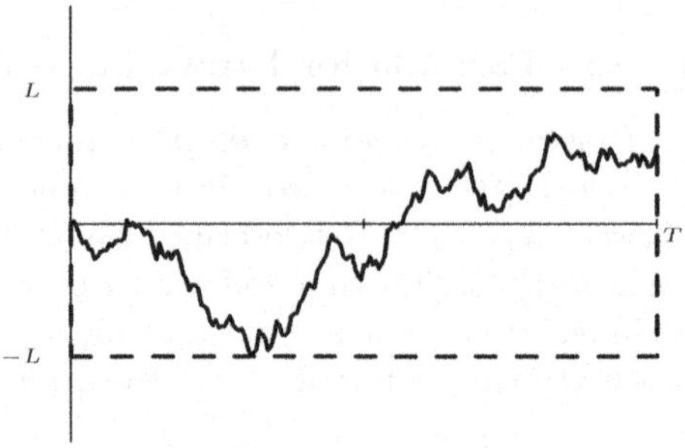

Fix $0 \le u < v \le T$ and let $\mathcal{T} = \{t_i\}$ be a partition of $[u, v]$. Using standard estimation techniques for functions of two variables (see §45), we get that

$$\begin{aligned}\Delta X_{t_i} &= f(t_{i+1}, B_{t_{i+1}}) - f(t_i, B_{t_i}) \\ &= f_t(t_i, B_{t_i})\Delta t_i + f_b(t_i, B_{t_i})\Delta B_{t_i} + \frac{1}{2} f_{bb}(t_i, B_{t_i})\Delta B_{t_i}^2 \\ &\quad + \text{Error}_i,\end{aligned}$$

where $|\textit{Error}_i| \le C \cdot (\Delta t_i^2 + \Delta t_i |\Delta B_{t_i}| + |\Delta B_{t_i}|^3)$. Here the term $\epsilon(t, \Delta t)$ as defined in (37.3) is given by

$$\begin{aligned}\epsilon(t_i, \Delta t_i) &= \Delta X_{t_i} - \left[\left(f_t(t_i, B_{t_i}) + \frac{1}{2} f_{bb}(t_i, B_{t_i})\right)\Delta t_i + f_b(t_i, B_{t_i})\Delta B_{t_i}\right] \\ &= \frac{1}{2} f_{bb}(t_i, B_{t_i})\Delta B_{t_i}^2 - \frac{1}{2} f_{bb}(t_i, B_{t_i})\Delta t_i + \text{Error}_i. \qquad (38.5)\end{aligned}$$

We will show that, as $||\mathcal{T}|| \to 0$,

$$\sum_i \frac{1}{2} f_{bb}(t_i, B_{t_i})\Delta B_{t_i}^2 \xrightarrow{\text{P}} \int_u^v \frac{1}{2} f_{bb}(t, B_t)\, dt, \qquad (38.6)$$

and

$$\sum_i |\text{Error}_i| \le C \cdot \sum_i \left(\Delta t_i^2 + \Delta t_i |\Delta B_{t_i}| + |\Delta B_{t_i}|^3\right) \xrightarrow{\text{P}} 0. \qquad (38.7)$$

Note that the right side of (38.6) is a perfectly legitimate Riemann integral, as, for each ω, the integrand is a continuous function of t. If we combine (38.6) with the fact that

$$\sum_i \frac{1}{2} f_{bb}(t_i, B_{t_i})\Delta t_i \to \int_u^v \frac{1}{2} f_{bb}(t, B_t)\, dt \quad \text{as} \quad ||\mathcal{T}|| \to 0,$$

we see that

$$\left|\sum_i \frac{1}{2} f_{bb}(t_i, B_{t_i}) \Delta B_{t_i}^2 - \sum_i \frac{1}{2} f_{bb}(t_i, B_{t_i}) \Delta t_i\right| \xrightarrow{\mathrm{P}} 0 \quad \text{as} \quad ||\mathcal{T}|| \to 0.$$

($X_m \xrightarrow{\mathrm{P}} X$, $Y_m \xrightarrow{\mathrm{P}} X \implies |X_m - Y_m| \xrightarrow{\mathrm{P}} 0$, see exercises.) This together with (38.7) shows that $\sum_i \epsilon(t_i, \Delta t_i) \xrightarrow{\mathrm{P}} 0$, as required (refer to (38.5)).

First we turn our attention to the error terms in (38.7). Note that

$$\sum_i \Delta t_i{}^2 \;\le\; ||\mathcal{T}|| \sum_i \Delta t_i \;=\; ||\mathcal{T}|| \cdot (v - u) \;\to\; 0, \quad \text{as } ||\mathcal{T}|| \to 0.$$

That was easy. As for the second error term,

$$\begin{aligned} \sum_i \Delta t_i |\Delta B_{t_i}| \;&\le\; \max\{|\Delta B_{t_i}| : 0 \le i < n\} \cdot \sum_i \Delta t_i \\ &=\; \max\{|\Delta B_{t_i}|\} \cdot (v - u), \end{aligned}$$

so it will suffice to show that

$$\max\{|\Delta B_{t_i}|\} \;\to\; 0, \quad \text{as } ||\mathcal{T}|| \to 0. \tag{38.8}$$

But B_t is a continuous function of t. Hence, on any closed and bounded interval $[u, v]$ it is *uniformly continuous* (see Appendix II). That is, for any $\epsilon > 0$ there is a $\delta > 0$ such that $|B_t - B_{t'}| < \epsilon$ whenever $t, t' \in [u, v]$ with $|t - t'| < \delta$. Note that $\delta = \delta(\omega)$ is random because it will depend on the Brownian path $B = B(\omega)$. Now fix any $\epsilon > 0$ and let $\delta > 0$ correspond to that ϵ. Then $||\mathcal{T}|| < \delta \implies \Delta t_i < \delta \implies |B_{t_i} - B_{t_{i+1}}| < \epsilon$. Because this holds for each individual i, we see that $||\mathcal{T}|| < \delta \implies \max\{|\Delta B_{t_i}|\} < \epsilon$. This establishes (38.8).

For the third error term, use Lemma 38.9 (stated below) as applied to $H_t = 1$ to get that $\sum_i \Delta B_{t_i}^2 \xrightarrow{\mathrm{P}} v - u$ as $||\mathcal{T}|| \to 0$. Then, as $||\mathcal{T}|| \to 0$,

$$\sum_i |\Delta B_{t_i}^3| \;\le\; \max\{|\Delta B_{t_i}|\} \cdot \sum_i \Delta B_{t_i}^2 \;\xrightarrow{\mathrm{P}}\; 0,$$

as $X_m \xrightarrow{\mathrm{P}} 0$, $Y_m \xrightarrow{\mathrm{P}} c \implies X_m Y_m \xrightarrow{\mathrm{P}} 0$ (see exercises). Collecting these three error terms yields that

$$\sum_i \left(\Delta t_i^2 + \Delta t_i |\Delta B_{t_i}| + |\Delta B_{t_i}|^3\right) \;\xrightarrow{\mathrm{P}}\; 0,$$

as $||\mathcal{T}|| \to 0$. This is enough to get (38.7), for $Y_n \xrightarrow{\mathrm{P}} 0 \implies XY_n \xrightarrow{\mathrm{P}} 0$ (see, e.g., Theorem 8.6).

Now we turn our attention to establishing (38.6), which is the crux of the matter. We have:

LEMMA 38.9. *If the process H_t satisfies* (36.1) *and has bounded second moment for $t \in [0,T]$ then*

$$\sum_{i=0}^{n-1} H_{t_i} \Delta B_{t_i}^2 \xrightarrow{\text{P}} \int_u^v H_t\, dt \quad \textit{as} \quad ||\mathcal{T}|| \to 0.$$

Statement (38.6) is simply this lemma as applied to $H_t = \frac{1}{2} f_{bb}(t, B_t)$, which has bounded second moment for $t \in [0,T]$ by virtue of (38.1).

PROOF OF LEMMA 38.9. For partition $\mathcal{T} = \{t_0, \ldots, t_n\}$ of $[u,v]$, let

$$\begin{aligned} Z_{\mathcal{T}} &= \sum_i H_{t_i} \Delta B_{t_i}^2 - \sum_i H_{t_i} \Delta t_i \\ &= \sum_i H_{t_i} [\Delta B_{t_i}^2 - \Delta t_i] \\ &= \sum_i H_i M_i, \end{aligned}$$

where $H_i = H_{t_i}$ and $M_i = \Delta B_{t_i}^2 - \Delta t_i$. We will show that $Z_{\mathcal{T}} \xrightarrow{\text{P}} 0$ as $||\mathcal{T}|| \to 0$. Since

$$\sum_{i=0}^{n-1} H_{t_i} \Delta t_i \to \int_u^v H_t\, dt \quad \text{as } ||\mathcal{T}|| \to 0,$$

the lemma will follow ($X_m - Y_m \xrightarrow{\text{P}} 0$, $Y_m \xrightarrow{\text{P}} Y \implies X_m \xrightarrow{\text{P}} Y$, see exercises). Note that this holds because the left side is a Riemann sum approximation of the integral on the right side. (Again, the integrand is a continuous function of t.) We will show that $EZ_{\mathcal{T}} = 0$ and that $\operatorname{Var} Z_{\mathcal{T}} \to 0$ as $||\mathcal{T}|| \to 0$. Then, by Chebyshev's inequality, $P[|Z_{\mathcal{T}}| \geq \epsilon] \leq \frac{\operatorname{Var} Z_{\mathcal{T}}}{\epsilon^2} \to 0$, establishing that $Z_{\mathcal{T}} \xrightarrow{\text{P}} 0$.

Now $E[\Delta B_{t_i}] = 0$, so $E[\Delta B_{t_i}^2] = \operatorname{Var} \Delta B_{t_i} = \Delta t_i$ and $EM_i = 0$. Furthermore, H_i depends on B_t only through time t_i while M_i depends on the Brownian increment on $[t_i, t_{i+1}]$, so H_i and M_i are independent and $E[H_i M_i] = EH_i \cdot EM_i = 0$. By linearity of expectation, $EZ_{\mathcal{T}} = 0$. Regarding the variance, if $0 \leq i < j$, then

$$\begin{aligned} E[H_i M_i H_j M_j] &= E[H_i M_i H_j] \cdot E[M_j] \\ &= E[H_i M_i H_j] \cdot 0 = 0 = E[H_i M_i] \cdot E[H_j M_j]. \end{aligned}$$

The first equality holds because $H_i M_i H_j$ depends on the Brownian path up to time t_j, while M_j depends on the Brownian increment over the interval $[t_j, t_{j+1}]$ — $H_i M_i H_j$ and M_j are therefore independent. We have that $H_i M_i$ and $H_j M_j$ are uncorrelated if $i \neq j$, so

$$\begin{aligned} \operatorname{Var} Z_{\mathcal{T}} = \sum_i \operatorname{Var}[H_i M_i] = \sum_i E[(H_i M_i)^2] &= \sum_i E[H_i^2] E[M_i^2] \\ &\leq h \sum_i \operatorname{Var}[M_i], \end{aligned} \tag{38.10}$$

where h is the hypothesized bound for EH_t^2. The second equality above holds because H_iM_i is mean 0; the third because H_i^2 and M_i^2 are independent. The inequality uses that M_i is mean 0. Now $\text{Var}[M_i] = \text{Var}[\Delta B_{t_i}^2 - \Delta t_i] = \text{Var}[\Delta B_{t_i}^2]$, and

$$\Delta B_{t_i}^2 \sim (\text{Normal}(0, \Delta t_i))^2 \sim \Delta t_i(\text{Normal}(0,1))^2,$$

so

$$\text{Var}\, M_i = \Delta t_i^2 \text{Var}[(\text{Normal}(0,1))^2] = 2\Delta t_i^2,$$

using that $(\text{Normal}(0,1))^2$ has the chi-squared distribution with 1 degree of freedom which has variance 2. Then

$$\sum_i \text{Var}[M_i] = \sum_i 2\Delta t_i^2 \le 2||\mathcal{T}|| \cdot \sum_i \Delta t_i = 2||\mathcal{T}|| \cdot (v-u) \to 0,$$

as $||\mathcal{T}|| \to 0$. Combining this with (38.10) gives that $\text{Var}\, Z_{\mathcal{T}} \to 0$ as $||\mathcal{T}|| \to 0$. This concludes the proof of Lemma 38.9 and Theorem 38.2. □

REMARK. Lemma 38.9 is sometimes informally paraphrased as "$dB^2 = dt$."

EXERCISES

• In problems 1 – 3, verify the assertions made in the proof of Theorem 38.2 and Lemma 38.9.

1. $X_m \xrightarrow{\text{P}} X,\ Y_m \xrightarrow{\text{P}} X \implies |X_m - Y_m| \xrightarrow{\text{P}} 0.$

2. $X_m \xrightarrow{\text{P}} 0,\ Y_m \xrightarrow{\text{P}} c \implies X_mY_m \xrightarrow{\text{P}} 0.$

3. $X_m - Y_m \xrightarrow{\text{P}} 0,\ Y_m \xrightarrow{\text{P}} Y \implies X_m \xrightarrow{\text{P}} Y.$

• In problems 4 – 7, compute the dynamics of $X_t = f(t,b)$.

4. $f(t,b) = b$.

5. $f(t,b) = tb$.

6. $f(t,b) = tb^2$.

7. $f(t,b) = e^b$ (geometric Brownian motion).

8. Determine a function $g(t)$ so that $f(t,b) = g(t)e^b$ generates a zero drift process.

§39. Application: The Black-Scholes PDE and Formula

Here we study a classic application of Itô's formula to develop the Black-Scholes formula which values a call option on an underlying stock. It is customary in finance to model the evolution of a stock price as a random process following *geometric Brownian motion*:

$$S_t = s_0 e^{\mu t + \sigma B_t}, \tag{39.1}$$

where B_t is a standard Brownian motion. Here, $s_0 > 0$ is the value of the stock today (i.e., at time 0), μ is a deterministic *drift* parameter and $\sigma \geq 0$ is the *volatility* of the stock price, which scales the amount of randomness that enters the process. Note that we are assured that $S_t > 0$, which is the sole purpose for exponentiation in (39.1). (Stock prices never go negative — although stocks can become worthless, something this model does not allow.) In this discussion we assume that the stock *pays no dividends.* Holding the stock over a period of time generates no cash flow until the stock is sold and its value at that time is captured in the form of cash from the sale. We ignore taxes and transaction costs.

Now put $s(t,b) = s_0 e^{\mu t + \sigma b}$, so $s(t,b)$ is just an ordinary function of two variables and S_t, as defined in (39.1), is equal to $s(t, B_t)$. Presently we use Itô's formula to describe the dynamics of the stochastic process S_t in terms of B_t. Taking the relevant partial derivatives of $s(t,b)$, we find that

$$s_t = \mu s, \quad s_b = \sigma s, \text{ and } \quad s_{bb} = \sigma^2 s.$$

Here the partial derivatives and the function s itself are all evaluated at (t,b). (The reader will verify that $s(t,b)$ satisfies the conditions in (38.1).) By Itô's formula, the dynamics of S_t in terms of B_t are then given by

$$\begin{aligned} dS_t &= \left[\mu s(t,B_t) + \frac{1}{2}\sigma^2 s(t,B_t)\right] dt + \sigma s(t,B_t)\, dB_t \\ &= r^* S_t\, dt + \sigma S_t\, dB_t, \end{aligned}$$

where $r^* = \mu + \frac{1}{2}\sigma^2$. Here, of course, we have used that $s(t,B_t) = S_t$. Over a small interval of time Δt these dynamics imply that

$$\Delta S_t \approx r^* S_t \Delta t + \sigma S_t \Delta B_t.$$

If the stock is trading at price s at time t,

$$\begin{aligned} E[\Delta S_t \mid S_t = s] &\approx E[r^* S_t \Delta t + \sigma S_t \Delta B_t \mid S_t = s] \\ &= r^* s \Delta t \; + \; \sigma s E[\Delta B_t \mid S_t = s] \\ &= s \cdot r^* \Delta t, \end{aligned}$$

where last equality uses that ΔB_t and S_t are independent so $E[\Delta B_t | S_t = s] = E[\Delta B_t] = 0$. Given that the stock's price at any particular time is s, the expected change in its value over the next small time increment Δt is given by $s \cdot r^* \Delta t$, i.e., the expected growth rate is r^*.

Let r denote the annual continuously compounding rate of interest as observed in the financial markets. For example, $r = .05$ corresponds to an interest rate of 5% per year and \$100 invested at this rate of interest for three years grows to $\$100 \cdot e^{0.05 \times 3} \approx \$116.18$. Conversely, \$100 to be paid 3 years from now is worth

$\$100 \cdot e^{-0.05\times 3} \approx \$86.07$ today. Let us agree that any security that does not generate cash flow is *priced fairly* if its market value's expected growth rate over the next infinitesimal increment of time is equal to the market rate of interest r. For the our stock S_t, this means that $r^* = r$, or, equivalently,

$$\mu = r - \frac{1}{2}\sigma^2. \tag{39.2}$$

With this choice of drift μ, the stock S_t is always priced fairly in the above sense.

Presently we show that, as a consequence, for any price s observed at time t:

$$\text{if } T > t, \text{ then } s = e^{-r(T-t)} \cdot E[S_T \,|\, S_t = s]. \tag{39.3}$$

Equation (39.3) says that, at any time t, the current value of the stock (s) can be obtained by computing the *expected* value of the stock at a future time (T) and discounting back over the elapsed time period $(T - t)$. The expected value is computed using the conditional distribution of S_T given that the stock's value at time t is s. To establish (39.3), note that if $S_t = s$, then

$$\begin{aligned} S_T &= s_0 e^{\mu T + \sigma B_T} \\ &= s_0 e^{\mu t + \sigma B_t} \cdot e^{\mu(T-t) + \sigma(B_T - B_t)} \\ &= s e^{\mu(T-t) + \sigma \Delta B}, \end{aligned}$$

where $\Delta B = B_T - B_t$. Now S_t depends on B_t, which is independent of the increment ΔB. Using that $\Delta B \sim \text{Normal}\,(0, T-t) \sim \sqrt{T-t} \cdot \text{Normal}\,(0,1)$, we get

$$\begin{aligned} E[S_T \,|\, S_t = s] &= E[s e^{\mu(T-t) + \sigma \Delta B} \,|\, S_t = s] \\ &= E[s e^{\mu(T-t) + \sigma \Delta B}] \\ &= s e^{\mu(T-t)} E[e^{\sigma \sqrt{T-t} \cdot \text{Normal}\,(0,1)}] \end{aligned}$$

The moment generating function for a standard normal is $M(\theta) = e^{\theta^2/2}$, so

$$\begin{aligned} e^{-r(T-t)} \cdot E[S_T \,|\, S_t = s] &= e^{-r(T-t)} s e^{\mu(T-t)} e^{\sigma^2 (T-t)/2} \\ &= s e^{(\mu - r + \sigma^2/2)(T-t)} = s, \ \checkmark \end{aligned}$$

where the final equality uses (39.2).

Derivative Securities. Now consider another security, call it C, that generates no cash flow until a future time $T > 0$. At that time, it generates a cash payment that depends on S_T, the value of the stock price at that time. Let $g(s)$ denote this *payoff function*, so at time T the security C generates its first and only cash payment in the amount of $g(S_T)$. Such a security C is called a *derivative security* because its value is *derived* from the value of S_T.

EXAMPLE 39.4. A *call option* gives the option holder the right, but not the obligation, to purchase a share of stock at a predetermined price k (the *strike* price) at a specified future time T (the *expiration*). If the stock is trading at a price S_T above the strike price at time T, the option holder can purchase the stock at price k and immediately sell it in the market at price S_T — the locked-in profit is $S_T - k$. If the stock is trading at a price below the strike, the option holder simply walks away from the option with no obligation — it makes no sense to buy the stock at price $k > S_T$ when the option holder can buy it in the market at price S_T. Since the option expires at time T, the payout in this situation is 0. Thus the call option payout at time T is

$$\begin{aligned} C_T &= \begin{cases} S_T - k & \text{if } S_T > k \\ 0 & \text{otherwise} \end{cases} \\ &= g_{\text{call}}(S_T), \end{aligned}$$

where $g_{\text{call}}(s) = \max(0, s - k)$. A *put option* gives the holder the right, but not the obligation, to *sell* the stock at the strike price. It has payoff function $g_{\text{put}}(s) = \max(0, k - s)$.

Returning to our general derivative with payoff function $g(s)$, let C_t denote the derivative security's value at time t. Here, we restrict attention to $0 \le t \le T$ since the derivative security makes its one and only cash payment at time T. We seek to find a function $c(t, s)$ such that $C_t = c(t, S_t)$. This function will tell us at any point in time what the value of the derivative security is in terms of the value of the underlying stock.

Bringing C_t back to the Brownian motion, we have

$$C_t = c(t, S_t) = c(t, s(t, B_t)),$$

so, with

$$f(t, b) = c(t, s(t, b)),$$

we have

$$C_t = f(t, B_t).$$

We again use Itô's formula to describe the dynamics of C_t with respect to the Brownian process. Computing the relevant partials, we have

$$\begin{aligned} f_t &= c_t + c_s s_t = c_t + c_s \mu s, \\ f_b &= c_s s_b = c_s \sigma s, \quad \text{and} \\ f_{bb} &= c_s s_{bb} + s_b^2 c_{ss} = c_s \sigma^2 s + c_{ss} \sigma^2 s^2. \end{aligned}$$

To be clear, the function s is evaluated at (t, b) and the various partials of c are evaluated at $(t, s(t, b))$. Itô's formula then gives that

$$dC_t = \left[c_t + c_s \mu s + \frac{1}{2} \left(c_s \sigma^2 s + c_{ss} \sigma^2 s^2 \right) \right] dt + c_s \sigma s \, dB_t.$$

Here the function s is evaluated at (t, B_t), so $s = S_t$, and the partials of c are evaluated at $(t, s(t, B_t)) = (t, S_t)$.

For the derivative to be priced fairly, we wish for its expected growth rate to equal the rate of interest r. This means that we must always have

$$c_t + c_s \mu s + \frac{1}{2}\left(c_s \sigma^2 s + c_{ss}\sigma^2 s^2\right) \;=\; C_t r \;=\; c(t, S_t) r \;=\; cr,$$

where, on the right and left sides above, c and its various partials are all evaluated at (t, S_t). This will hold provided the function $c = c(t, s)$ solves the PDE:

$$c_t + \frac{1}{2}\sigma^2 s^2 c_{ss} + rsc_s - rc \;=\; 0. \qquad (39.5)$$

Here again we have used (39.2): $c_s \mu s + \frac{1}{2} c_s \sigma^2 s = rsc_s$. Equation (39.5) is the much celebrated Black-Scholes PDE.

Solving the Black-Scholes PDE. Motivated by (39.3), let's hope that we can value the derivative at time t if the underlying stock price is $S_t = s$, by computing its expected payout at time T and discounting back to time t at rate r. As in (39.3), the expected value will be computed using the conditional distribution of S_T given that s is the stock's value at time t. This reasoning leads us to hope that

$$c(t, s) \;=\; e^{-r(T-t)} E[g(S_T) | S_t = s] \qquad (39.6)$$

solves the Black-Scholes PDE. Note that it has the correct boundary conditions, as

$$c(T, s) \;=\; e^{-r(T-T)} E[g(S_T) | S_T = s] \;=\; g(s),$$

which produces $C_T = c(T, S_T) = g(S_T)$ — the derivative's payout at time T.

Next we verify that (39.6) is indeed a solution to (39.5). Warning: This is tedious! Given that $S_t = s$, we have seen that $S_T \sim s e^{\mu(T-t)+\sigma\sqrt{T-t}\,\text{Normal}(0,1)}$. If we let

$$y \;=\; y(s, t, x) \;=\; s e^{\mu(T-t)+\sigma\sqrt{T-t}\cdot x} \quad \text{and} \quad \psi(x) \;=\; \frac{1}{\sqrt{2\pi}} e^{-x^2/2},$$

then

$$c \;=\; c(t, s) \;=\; e^{-r(T-t)} \int_{-\infty}^{\infty} g(y)\psi(x)\, dx. \qquad (39.7)$$

Therefore

$$\begin{aligned} c_t \;&=\; rc + e^{-r(T-t)} \int_{-\infty}^{\infty} g'(y) \frac{\partial y}{\partial t} \psi(x)\, dx \\ &=\; rc - e^{-r(T-t)} \int_{-\infty}^{\infty} y g'(y)(\mu + \frac{1}{2}\sigma(T-t)^{-1/2} x)\psi(x)\, dx, \end{aligned}$$

while

$$\begin{aligned} c_s &= e^{-r(T-t)} \int_{-\infty}^{\infty} g'(y) \frac{\partial y}{\partial s} \psi(x)\, dx \\ &= e^{-r(T-t)} \int_{-\infty}^{\infty} \frac{y}{s} g'(y) \psi(x)\, dx, \end{aligned} \tag{39.8}$$

and

$$c_{ss} = e^{-r(T-t)} \int_{-\infty}^{\infty} \left(\frac{y}{s}\right)^2 g''(y)\psi(x)\, dx. \tag{39.9}$$

Equation (39.9) follows from (39.8), since $\frac{\partial}{\partial s}\left(\frac{y}{s}\right) = 0$. In the three computations above, the passing of the derivative through the integral must be justified. This will work for any "reasonable" payoff function $g()$. Presently we assemble all this to verify that (39.6) is satisfied by our specification of $c(t,s)$. Now

$$\begin{aligned} &\textit{Left Side of } (39.5) = \\ &rc - e^{-r(T-t)} \int_{-\infty}^{\infty} y g'(y)(\mu + \frac{1}{2}\sigma(T-t)^{-1/2}x)\psi(x)\, dx \\ &+ e^{-r(T-t)} \int_{-\infty}^{\infty} \frac{1}{2}\sigma^2 y^2 g''(y)\psi(x)\, dx + e^{-r(T-t)} \int_{-\infty}^{\infty} r y g'(y)\psi(x)\, dx \\ &- rc. \end{aligned}$$

After some cancellation and rearrangement using (39.2), we see that we must check that

$$\int_{-\infty}^{\infty} y g'(y)(T-t)^{-1/2} x\psi(x)\, dx = \int_{-\infty}^{\infty} \sigma y^2 g''(y)\psi(x)\, dx + \int_{-\infty}^{\infty} \sigma y g'(y)\psi(x)\, dx.$$

This can be confirmed by an integration by parts setting $v = yg'(y)(T-t)^{-1/2}$ and $u = -\psi(x)$ (so $du = x\psi(x)dx$). The details of this are left to the reader. Note that the boundary term requires that

$$0 = uv\Big|_{x=-\infty}^{x=\infty} = -(T-t)^{-1/2}\left[yg'(y)\psi(x)\right]_{x=-\infty}^{x=\infty}.$$

This will hold for any reasonable choice of payoff function $g()$ since $\psi(x) \to 0$ very rapidly as $x \to \pm\infty$.

Black-Scholes PDE as Applied to Call Options. Fischer Black and Myron Scholes (1973) applied this to European call and put options on the underlying stock S_t. When (39.6) (or, equivalently, (39.7)) is applied to the call option, the result is the following famous formula:

$$c(t,s) = s\Psi(d_+) - ke^{-r(T-t)}\Psi(d_-), \tag{39.10}$$

where

$$d_\pm = \frac{\ln(s/k) + r(T-t)}{\sigma\sqrt{T-t}} \pm \frac{1}{2}\sigma\sqrt{T-t}, \text{ and } \Psi(x) = P[\text{Normal}\,(0,1) \le x].$$

We have seen that, conditioned on $S_t = s$, $S_T \sim se^{\mu(T-t)+\sigma\sqrt{T-t}\,N}$, where $N \sim \text{Normal}\,(0,1)$. Hence

$$E[g(S_T)\,|\,S_t = s] = \int_{-\infty}^{\infty} g_{\text{call}}\left(se^{\mu(T-t)+\sigma\sqrt{T-t}\cdot x}\right)\psi(x)\,dx.$$

Now $g_{\text{call}}(s) = 0$ if $s < k$ and $g_{\text{call}}(s) = s - k$ if $s \ge k$, hence

$$E[g(S_T)\,|\,S_t = s] = \int_{x^*}^{\infty} \left(se^{\mu(T-t)+\sigma\sqrt{T-t}\cdot x} - k\right)\psi(x)\,dx, \qquad (39.11)$$

where x^* solves $se^{\mu(T-t)+\sigma\sqrt{T-t}\cdot x^*} = k$, i.e., when $x^* = -d_-$. We break the integral in (39.11) into two pieces. First, note that

$$\int_{x^*}^{\infty} -k\psi(x)\,dx = -kP[N \ge x^*] = -kP[N \le -x^*] = -k\Psi(d_-).$$

Next, a little algebra (using (39.2) again) verifies that

$$se^{\mu(T-t)+\sigma\sqrt{T-t}\cdot x}\psi(x) = se^{r(T-t)}\psi\left(x - \sigma\sqrt{T-t}\right),$$

so

$$\begin{aligned}
\int_{x^*}^{\infty} se^{\mu(T-t)+\sigma\sqrt{T-t}\cdot x}\psi(x)\,dx &= se^{r(T-t)}\int_{x^*}^{\infty}\psi\left(x - \sigma\sqrt{T-t}\right)dx \\
&= se^{r(T-t)}P[N \ge x^* - \sigma\sqrt{T-t}] \\
&= se^{r(T-t)}P[N \le -x^* + \sigma\sqrt{T-t}] \\
&= se^{r(T-t)}\Psi(d_+).
\end{aligned}$$

Combining the two pieces of the integral gives that

$$E[g(S_T)\,|\,S_t = s] = se^{r(T-t)}\Psi(d_+) - k\Psi(d_-),$$

so

$$c(t,s) = e^{-r(T-t)}E[g(S_T)\,|\,S_t = s] = s\Psi(d_+) - ke^{-r(T-t)}\Psi(d_-). \quad \square$$

EXERCISES

1. What is the value of a call option expiring in 6 months struck at \$110 if: (i) the stock is currently trading at \$100; (ii) the market interest rate is 5%; and (iii) the stock's volatility is $\sigma = 40\%$.

2. Verify that the function $s(t, b) = s_0 e^{\mu t + \sigma b}$ satisfies the conditions in (38.1).

3. Use (39.7) and mimic the calculations above to derive a formula for the value (call it $p(t, s)$) of a put option. Recall that $g_{\text{put}}(S_T) = \max(0, k - S_T)$.

4. Verify that call and put options satisfy *put-call parity*:

$$c(t, s) - p(t, s) + e^{-r(T-t)}k = s.$$

5. Show that $g_{\text{call}}(S_T) - g_{\text{put}}(S_T) = S_T - K$ and explain why this also yields put-call parity.

6. A certain *binary* option pays \$1 at time T if $S_T \leq k$, and nothing otherwise. Derive a formula for the value (call it $b(t, s)$) of this option.

7. Simplifying to the case where $r = 0$, derive formulas for $b_t(t, s)$, $b_s(t, s)$, and $b_{ss}(t, s)$.

8. The Black-Scholes PDE (39.5) for the binary option with $r = 0$ says that $b_t = -\frac{1}{2}\sigma^2 s^2 b_{ss}$. Verify using your formulas in problem 7 that this is true. Hint: at some point you may wish to use that

$$\frac{\Psi''(x)}{\Psi'(x)} = \frac{d}{dx} \ln \Psi'(x) = \frac{d}{dx}\left(\ln \frac{1}{\sqrt{2\pi}} - \frac{1}{2}x^2\right) = -x.$$

§40. The Ornstein-Uhlenbeck Process

The Ornstein-Uhlenbeck (O-U) process is a normal mean reverting stochastic process whose dynamics and initial condition are given by

$$dU_t = -\alpha U_t dt + dB_t, \tag{40.1}$$
$$U_0 = 0, \tag{40.2}$$

where $\alpha \geq 0$ is called the *mean reversion* parameter. It is apparent from the dynamics that when $U_t > 0$, the drift term tends to pull U_t "downward" back toward 0. The reverse holds when $U_t < 0$. Think of U_t as constructed from B_t with a rubber band attached at one end to U_t and fixed at the other end at 0. The rubber band acts to pull U_t back toward 0 — the parameter α measures the strength of the rubber band. Here we start the process off at $U_0 = 0$, shortly we will allow for a random initial value.

Figure 40.4 shows four O-U paths generated from the same Brownian path. The first path, where $\alpha = 0$, actually is the Brownian path. (With $\alpha = 0$, we have $U_0 = 0 = B_0$ and $dU_t = dB_t$, so $U_t = B_t$ for all t.) The other three paths show the U_t path resulting from increasing strengths of mean reversion. If you look carefully at the $\alpha = 1.0$ case, you can still see the random "bumps" arising from the dB_t term in (40.1). This gets harder to see in for larger α as the drift term plays a larger role.

Perhaps the most straightforward way to *define* the O-U process is as follows:

$$U_t = B_t - \alpha \int_0^t e^{-\alpha(t-s)} B_s ds. \tag{40.3}$$

Note that B_s is a continuous function in s, so with a particular Brownian path in hand the integral in (40.3) is just an ordinary Riemann integral. Since both B_t and the integral in (40.3) vary continuously in t so does U_t.

Figure 40.4. U_t with various values of α.

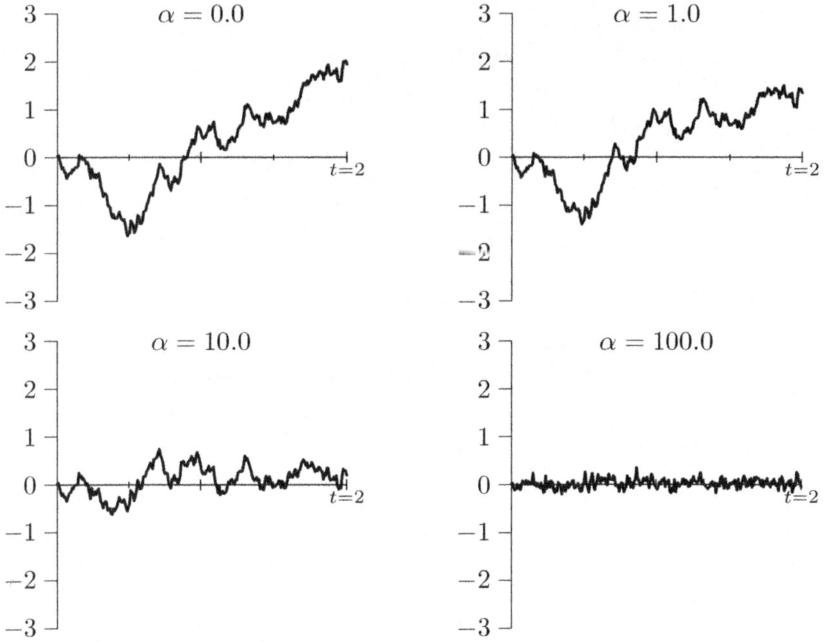

Note that $U_0 = B_0 = 0$, so (40.2) holds. To confirm that the process defined in (40.3) satisfies the dynamics in (40.1) on any time interval $[0,T]$, we compute

$$\begin{aligned}\Delta U_t &= U_{t+\Delta t} - U_t \\ &= \left(B_{t+\Delta t} - \alpha\int_0^{t+\Delta t} e^{-\alpha(t+\Delta t-s)}B_s ds\right) \\ &\quad - \left(B_t - \alpha\int_0^t e^{-\alpha(t-s)}B_s ds\right) \\ &= \Delta B_t - \alpha\Delta t\left[\begin{array}{l}\frac{1}{\Delta t}\int_t^{t+\Delta t} e^{-\alpha(t+\Delta t-s)}B_s ds \\ +\int_0^t \frac{e^{-\alpha(t+\Delta t-s)} - e^{-\alpha(t-s)}}{\Delta t}B_s ds\end{array}\right].\end{aligned}$$

Now

$$\begin{aligned}\frac{1}{\Delta t}\int_t^{t+\Delta t} e^{-\alpha(t+\Delta t-s)}B_s ds &= B_t + \boxed{\frac{1}{\Delta t}\int_t^{t+\Delta t}\left(e^{-\alpha(t+\Delta t-s)}B_s - B_t\right)ds} \\ &= B_t + C_1(t,\Delta t),\end{aligned} \tag{40.5}$$

and

$$\begin{aligned}&\int_0^t \frac{e^{-\alpha(t+\Delta t-s)} - e^{-\alpha(t-s)}}{\Delta t}B_s ds \\ &\qquad = -\alpha\int_0^t e^{-\alpha(t-s)}B_s ds + \\ &\qquad\qquad \boxed{\int_0^t\left(\frac{e^{-\alpha(t+\Delta t-s)} - e^{-\alpha(t-s)}}{\Delta t} + \alpha e^{-\alpha(t-s)}\right)B_s\, ds} \\ &\qquad = -\alpha\int_0^t e^{-\alpha(t-s)}B_s ds + C_2(t,\Delta t),\end{aligned} \tag{40.6}$$

where C_1 and C_2 are the boxed quantities. Together this produces

$$\begin{aligned}\Delta U_t &= \Delta B_t - \alpha\Delta t\left[\boxed{B_t} + C_1(t,\Delta t) + \boxed{-\alpha\int_0^t e^{-\alpha(t-s)}B_s ds} + C_2(t,\Delta t)\right] \\ &= -\alpha U_t\Delta t + \Delta B_t + \epsilon(t,\Delta t),\end{aligned}$$

where

$$\epsilon(t,\Delta t) = -\alpha\Delta t\cdot(C_1(t,\Delta t) + C_2(t,\Delta t)). \tag{40.7}$$

Here we have used that the sum of the two boxed terms immediately above is U_t.

First we bound C_1. We measure B's *oscillation* on $[0,T]$ on subintervals of length δ by

$$\mathrm{Osc}_B(\delta) \;=\; \sup\{|B_{t'} - B_t| : 0 \le t < t' \le T \text{ and } t' - t \le \delta\},$$

and put $B_{\max} = \sup\{|B_t| : 0 \le t \le T\}$. These are random quantities as they depend on the Brownian path, but since B_t is uniformly continuous on $[0,T]$ we see that $\mathrm{Osc}_B(\delta) \to 0$ as $\delta \to 0$ and also $B_{\max} < \infty$. Note also that $\delta_2 < \delta_1 \implies \mathrm{Osc}_B(\delta_2) \le \mathrm{Osc}_B(\delta_1)$, for the condition $t' - t \le \delta_2$ is more restrictive than $t' - t \le \delta_1$. Using that $|e^{-x} - 1| \le x$ for $x \ge 0$ (see Appendix II, Example 45.13), we get for $t \le s \le t + \Delta t$ that

$$\begin{aligned} |e^{-\alpha(t+\Delta t-s)}B_s - B_t| \;&\le\; |e^{-\alpha(t+\Delta t-s)} - 1| \cdot |B_s| + |B_s - B_t| \\ &\le\; \alpha\Delta t B_{\max} + \mathrm{Osc}_B(\Delta t), \end{aligned}$$

which, combined with (40.5) — where C_1 is defined — gives that

$$|C_1(t,\Delta t)| \;\le\; \alpha\Delta t B_{\max} + \mathrm{Osc}_B(\Delta t).$$

As for C_2, note that for $0 \le s \le t$

$$\begin{aligned} \left|\frac{e^{-\alpha(t+\Delta t-s)} - e^{-\alpha(t-s)}}{\Delta t} + \alpha e^{-\alpha(t-s)}\right| \;&=\; \left|e^{-\alpha(t-s)} \cdot \frac{e^{-\alpha\Delta t} - (1 - \alpha\Delta t)}{\Delta t}\right| \\ &\le\; \alpha^2\Delta t, \end{aligned}$$

as $e^{-\alpha(t-s)} \le 1$ and $|e^{-x} - (1-x)| \le \frac{1}{2}x^2 \le x^2$ for $x \ge 0$ (Example 45.13 again). Combining this with (40.6) — where C_2 is defined — gives that

$$|C_2(t,\Delta t)| \;\le\; \alpha^2\Delta t \int_0^t |B_s|\, ds \;\le\; \alpha^2\Delta t \cdot T \cdot B_{\max}.$$

Now fix u and v with $0 \le u < v \le T$ and let $\mathcal{T} = \{t_i\}$ partition $[u,v]$. Then

$$\sum_i |\epsilon(t_i,\Delta t_i)| \;\le\; \sum_i \left(\alpha^2\Delta t_i^2 B_{\max} + \alpha\Delta t_i \mathrm{Osc}_B(\Delta t_i) + \alpha^3 T\Delta t_i^2 B_{\max}\right).$$

Since each $\Delta t_i \le ||\mathcal{T}||$ and $\sum_i \Delta t_i = v - u \le T$, we have

$$\begin{aligned} \sum_i |\epsilon(t_i,\Delta t_i)| \;&\le\; \left(\alpha^2||\mathcal{T}||B_{\max} + \alpha\mathrm{Osc}_B(||\mathcal{T}||) + \alpha^3 T||\mathcal{T}||B_{\max}\right) \cdot \sum_i \Delta t_i \\ &\le\; \alpha^2 T||\mathcal{T}||B_{\max} + \alpha T\mathrm{Osc}_B(||\mathcal{T}||) + \alpha^3 T^2||\mathcal{T}||B_{\max} \\ &\to\; 0 \text{ as } ||\mathcal{T}|| \to 0, \end{aligned}$$

proving:

THEOREM 40.8. *The Ornstein-Uhlenbeck process, as defined in* (40.3), *has dynamics* $dU_t = -\alpha U_t dt + dB_t$ *on any interval* $[0,T]$ *with initial value* $U_0 = 0$.

REMARK. We have demonstrated that $\sum_i |\epsilon(t_i, \Delta t_i)| \to 0$ as $||T|| \to 0$, which certainly implies that $\sum_i \epsilon(t_i, \Delta t_i) \to 0$. This is almost sure convergence, more than the required convergence in probability.

Another Formulation of O-U. We formally defined the O-U process in (40.3). Here we establish an alternative formulation which is useful to establish some of the properties of the process. Our second formulation of the O-U process, which we will show agrees with (40.3), is

$$\widetilde{U}_t = \int_0^t e^{-\alpha(t-s)}\, dB_s, \tag{40.9}$$

where $0 \le t \le T$. A sort of symbolic "proof" goes as follows. Let $u = e^{-\alpha(t-s)}$ and $v = B_s$, so $du = \alpha e^{-\alpha(t-s)} ds$ and $dv = dB_s$. Then, in a stochastic rendition of integration by parts, we have

$$\begin{aligned}
\widetilde{U}_t &= \int_0^t u\, dv = uv\Big|_{s=0}^{s=t} - \int_0^t v\, du \\
&= e^{-\alpha(t-s)} B_s\Big|_{s=0}^{s=t} - \int_0^t B_s \alpha e^{-\alpha(t-s)}\, ds \\
&= B_t - \alpha \int_0^t B_s e^{-\alpha(t-s)}\, ds = U_t.
\end{aligned}$$

Unfortunately, these steps must be justified — an endeavor to which we presently turn our attention.

First we prove a result that is useful in many contexts, called *summation by parts*. Specifically, for any two sequences of numbers (a_i) and (b_i), $i \ge 0$, we have

$$\sum_{i=0}^{n-1} a_i(b_{i+1} - b_i) = a_{n-1}b_n - a_0 b_0 - \sum_{i=1}^{n-1} b_i(a_i - a_{i-1}). \tag{40.10}$$

Proof:

$$\begin{aligned}
\text{LHS} &= \sum_{i=0}^{n-1} a_i b_{i+1} - \sum_{i=0}^{n-1} a_i b_i \\
&= a_{n-1}b_n - a_0 b_0 + \sum_{i=0}^{n-2} a_i b_{i+1} - \sum_{i=1}^{n-1} a_i b_i \\
&= a_{n-1}b_n - a_0 b_0 + \sum_{i=1}^{n-1} a_{i-1} b_i - \sum_{i=1}^{n-1} a_i b_i = \text{RHS}.
\end{aligned}$$

Here LHS and RHS refer to the left and right hand sides of (40.10), respectively.

Let $\mathcal{T} = \{t_0, \ldots, t_n\}$ partition the interval $[0, t]$ and put

$$S(\mathcal{T}) = \sum_{i=0}^{n-1} e^{-\alpha(t-t_i)} \Delta B_{t_i}. \tag{40.11}$$

We must show that $S(\mathcal{T}) \to U_t$ as $||\mathcal{T}|| \to 0$. Applying (40.10) to (40.11) with $a_i = e^{-\alpha(t-t_i)}$ and $b_i = B_{t_i}$ we get

$$S(\mathcal{T}) = e^{-\alpha(t-t_{n-1})} B_{t_n} - e^{-\alpha(t-t_0)} B_{t_0} - \sum_{i=1}^{n-1} B_{t_i} \big(e^{-\alpha(t-t_i)} - e^{-\alpha(t-t_{i-1})}\big). \tag{40.12}$$

Now $t_n = t$, $t_{n-1} \to t$ as $||\mathcal{T}|| \to 0$, and $B_{t_0} = B_0 = 0$, so

$$e^{-\alpha(t-t_{n-1})} B_{t_n} - e^{-\alpha(t-t_0)} B_{t_0} \to B_t \quad \text{as } ||\mathcal{T}|| \to 0.$$

Also,

$$\begin{aligned} e^{-\alpha(t-t_i)} - e^{-\alpha(t-t_{i-1})} &= e^{-\alpha(t-t_i)}(1 - e^{-\alpha \Delta t_{i-1}}) \\ &= \alpha e^{-\alpha(t-t_i)} \Delta t_{i-1} + E_i, \end{aligned}$$

where $|E_i| \le \alpha^2 \Delta t_{i-1}^2 \le \alpha^2 ||\mathcal{T}|| \Delta t_{i-1}$. (This estimate follows from the fact that $e^{-\alpha(t-t_i)} \le 1$ and $|(1 - e^{-x}) - x| \le x^2$ for $x \ge 0$ — Example 45.13 again.) Then

$$\sum_{i=1}^{n-1} B_{t_i} \big(e^{-\alpha(t-t_i)} - e^{-\alpha(t-t_{i-1})}\big) = \alpha \sum_{i=1}^{n-1} e^{-\alpha(t-t_i)} B_{t_i} \Delta t_{i-1} + E,$$

where

$$\begin{aligned} |E| \le \sum_{i=1}^{n-1} |B_{t_i} E_i| &\le B_{\max} \alpha^2 ||\mathcal{T}|| \sum_{i=1}^{n-1} \Delta t_{i-1} \\ &\le B_{\max} \alpha^2 ||\mathcal{T}|| t \to 0 \text{ as } ||\mathcal{T}|| \to 0. \end{aligned}$$

By ordinary calculus (using that B_s is continuous in s),

$$\alpha \sum_{i=1}^{n-1} e^{-\alpha(t-t_i)} B_{t_i} \Delta t_{i-1} \to \alpha \int_0^t e^{-\alpha(t-s)} B_s \, ds \quad \text{as } ||\mathcal{T}|| \to 0.$$

Putting (40.12) together with these subsequent estimates gives that

$$S(\mathcal{T}) \to B_t - \alpha \int_0^t e^{-\alpha(t-s)} B_s \, ds = U_t \quad \text{as } ||\mathcal{T}|| \to 0.$$

Some Properties of the O-U Process. Here we derive some basic facts about the O-U process. They will be needed for our application as developed in the next section. First we have:

THEOREM 40.13. $U_t \sim \text{Normal}\left(0, \frac{1-e^{-2\alpha t}}{2\alpha}\right)$.

REMARK. The limiting variance of the O-U process with $\alpha > 0$ as $t \to \infty$ is $\frac{1}{2\alpha} < \infty$. This is in contrast with the Brownian process B_t whose variance at time t is t.

PROOF. Fix $t > 0$ and let $U = \int_0^t e^{-\alpha(t-s)} dB_s$. For $n \geq 1$, put $\Delta t = \frac{t}{n}$ and $t_i = i\Delta t$ so $\mathcal{T} = \{t_0, \ldots, t_n\}$ is a partition of $[0, t]$ with mesh Δt. Now set

$$U_n = \sum_{i=0}^{n-1} e^{-\alpha(t-t_i)} \Delta B_{t_i},$$

so, by the previous section, $U_n \to U$ as $n \to \infty$. This is almost sure convergence, which implies convergence in probability. Now the increments ΔB_{t_i}, $0 \leq i < n$, are independent $\text{Normal}(0, \Delta t)$ random variables so U_n is normal with mean 0 and variance

$$\sigma_n^2 = \operatorname{Var} U_n = \sum_{i=0}^{n-1} e^{-2\alpha(t-t_i)} \Delta t = e^{-2\alpha t} \Delta t \sum_{i=0}^{n-1} e^{2\alpha \Delta t \cdot i}.$$

Noting that the right side is a geometric series with ratio $e^{2\alpha\Delta t}$ and using that $\Delta t \cdot n = t$, we get

$$\sigma_n^2 = e^{-2\alpha t} \Delta t \frac{1 - e^{2\alpha t}}{1 - e^{2\alpha \Delta t}} \to \frac{1 - e^{-2\alpha t}}{2\alpha},$$

as $n \to \infty$. (This is an application of L'Hôpital's rule since $\Delta t \to 0$ as $n \to \infty$.) The result then follows from Lemma 35.10. □

Next, for $t > t'$, we investigate what the conditional distribution of U_t is given that $U_{t'} = u$. Now

$$\begin{aligned}
U_t &= \int_0^t e^{-\alpha(t-s)}\, dB_s \\
&= \int_0^{t'} e^{-\alpha(t-s)}\, dB_s + \int_{t'}^t e^{-\alpha(t-s)}\, dB_s \\
&= e^{-\alpha(t-t')} \int_0^{t'} e^{-\alpha(t'-s)}\, dB_s + \int_{t'}^t e^{-\alpha(t-s)}\, dB_s \\
&= e^{-\alpha(t-t')} U_{t'} + \int_{t'}^t e^{-\alpha(t-s)}\, dB_s.
\end{aligned}$$

Now the two summands above are independent as the first depends on the Brownian path up to time t' and the second depends on the Brownian increments on the time interval $[t', t]$. Therefore, conditioned on $U_{t'} = u$, we get that

$$\begin{aligned} U_t &\sim e^{-\alpha(t-t')}u + \int_{t'}^{t} e^{-\alpha(t-s)}\, dB_s \\ &\sim e^{-\alpha(t-t')}u + \text{Normal}\left(0, \frac{1-e^{-2\alpha(t-t')}}{2\alpha}\right). \end{aligned}$$

The second "$\sim$" above follows exactly as in the proof of Theorem 40.13. In particular, we see that

$$E[U_t \mid U_{t'} = u] = e^{-\alpha(t-t')}u, \quad \text{and} \tag{40.14}$$

$$\text{Var}\,[U_t \mid U_{t'} = u] = \frac{1-e^{-2\alpha(t-t')}}{2\alpha}. \tag{40.15}$$

The Invariant Distribution for the O-U Process. Up to now we have started the O-U process at $U_0 = 0$. This is not necessary — the process can begin with any value. Here we start the process at a *random* value $V_0(\omega)$. We require that V_0 is independent of the underlying Brownian process B. For $t > 0$ we put

$$V_t = V_0 e^{-\alpha t} + U_t, \tag{40.16}$$

where U_t is the usual O-U process starting at 0. Then we have

THEOREM 40.17. *The process V_t defined in* (40.16) *has the O-U dynamics $dV_t = -\alpha V_t dt + dB_t$ on any interval $[0, T]$.*

PROOF. We compute

$$\Delta V_t = V_{t+\Delta t} - V_t = V_0(e^{-\alpha(t+\Delta t)} - e^{-\alpha t}) + U_{t+\Delta t} - U_t.$$

Noting that

$$\begin{aligned} |(e^{-\alpha(t+\Delta t)} - e^{-\alpha t}) - (-\alpha e^{-\alpha t}\Delta t)| &= e^{-\alpha t} \cdot |e^{-\alpha \Delta t} - (1 - \alpha\Delta t)| \\ &\leq 1 \cdot \alpha^2 \Delta t^2, \end{aligned}$$

we get that $V_0(e^{-\alpha(t+\Delta t)} - e^{-\alpha t}) = -\alpha V_0 e^{-\alpha t}\Delta t + \epsilon_1(t, \Delta t)$, where $|\epsilon_1(t, \Delta t)| \leq |V_0|\alpha^2\Delta t^2$. Now define $\epsilon_2(t, \Delta t)$ by

$$U_{t+\Delta t} - U_t = \Delta U_t = -\alpha U_t \Delta t + \Delta B_t + \epsilon_2(t, \Delta t),$$

so

$$\begin{aligned} \Delta V_t &= -\alpha V_0 e^{-\alpha t}\Delta t + \epsilon_1(t, \Delta t) - \alpha U_t \Delta t + \Delta B_t + \epsilon_2(t, \Delta t) \\ &= -\alpha(V_0 e^{-\alpha t} + U_t)\Delta t + \Delta B_t + \epsilon_1(t, \Delta t) + \epsilon_2(t, \Delta t) \\ &= -\alpha V_t \Delta t + \Delta B_t + \epsilon_1(t, \Delta t) + \epsilon_2(t, \Delta t). \end{aligned}$$

Let $0 \le u < v \le T$ and suppose $\{t_0, \ldots, t_n\}$ partitions $[u, v]$. To establish the Theorem we must show that $\sum_i [\epsilon_1(t_i, \Delta t_i) + \epsilon_2(t_i, \Delta t_i)] \xrightarrow{\text{P}} 0$ as $||\mathcal{T}|| \to 0$. We know from Theorem 40.8 that $\sum_i |\epsilon_2(t_i, \Delta t_i)| \to 0$ (see the remark following the theorem). Also,

$$\begin{aligned} \sum_i |\epsilon_1(t_i, \Delta t_i)| &\le \sum_i |V_0| \alpha^2 \Delta t_i^2 \\ &\le |V_0| \alpha^2 ||\mathcal{T}|| \sum_i \Delta t_i \le |V_0| \alpha^2 ||\mathcal{T}|| T \to 0, \end{aligned}$$

as $\mathcal{T} \to 0$. □

Since V_0 and the Brownian path B are independent, so are V_0 and the O-U path U. It follows that the conditional distribution of U_t given $V_0 = v_0$ is the same as its unconditional distribution. Hence, given $V_0 = v_0$ we have via (40.16) that

$$V_t \sim v_0 e^{-\alpha t} + U_t \sim \text{Normal}\left(v_0 e^{-\alpha t}, \frac{1 - e^{-2\alpha t}}{2\alpha}\right).$$

We see that, regardless of the value of V_0, the distribution of V_t converges to a "long-run" distribution that is Normal $\left(0, \frac{1}{2\alpha}\right)$. Here we show that this long-run distribution is, in fact, an invariant distribution precisely in the Markovian sense of Chapter III. Specifically,

THEOREM 40.18. *If $V_0 \sim$ Normal $\left(0, \frac{1}{2\alpha}\right)$ and V_0 and B are independent then $V_t \sim$ Normal $\left(0, \frac{1}{2\alpha}\right)$ for all $t \ge 0$.*

PROOF. Since V_0 and U_t are independent mean-zero normals, V_t is also a mean-zero normal. Furthermore,

$$\begin{aligned} \operatorname{Var} V_t &= e^{-2\alpha t} \operatorname{Var} V_0 + \operatorname{Var} U_t \\ &= \frac{e^{-2\alpha t}}{2\alpha} + \frac{1 - e^{-2\alpha t}}{2\alpha} = \frac{1}{2\alpha}. \end{aligned}$$ □

EXERCISES

1. Use L'Hôpital's rule to evaluate $\lim_{\alpha \to 0} \operatorname{Var} U_t$ as given in Theorem 40.13. What do you observe?

• In the following exercises we develop a random walk $(W_n : n \ge 0)$ whose distributional properties when scaled properly mimic the continuous-time O-U process. The state space for the walk is $S = \{i \in \mathbb{Z} : |i| \le i^*\}$, where the states $\pm i^*$ are retaining. For fixed time increment $\Delta t > 0$, put $t_n = n\Delta t$ and for an appropriate scaling factor h we let

$$\widehat{U}_{t_n} = h \cdot W_n,$$

all for $n = 0, 1, 2, \ldots$. With the $\widehat{U}$ process defined at the discrete times $\{t_n\}$, we define $\widehat{U}_t$ in general by linear interpolation. Here Δt, i^*, h, and mean reversion strength $\alpha > 0$ are inter-related parameters of the process. Our goal is to construct the process in such a way that (40.14) and (40.15) are satisfied for $t' = t_n$ and $t = t_{n+1}$, i.e., so that

$$E[\widehat{U}_{t_{n+1}} \,|\, \widehat{U}_{t_n} = u] = e^{-\alpha\Delta t}u, \quad \mathrm{Var}\,[\widehat{U}_{t_{n+1}} \,|\, \widehat{U}_{t_n} = u] = \frac{1 - e^{-2\alpha\Delta t}}{2\alpha}. \qquad (*)$$

In 1994, John Hull and Alan White introduced a discrete model of short-term interest rate behavior using this process.

1. Our walk starts at $W_0 = 0$. In general, when $W_n = 0$, then $W_{n+1} = -1$, 0, or $+1$, each with probability 1/3. Now $W_n = 0 \implies u = 0$ in $(*)$. Verify here that $(*)$ holds if and only if we take

$$h = \sqrt{\frac{3(1 - e^{-2\alpha\Delta t})}{4\alpha}}.$$

2. If $W_n = i$, where $0 < i < i^*$, then $W_{n+1} = i - 1$, i, or $i + 1$, with probabilities $q(i)$, $r(i)$, and $p(i)$ respectively. Now $W_n = i \implies u = ih$ in $(*)$. Show that the first moment condition says that $ih + (p(i) - q(i))h = ihe^{-\alpha\Delta t}$ and the variance condition says that $(p(i) + q(i))h^2 - (p(i) - q(i))^2h^2 = \frac{2}{3}h^2$. Variance hint: If a random variable X takes on values $(i - 1)h$, ih, and $(i + 1)h$ with respective probabilities q, r, and p, then $\mathrm{Var}\,X$ does not depend on i (why?); the variance is easiest to compute when $i = 0$.

3. Use problem 2 to solve for $q(i)$, $r(i)$, and $p(i)$. With $x = (e^{-\alpha\Delta t} - 1)i$, show that

$$\begin{aligned} p(i) &= \frac{x^2}{2} + \frac{x}{2} + \frac{1}{3}, \\ q(i) &= \frac{x^2}{2} - \frac{x}{2} + \frac{1}{3}, \quad \text{and} \\ r(i) &= -x^2 + \frac{1}{3}. \end{aligned} \qquad (**)$$

4. Here we verify that these are legitimate probabilities. Since they sum to one, we must only verify that they are all non-negative. Verify by calculus that the quadratics for $p(i)$ and $q(i)$ have minimum value 5/24. Verify that $r(i) > 0$ provided

$$i \le \frac{1}{2(1 - e^{-\alpha\Delta t})}.$$

To insure that our transition probabilities are legitimate for states $0 < i < i^*$, we therefore take

$$i^* = \left\lfloor \frac{1}{2(1 - e^{-\alpha\Delta t})} \right\rfloor + 1,$$

where $\lfloor y \rfloor$ is the largest integer less than or equal to y.

5. Here we compute the transition probabilities for state i^*. If $W_n = i^*$, then $W_{n+1} = i^* - 2$, $i^* - 1$, or i^*, with probabilities $q(i^*)$, $r(i^*)$, and $p(i^*)$ respectively. Now $W_n = i^* \implies u = i^* h$ in $(*)$. Show that the first moment condition says that $(i^* - 1)h + (p(i^*) - q(i^*))h = i^* h e^{-\alpha \Delta t}$ and the variance condition says that $(p(i^*) + q(i^*))h^2 - (p(i^*) - q(i^*))^2 h^2 = \frac{2}{3}h^2$.

6. Show that with $x = (e^{-\alpha \Delta t} - 1)i^* + 1$ the transition probabilities for state i^* are given by $(**)$, with $i = i^*$. Verify that these are legitimate probabilities.

7. The distribution of the continuous-time O-U process is symmetric about 0. Use this symmetry to derive formulas for the transition probabilities for states i with $-i^* \leq i < 0$.

8. As a specific example, with $\Delta t = 1$ and $\alpha = 0.05$, verify that $h = 1.195$ and $i^* = 11$.

9. Compute the transition probabilities for states 1, 5, 10, and 11.

• The figure below shows the values attainable by the $\widehat{U}$ process (the dots) out to time 20 with the parameters of problem 8. Several transition arrangements and a randomly sampled path are shown. The long-run standard deviation of the process, $\sqrt{1/2\alpha}$, is approximately 3.16 and the process usually is within ± 5.

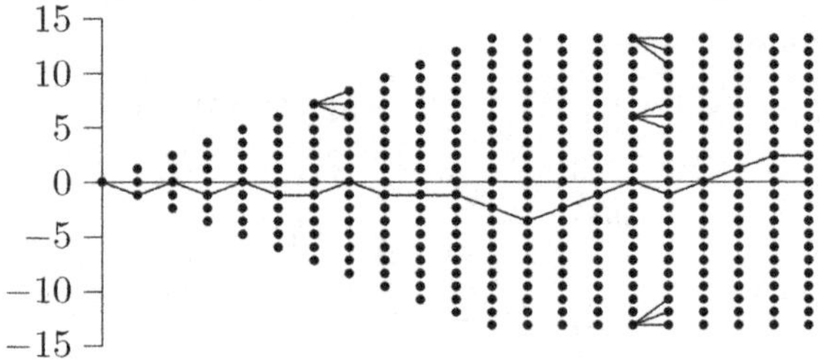

§41. Application: Modeling the Index of Consumer Sentiment

Our US economy has various built-in features that create what the economists call *countercyclical* forces. When the economy is weak these forces tend to stimulate it; they subside when the economy is strong. Some examples include: welfare and unemployment benefits that kick in when incomes fall and jobs are lost; lower tax collections when incomes fall; lower interest rates to stimulate investment (the Fed controls short-term interest rates); Federal legislation funding stimulative job-creating projects (when Congress is functioning properly). Due to these self-correcting restorative forces, it seems reasonable to view economic activity broadly speaking as a mean reverting process. In this final section we explore this idea.

The Index of Consumer Sentiment. Every month, the Survey Research Center at the University of Michigan interviews a minimum of 500 randomly selected American households. The Index of Consumer Sentiment (ICS) is based on the responses to five of their questions. (One of these five questions, for example, is: Are you better or worse off financially than you were a year ago?) Here we view the ICS as a measure of the overall health of the US economy. (Admittedly, this is a bit like a doctor asking a patient how he feels instead of directly measuring the patient's vitals.) Figure 41.1 shows the ICS for a 38 year period starting January 1978, which is the earliest data published on their web site (www.sca.isr.umich.edu). The index is published monthly and the figure plots $38 \times 12 = 456$ individual items of data. Note that the index can exceed 100, as it did in the late 1990s. The horizontal line is at 85, which is roughly the 38 year average value of the index.

We will model the ICS as $C_t = \mu + \sigma U_t$, where μ is the long-run mean of the process, σ is a scaling factor, and U_t is an O-U process with mean reversion α. The process therefore has three parameters, μ, σ, and α, which we seek to estimate using the actual data.

Figure 41.1. The ICS from January 1978 to December 2015.

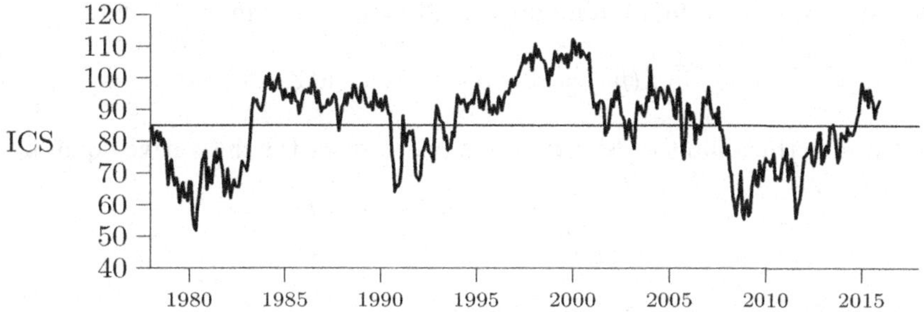

Maximum Likelihood Estimation. There are numerous statistical techniques for using data to estimate the value of parameters that govern the probability distribution of that data. *Maximum likelihood estimation* is one such technique. In a nutshell, the general idea is to determine the value of the parameters that make the data most likely to have been observed. Before applying this to the problem at hand, we illustrate the methodology in two simpler settings.

Batting Averages. Suppose each Major League baseball player has his own hidden (i.e., not directly observable) batting probability p. Each time he comes up to bat he gets a hit with probability p, independent of outcomes for previous at-bats. The value of p is fixed and does not change with time. (This model clearly oversimplifies things.) Let's say that after n at-bats he has produced k hits. How should we estimate p?

The maximum likelihood estimator (MLE) works as follows. Define the random variable X_i to be 1 if he gets a hit the ith time at bat and $X_i = 0$ otherwise.

He has been at bat n times yielding k hits. Suppose $(x_1, x_2, \ldots, x_n)$ is the realization of $(X_1, X_2, \ldots, X_n)$ that we observe, so the x_i's contain k ones and $n-k$ zeros. For any particular at-bat, we have

$$p_X(x) = \begin{cases} p & \text{if } x = 1 \\ 1-p & \text{if } x = 0. \end{cases}$$

With independent at-bats, the probability of observing the data $(x_1, x_2, \ldots, x_n)$ is given by

$$p_X(x_1) \cdot p_X(x_2) \cdot \ldots \cdot p_X(x_n) = p^k(1-p)^{n-k}.$$

Usually we think of p as fixed and view the right side as a function of $(x_1, \ldots, x_n)$. (The value of k depends on the data.) Maximum likelihood estimation turns this around. We take the data $(x_1, \ldots, x_n)$ as fixed and think of the right side as a function of p

$$L(p) = p^k(1-p)^{n-k},$$

known as the *likelihood function.* The maximum likelihood estimate of p is the value of p that maximizes $L(p)$ — it maximizes the probability of observing the data. It is easier to work with the logarithm of L, which is maximized for the same value of p since $\ln(x)$ is monotonically increasing in x. We have

$$\ln L(p) = k \ln p + (n-k)\ln(1-p).$$

To maximize this quantity let's differentiate and set the derivative equal to zero:

$$\frac{d}{dp} \ln L(p) = \frac{k}{p} - \frac{n-k}{1-p},$$

which is 0 when $p = \frac{k}{n}$. This is perfectly reasonable — we estimate his probability of getting a hit by the fraction of observed at-bats that have produced hits.

REMARK. This is not always as reasonable as it seems. Suppose our baseball player is a rookie with only 10 at-bats that have produced 6 hits. Would you believe that his p is 0.6? By the Law of Large Numbers that would produce a career batting average near .600, which is unlikely, to say the least, as major leaguers have a mean batting average of roughly .260 with a standard deviation of roughly .030. *Bayesian estimation* is a technique that considers both the *prior* distribution that the rookie is drawn from (e.g., Normal$(.260, .030^2)$) and the observed data to produce an updated *posterior* distribution. That methodology would be more appropriate with only 10 at-bats.

Quality Control. In a second example a little closer to our ICS application, suppose we manufacture and package boxes of crackers. We seek to test our production facility to determine the probability distribution of the weight of the boxes produced. Let us assume that the underlying distribution is normal, but of unknown mean and variance. If X is the weight of a randomly selected box of crackers, we have $X \sim \text{Normal}(\mu, \sigma^2)$ — but we do not know what the values of μ and σ^2 are. To determine these values, we randomly sample n boxes of crackers and we measure the weights $x_1, x_2, \ldots, x_n$. We wish to use this data to estimate μ and σ^2.

In the batting average application, we were able to compute the probability of observing the data (as a function of p). The result was simply the joint mass function evaluated at the data. Here the data is jointly continuously distributed so the probability of observing any particular sequence of data is 0. Rather than computing the joint mass function, we compute the joint density function and evaluate that at the data. This is the quantity we will maximize in μ and σ^2.

The probability density function for each individual box weight is

$$f(x) = \frac{1}{\sqrt{2\pi\sigma^2}} e^{-(x-\mu)^2/2\sigma^2},$$

so joint density function evaluated at the data $(x_1, x_2, \ldots, x_n)$ is

$$L(\mu, \sigma^2) = f(x_1) \cdot f(x_2) \cdot \ldots \cdot f(x_n) = \prod_{i=1}^{n} \frac{1}{\sqrt{2\pi\sigma^2}} e^{-(x_i-\mu)^2/2\sigma^2};$$

this is our likelihood function in this setting. We have used that the individual observations are independent (that's what a random sample is). As before, our objective is to determine the values of μ and σ^2 that maximize the likelihood function.

In this setting we can accomplish this analytically. Again it is easier to work with the logarithm of L:

$$\ln L(\mu, \sigma^2) = -\frac{n}{2}\ln(2\pi) - \frac{n}{2}\ln\sigma^2 - \sum_{i=1}^{n} \frac{(x_i-\mu)^2}{2\sigma^2}$$

Now

$$\frac{\partial}{\partial\mu}\ln L = \sum_{i=1}^{n} \frac{x_i-\mu}{\sigma^2},$$

which is zero when $\mu = \frac{1}{n}\sum_i x_i$ — the maximum likelihood estimate (MLE) of μ is the sample mean. As for the variance,

$$\frac{\partial}{\partial\sigma^2}\ln L = -\frac{n}{2}\frac{1}{\sigma^2} + \sum_{i=1}^{n} \frac{(x_i-\mu)^2}{2\sigma^4} = \frac{-n\sigma^2 + \sum_i (x_i-\mu)^2}{2\sigma^4}.$$

(If it bothers you to differentiate with respect to σ^2, just re-label it v.) By setting this equal to zero, we see that $\ln L$ is maximized when $\sum_i (x_i-\mu)^2 = n\sigma^2$, i.e., when $\sigma^2 = \frac{1}{n}\sum_i (x_i-\mu)^2$. Since μ is the sample mean, the MLE variance estimator is the sample variance.

Back to the ICS. Now that we understand maximum likelihood estimation and see that it produces reasonable results, let us apply it to our setting. Here the solution will not yield to analytical techniques, but numerical methods come to the rescue. Recall our model is $C_t = \mu + \sigma U_t$, where U_t is an O-U process with mean reversion α. There are three parameters that we must estimate: μ, σ, and α. The data we observe is $(c_t : 0 \leq t \leq 455)$. Time t is in months and the ICS data is observed at integer times t.

Since U_t has a long-run (invariant) distribution given by Normal $(0, \frac{1}{2\alpha})$, the C_t process has long-run distribution given by Normal $(\mu, \frac{\sigma^2}{2\alpha})$. Let us assume that c_0, the first value of the C_t process that we observe, is drawn from this distribution:

$$f_{C_0}(x) = \sqrt{\frac{\alpha}{\pi\sigma^2}} e^{-\alpha(x-\mu)^2/\sigma^2}.$$

(No, there should not be a 2 in the denominator of the exponent!) Now suppose that $(C_0, \ldots, C_t)$ has been observed and the ICS data is $(c_0, \ldots, c_t)$. Presently we compute the conditional distribution of C_{t+1}, given this data. First we observe that

$$C_i = c \iff \mu + \sigma U_i = c \iff U_i = \frac{c - \mu}{\sigma}.$$

With

$$\rho = e^{-\alpha} \quad \text{and} \quad v = \frac{1 - e^{-2\alpha}}{2\alpha},$$

we have

$$\begin{aligned} C_t = c_t \implies U_t &= \frac{c_t - \mu}{\sigma} \\ \implies U_{t+1} &\sim \text{Normal}\left(\rho \frac{c_t - \mu}{\sigma}, v\right) \text{ (see (40.14) and (40.15),} \\ &\qquad \text{here } t - t' = 1) \\ \implies \sigma U_{t+1} &\sim \text{Normal}\left(\rho(c_t - \mu), \sigma^2 v\right) \\ \implies C_{t+1} &= \mu + \sigma U_{t+1} \sim \text{Normal}\left(\mu + \rho(c_t - \mu), \sigma^2 v\right). \end{aligned}$$

Letting $H_t = \{(C_0, \ldots, C_t) = (c_0, \ldots, c_t)\}$ denote the history of the ICS up to month t, we see that the conditional density function of C_{t+1} given H_t is

$$f_{(C_{t+1}|H_t)}(x) = \frac{1}{\sqrt{2\pi\sigma^2 v}} e^{-(x-(\mu+\rho(c_t-\mu)))^2/2\sigma^2 v}.$$

Multiplying the marginal distribution of C_0 and the subsequent conditional distributions of the C_t, $t \geq 1$ we see that the joint density function for $(C_0, \ldots, C_{455})$ evaluated at the data $(c_0, \ldots, c_{455})$ is given by:

$$f(c_0, \ldots, c_{455}) = \sqrt{\frac{\alpha}{\pi\sigma^2}} e^{-\alpha(c_0-\mu)^2/\sigma^2} \cdot \prod_{t=0}^{454} \frac{1}{\sqrt{2\pi\sigma^2 v}} e^{-(c_{t+1}-(\mu+\rho(c_t-\mu)))^2/2\sigma^2 v}. \tag{41.2}$$

As in our previous example, to compute the MLEs of μ, σ, and α, we think of $(c_0, \ldots, c_{455})$ as fixed (they are the actual data) and view the right side of (41.2) as a function of μ, σ, and α — call it $L(\mu, \sigma, \alpha)$. Note that the expression in (41.2) also involves ρ and v, but these quantities are functions of α.

A Numerical Approach. The expression in (41.2) is formidable, and finding the values of μ, σ, and α that maximize it is not amenable to simple calculus. Instead we employ a numerical (computer-based) search. Our algorithm works as follows:

(i) Pick initial trial values for μ, σ, and α.

(ii) Compute $L(\tilde{\mu}, \tilde{\sigma}, \tilde{\alpha})$ for the 27 combinations where $\tilde{\mu} \in \{\mu - \delta, \mu, \mu + \delta\}$, $\tilde{\sigma} \in \{\sigma - \delta, \sigma, \sigma + \delta\}$, and $\tilde{\alpha} \in \{\alpha - \delta, \alpha, \alpha + \delta\}$. Here δ is a specified step size (taken to be 0.001 in our application). Let L^* denote the largest value of L observed among the 27 values computed, and let μ^*, σ^*, and α^* denote the values of $\tilde{\mu}$, $\tilde{\sigma}$, and $\tilde{\alpha}$ that realize L^*.

(iii) If $(\mu^*, \sigma^*, \alpha^*) = (\mu, \sigma, \alpha)$, stop — these are the MLE results. Otherwise, put $\mu = \mu^*$, $\sigma = \sigma^*$, and $\alpha = \alpha^*$ and go back to step (ii).

REMARK. As in the iid normal case where we solved for μ and σ analytically, we actually work with $\ln L$ rather than L directly. Here we do so for a different reason. Multiplying together 456 numbers on a computer is problematic. If they are small, the result will be *very* small; if they are large, the result will be *very* large. An unfortunate consequence could be what computer scientists call *underflow* or *overflow*. By taking logarithms and adding rather than multiplying, the order of magnitude of the result is manageable.

This procedure is a sort of numerical steepest ascent algorithm, and is guaranteed to yield a local maximum, at least on our 3 dimensional numerical grid. Running this algorithm produces $\mu = 85.542$, $\sigma = 4.004$, and $\alpha = 0.050$. That it produces the same result over an assortment of initial trial values in step (i) gives some comfort that we are actually obtaining a global maximum. As an additional "sanity check", we have computed the sample mean and standard deviation for the ICS data $(c_0, \ldots, c_{455})$. They are 85.319 and 12.830, respectively. The C_t process given by $C_t = 85.542 + 4.004U_t$ (where U_t has mean reversion $\alpha = 0.050$) has a long-run mean of 85.542 and a long-run standard deviation of $\sqrt{\frac{\sigma^2}{2\alpha}} = 12.662$. These numbers correspond quite closely to the data. Our model indicates that, on average, consumer sentiment reverts back toward its long-run average at a rate of about 5% per month. For your viewing amusement, Figure 41.3 displays a C_t path computer-simulated over the same time period (38 years) using our MLE model. You should visually compare this to the actual path in Figure 41.1 — qualitatively they should look similar.

Figure 41.3. A simulated ICS path using $C_t = 85.542 + 4.004\,U_t$.

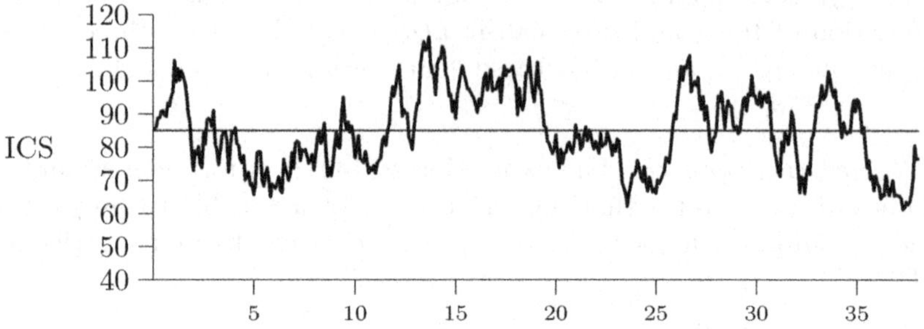

EXERCISES

1. Use our model of the ICS to estimate the probability that the ICS will be at or above 85 after two years if it is currently at 50.

2. How about after five years?

3. Formulate other questions of interest about the ICS that could be answered via computer simulation with the model developed above.

APPENDIX I

SOME COMMON DISTRIBUTIONS

Here we present some features of some of the more common probability distributions. In particular, for each distribution we specify the range and probability mass (or density) function and compute the mean, variance, and moment generating function. Our list omits several important distributions, but includes the distributions appearing in this text.

§42. Some Common Discrete Distributions

The Bernoulli Distribution. $X \sim \text{Bernoulli}\,(p)$, where $0 < p < 1$, if X takes on the values $\{0, 1\}$ with $P[X = 1] = p$ and $P[X = 0] = 1 - p$. The quantity $1 - p$ is usually called 'q', as we shall here.

- $EX = q \cdot 0 + p \cdot 1 = p$.
- $EX^2 = q \cdot 0^2 + p \cdot 1^2 = p$ again, so $\operatorname{Var} X = EX^2 - (EX)^2 = p - p^2 = p(1-p) = pq$.
- $M_X(\theta) = Ee^{\theta X} = qe^{\theta \cdot 0} + pe^{\theta \cdot 1} = q + pe^{\theta}$.

The Binomial Distribution. $X \sim \text{Binomial}\,(n, p)$ where $n \geq 1$ and $0 < p < 1$ if X takes on the values $\{0, 1, \ldots, n\}$ with, for such a number k, $P[X = k] = \binom{n}{k} p^k q^{n-k}$. These numbers sum to 1 because, by the binomial theorem, $\sum_{k=0}^{n} \binom{n}{k} p^k q^{n-k} = (p+q)^n = 1^n = 1$.

- The expected value of X can be calculated as follows.

$$EX = \sum_{k=0}^{n} k \cdot P[X = k] = \sum_k k \binom{n}{k} p^k q^{n-k} = q^n \sum_k \binom{n}{k} k t^k,$$

where the right hand side is evaluated at $t = \frac{p}{q}$. Now, viewed as a function of t,

$$\begin{aligned} q^n \sum_k \binom{n}{k} k t^k &= q^n t \sum_k \binom{n}{k} k t^{k-1} \\ &= q^n t \sum_k \binom{n}{k} \frac{d}{dt} t^k \quad \text{nice trick!} \\ &= q^n t \frac{d}{dt} \sum_k \binom{n}{k} t^k \end{aligned}$$

Continuing,

$$\begin{aligned} q^n \sum_k \binom{n}{k} kt^k &= q^n t \frac{d}{dt} \sum_k \binom{n}{k} t^k 1^{n-k} \\ &= q^n t \frac{d}{dt}(t+1)^n = q^n tn(t+1)^{n-1}. \end{aligned}$$

Evaluating this at $t = \frac{p}{q}$ and using that $p + q = 1$ gives

$$EX = q^n \frac{p}{q} n \left(\frac{p}{q} + 1\right)^{n-1} = npq^{n-1} \left(\frac{1}{q}\right)^{n-1} = np.$$

That was hard. A much easier way to see this is to observe that if $Y_1, \ldots, Y_n$ are independent and distributed like Bernoulli (p), then $X = Y_1 + \cdots + Y_n$ is distributed like Binomial (n, p). ($X = k$ if and only if exactly k of the n Y's are 1 and $n - k$ are 0. The probability that any particular selection of k of the n Y's are 1 and $n - k$ are 0 is $p^k q^{n-k}$ and there are $\binom{n}{k}$ ways to select k of the n Y's.) Therefore

$$EX = EY_1 + \cdots + EY_n = p + \cdots + p = np,$$

much easier!

One often interprets this latter representation of a binomial random variable as the number "successes" in n independent trials of an experiment if each individual experiment is successful with probability p.

• One may compute EX^2 and then $\operatorname{Var} X$ using methodology similar to the first approach above for computing EX. We leave that to the interested reader in favor of the easier second approach. Since the Y's are independent they are uncorrelated and

$$\operatorname{Var} X = \operatorname{Var} Y_1 + \cdots + \operatorname{Var} Y_n = pq + \cdots + pq = npq.$$

• $M_X(\theta) = \sum_{k=0}^n e^{k\theta} \binom{n}{k} p^k q^{n-k} = \sum_{k=0}^n \binom{n}{k} (pe^\theta)^k q^{n-k} = (pe^\theta + q)^n$, again by the binomial theorem. This result should not be surprising as we may view X as the sum of n independent Bernoulli (p)'s each with MGF $pe^\theta + q$.

The Poisson Distribution. $X \sim$ Poisson (λ) where $\lambda > 0$ if X takes on values in $\{0, 1, 2, \ldots\}$ with $P[X = k] = e^{-\lambda} \frac{\lambda^k}{k!}$. That these numbers sum to 1 follows from the fact that $\sum_{k=0}^\infty \frac{\lambda^k}{k!} = e^\lambda$.

• The mean may be computed as

$$\begin{aligned} EX &= \sum_{k=0}^{\infty} k e^{-\lambda}\frac{\lambda^k}{k!} \\ &= \sum_{k=1}^{\infty} k e^{-\lambda}\frac{\lambda^k}{k!} \quad \text{(because the first summand above is 0)} \\ &= \lambda \sum_{k=1}^{\infty} e^{-\lambda}\frac{\lambda^{k-1}}{(k-1)!} \\ &= \lambda \sum_{k=0}^{\infty} e^{-\lambda}\frac{\lambda^k}{k!} \;=\; \lambda. \end{aligned}$$

• The second moment is similar.

$$\begin{aligned} EX^2 &= \sum_{k=0}^{\infty} k^2 e^{-\lambda}\frac{\lambda^k}{k!} \\ &= \sum_{k=1}^{\infty} k e^{-\lambda}\frac{\lambda^k}{(k-1)!} \\ &= \lambda \sum_{k=1}^{\infty} k e^{-\lambda}\frac{\lambda^{k-1}}{(k-1)!} \\ &= \lambda \sum_{k=0}^{\infty} (k+1) e^{-\lambda}\frac{\lambda^k}{k!} \\ &= \lambda E[X+1] \;=\; \lambda(\lambda+1) = \lambda^2 + \lambda. \end{aligned}$$

Hence $\operatorname{Var} X = EX^2 - (EX)^2 = \lambda^2 + \lambda - \lambda^2 = \lambda$. For the Poisson random variable, the parameter λ is both the mean and the variance of the distribution.

• As for the moment generating function, we have

$$\begin{aligned} M_X(\theta) &= \sum_{k=0}^{\infty} e^{\theta k} e^{-\lambda}\frac{\lambda^k}{k!} \\ &= e^{-\lambda} \sum_{k} \frac{(\lambda e^\theta)^k}{k!} \\ &= e^{-\lambda} e^{\lambda e^\theta} \;=\; e^{\lambda e^\theta - \lambda}. \end{aligned}$$

The Geometric Distribution. $X \sim \text{Geometric}\,(p)$ where $0 < p < 1$ if X takes on the values $\{1, 2, \dots\}$ and $P[X = k] = q^{k-1}p$. These numbers sum to 1 since $q < 1$ and $\sum_{k=1}^{\infty} q^{k-1} = \frac{1}{1-q} = \frac{1}{p}$. One often interprets X as the number of independent trials of an experiment it takes to get the first success if each experiment is successful with probability p. If $X = k$, one must first obtain $k-1$ consecutive failures (the factor q^{k-1}) followed by a success (the factor p).

• As for the mean:

$$EX = \sum_{k=1}^{\infty} kq^{k-1}p = p\frac{d}{dq}\sum_{k=1}^{\infty} q^k = p\frac{d}{dq}\frac{q}{1-q} = p\frac{1}{(1-q)^2} = \frac{p}{p^2} = \frac{1}{p}.$$

Another way to see this is through the experiment interpretation. Let A be the event that the first experiment is successful. Then by conditioning we obtain:

$$EX = E[X|A]PA + E[X|A^c]PA^c = 1 \cdot p + (EX+1) \cdot (1-p).$$

Solving for EX gives $EX = \frac{1}{p}$. Note that $E[X|A^c] = EX + 1$ because, conditioned on A^c, there is one failure and after that failure the expected number of experiments until success is exactly EX — by independence you start from scratch after the first failed experiment.

• Regarding the variance, we first compute the second moment (as usual). Using our experiment interpretation, X conditioned on A^c is distributed like $1+X$ — as noted above. So $E[X^2|A^c] = E[(1+X)^2] = 1 + 2EX + EX^2$. Using also that $E[X^2|A] = 1^2 = 1$ and $EX = \frac{1}{p}$, we get that

$$EX^2 = E[X^2|A]PA + E[X^2|A^c]PA^c = 1 \cdot p + \left(1 + \frac{2}{p} + EX^2\right) q.$$

Solving this for EX^2 gives $EX^2 = \frac{2-p}{p^2}$. Then

$$\operatorname{Var} X = \frac{2-p}{p^2} - \frac{1}{p^2} = \frac{1-p}{p^2}.$$

• The moment generating function is given by

$$M_X(\theta) = \sum_{k=1}^{\infty} e^{\theta k} q^{k-1} p = \frac{p}{q}\sum_{k=1}^{\infty} (e^{\theta} q)^k.$$

If $e^{\theta} q < 1$ or, equivalently, if $\theta < -\ln q$, then this latter sum is a convergent geometric series giving $M_X(\theta) = \frac{p}{q}\frac{e^{\theta}q}{1-e^{\theta}q} = \frac{pe^{\theta}}{1-qe^{\theta}}$. If $\theta \geq -\ln q$ then the sum diverges and $M_X(\theta)$ is undefined. Note that $-\ln q > 0$ as $q < 1$.

The Negative Binomial Distribution. $X \sim \text{NegBin}\,(n, p)$ where $n \geq 1$ and $0 < p < 1$ if X takes on values in $\{n, n+1, n+2, \ldots\}$ with $P[X = k] = \binom{k-1}{n-1} p^n q^{k-n}$. This random variable may be interpreted as the number of independent trials of an experiment that are necessary to get n successes if each experiment is successful with probability p. For example, Geometric (p)

and NegBin $(1, p)$ are the same distribution. It follows from this characterization that if $Y_1, \ldots, Y_n$ are n independent Geometric (p) random variables, then $X = Y_1 + \cdots + Y_n \sim$ NegBin (n, p). The quantity Y_k is interpreted as the number of additional trials after the $k - 1^{\text{st}}$ success that are necessary to produce the kth success. First we derive the probability mass function of such an X, i.e., we compute $P[X = k]$. For this latter quantity to be positive we clearly must have $k \geq n$, for at least n trials are necessary to obtain n successes. Fix $k \geq n$ and let

$$\begin{aligned} A &= \{\text{there are exactly } n-1 \text{ successes among the first } k-1 \text{ trials}\}, \text{ and} \\ B &= \{\text{the } k^{\text{th}} \text{ trial is a success}\}. \end{aligned}$$

Since the trials are independent, A and B are independent and

$$P[X = k] \;=\; P[A \cap B] \;=\; PA \cdot PB.$$

Now clearly $PB = p$, but also $PA = \binom{k-1}{n-1} p^{n-1} q^{k-n}$. To see this latter fact, note that $p^{n-1} q^{k-n}$ is the probability of observing any particular arrangement of $n-1$ successes and $k-n$ failures among the first $k-1$ trials. The combinatoric factor is there because there are $\binom{k-1}{n-1}$ such arrangements. Hence the mass function for NegBin (n, p) is given by $P[X = k] = \binom{k-1}{n-1} p^n q^{k-n}$ for $k \geq n$.

Using now that $X = Y_1 + \cdots + Y_n \sim$ NegBin (n, p) when the Y's are iid Geometric (p)'s, we see that

- $X = nEY = \frac{n}{p}$.
- $\operatorname{Var} X = n \operatorname{Var} Y = \frac{n(1-p)}{p^2}$.
- $M_X(\theta) = \left(\frac{pe^\theta}{1-qe^\theta}\right)^n$, for $\theta < -\ln q$.

The Discrete Uniform Distribution. A discrete random variable X is *uniformly distributed* if its range is finite and each number in its range is observed with equal probability. A special case of this is when X is uniformly distributed on the numbers $\{1, 2, \ldots, n\}$. In this case, we will write $X \sim$ Uniform $\{1, 2, \ldots, n\}$ and, for $1 \leq k \leq n$, we have $P[X = k] = \frac{1}{n}$. Then

- $EX = \frac{1}{n}(1 + 2 + \cdots + n) = \frac{1}{n}\frac{n(n+1)}{2} = \frac{n+1}{2}$.
- $EX^2 = \frac{1}{n}(1^2 + 2^2 + \cdots + n^2) = \frac{1}{n}\frac{n(n+1)(2n+1)}{6} = \frac{(n+1)(2n+1)}{6}$, so

$$\operatorname{Var} X \;=\; \frac{(n+1)(2n+1)}{6} - \left(\frac{n+1}{2}\right)^2 \;=\; \frac{n^2-1}{12}.$$

- $M_X(\theta) = \frac{1}{n}(e^\theta + e^{2\theta} + \cdots + e^{n\theta}) = \begin{cases} 1 & \text{if } \theta = 0 \\ \frac{e^\theta - e^{(n+1)\theta}}{n(1-e^\theta)} & \text{otherwise.} \end{cases}$

§43. Some Common Continuous Distributions

The Uniform Distribution. A continuous random variable X is uniformly distributed if its density function is constant on its range. The random variable X is uniformly distributed on the interval $[a,b]$ (or (a,b) as particular values are assumed with probability 0), if

$$f(x) = \begin{cases} \frac{1}{b-a} & \text{if } a < x < b \\ 0 & \text{otherwise.} \end{cases}$$

We will write $X \sim \text{Uniform}\,(a,b)$. The most common application in this text involves $U \sim \text{Uniform}\,(0,1)$. A very useful fact for such a U is that $P[U < x] = x$ when $0 \leq x \leq 1$. We have

- $EU = \int_0^1 x \cdot 1\, dx = \frac{1}{2}x^2\big|_0^1 = \frac{1}{2}$.
- $EU^2 = \int_0^1 x^2\, dx = \frac{1}{3}x^3\big|_0^1 = \frac{1}{3}$, so $\text{Var}\, X = \frac{1}{3} - \left(\frac{1}{2}\right)^2 = \frac{1}{12}$.
- For $\theta \neq 0$, $M_U(\theta) = \int_0^1 e^{x\theta}\, dx = \frac{1}{\theta}e^{x\theta}\big|_0^1 = \frac{e^\theta - 1}{\theta}$. $M_U(0) = 1$.

More generally, if $U \sim \text{Uniform}\,(0,1)$ and $X = a + (b-a)U$ then $X \sim \text{Uniform}\,(a,b)$. To see this, note that $(b-a)U$ is uniformly distributed on the interval $[0, b-a]$ and adding a shifts the interval over to $[a,b]$. Then

- $EX = a + (b-a)EU = a + (b-a)\frac{1}{2} = \frac{a+b}{2}$.
- $\text{Var}\, X = (b-a)^2 \text{Var}\, U = \frac{(b-a)^2}{12}$.
- $M_X(\theta) = e^{a\theta} M_U((b-a)\theta) = \begin{cases} \frac{e^{b\theta} - e^{a\theta}}{(b-a)\theta} & \text{if } \theta \neq 0 \\ 1 & \text{if } \theta = 0. \end{cases}$

The Exponential Distribution. $X \sim \text{Exponential}\,(\lambda)$ where $\lambda > 0$ if $X > 0$ with density function

$$f(x) = \begin{cases} \frac{1}{\lambda}e^{-x/\lambda} & \text{if } x > 0 \\ 0 & \text{otherwise.} \end{cases}$$

A useful fact for such an exponential is the "tail" mass

$$P[X > x] = \int_x^\infty \frac{1}{\lambda}e^{-t/\lambda}\, dt = -e^{-t/\lambda}\big|_x^\infty = e^{-x/\lambda}. \tag{43.1}$$

A common special case is $Y \sim \text{Exponential}\,(1)$. For such a Y we have (by parts) that

- $EY = \int_0^\infty xe^{-x}\, dx = \left[-(x+1)e^{-x}\right]_0^\infty = 1$.
- $EY^2 = \int_0^\infty x^2 e^{-x}\, dx = \left[-(x^2+2x+2)e^{-x}\right]_0^\infty = 2$, so $\text{Var}\, Y = 2 - 1^2 = 1$.
- $M_Y(\theta) = \int_0^\infty e^{\theta x}e^{-x}\, dx = \int_0^\infty e^{(\theta-1)x}\, dx = \frac{1}{\theta - 1}$ if $\theta < 1$; the integral diverges and $M_Y(\theta)$ is undefined if $\theta \geq 1$.

In general, if $Y \sim$ Exponential$\,(1)$ and $X = \lambda Y$ then $X \sim$ Exponential$\,(\lambda)$. To see this, we compute

$$P[X > x] \;=\; P[\lambda Y > x] \;=\; P[Y > x/\lambda] \;=\; e^{-x/\lambda},$$

as required (see (43.1)). Then, for $X \sim$ Exponential$\,(\lambda)$ we have

- $EX = \lambda EY = \lambda$.
- $\operatorname{Var} X = \lambda^2 \operatorname{Var} Y = \lambda^2$.
- $M_X(\theta) = M_Y(\lambda\theta) = \frac{1}{\lambda\theta - 1}$ if $\theta < \frac{1}{\lambda}$, undefined otherwise.

The Normal Distribution. The range of the normal distribution consists of all real numbers and its density function is given by

$$f(x) \;=\; \frac{1}{\sqrt{2\pi}\sigma} e^{-(x-\mu)^2/2\sigma^2}$$

where $\sigma > 0$ and μ can be any real number. We write $X \sim$ Normal$\,(\mu, \sigma^2)$. A common special case is the "standard normal", $N \sim$ Normal$\,(0, 1)$. Its density function is

$$\psi(x) \;=\; \frac{1}{\sqrt{2\pi}} e^{-x^2/2}. \tag{43.2}$$

In this special case

- $EN = 0$, because the distribution is symmetric about 0 (i.e., $\psi(x) = \psi(-x)$).
- By parts we get

$$\begin{aligned} EN^2 &= \frac{1}{\sqrt{2\pi}} \int_{-\infty}^{\infty} x^2 e^{-x^2/2}\, dx \\ &= -\frac{1}{\sqrt{2\pi}} x e^{-x^2/2} \Big|_{-\infty}^{\infty} + \frac{1}{\sqrt{2\pi}} \int_{-\infty}^{\infty} e^{-x^2/2}\, dx \\ &= 0 + 1 \;-\; 1, \end{aligned}$$

so $\operatorname{Var} N = 1 - 0^2 = 1$.

- As for the moment generating function, we have

$$\begin{aligned} M_N(\theta) &= \int_{-\infty}^{\infty} e^{\theta x} \frac{1}{\sqrt{2\pi}} e^{-x^2/2}\, dx \\ &= \int_{-\infty}^{\infty} \frac{1}{\sqrt{2\pi}} e^{-(x-\theta)^2/2} e^{\theta^2/2}\, dx \\ &= e^{\theta^2/2} \int_{-\infty}^{\infty} \frac{1}{\sqrt{2\pi}} e^{-(x-\theta)^2/2}\, dx \;=\; e^{\theta^2/2}, \end{aligned}$$

where the last equality holds because $\frac{1}{\sqrt{2\pi}} e^{-(x-\theta)^2/2}$ is the density function of a Normal$\,(\theta, 1)$ random variable and hence integrates to 1. This holds for any $\theta \in (-\infty, \infty)$.

If $N \sim \text{Normal}(0,1)$ and $X = \mu + \sigma N$, then $X \sim \text{Normal}(\mu, \sigma^2)$. To see this, we compute

$$P[X \le x] \;=\; P[\mu + \sigma N \le x] \;=\; P\left[N \le \frac{x-\mu}{\sigma}\right],$$

so

$$\begin{aligned} f_X(x) &= \frac{d}{dx} P[X \le x] \;=\; \frac{d}{dx} P\left[N \le \frac{x-\mu}{\sigma}\right] \\ &= \psi\left(\frac{x-\mu}{\sigma}\right) \cdot \frac{1}{\sigma} \;=\; \frac{1}{\sqrt{2\pi}\sigma} e^{-(x-\mu)^2/2\sigma^2}. \;\checkmark \end{aligned}$$

This produces:

- $EX = \mu + \sigma EN = \mu$.
- $\text{Var}\, X = \text{Var}\,(\mu + \sigma N) = \sigma^2 \text{Var}\, N = \sigma^2$.
- $M_X(\theta) = E[e^{(\mu+\sigma N)\theta}] = e^{\mu\theta} E e^{\sigma\theta N} = e^{\mu\theta} e^{\sigma^2\theta^2/2} = e^{\mu\theta + \sigma^2\theta^2/2}$.

For completeness we confirm that $\psi(x)$, defined in (43.2), is a bona-fide density function. As $\psi(x) > 0$ for all x, we must show that $I = \int_{-\infty}^{\infty} \psi(x)\, dx$ satisfies $I = 1$. We do this by showing that $I^2 = 1$ — since $I > 0$ we must therefore have $I = 1$. To wit:

$$\begin{aligned} I^2 &= \frac{1}{\sqrt{2\pi}} \int_{-\infty}^{\infty} e^{-x^2/2}\, dx \cdot \frac{1}{\sqrt{2\pi}} \int_{-\infty}^{\infty} e^{-y^2/2}\, dy \\ &= \frac{1}{2\pi} \iint_{\mathbb{R}^2} e^{-(x^2+y^2)/2}\, dx\, dy \\ &= \frac{1}{2\pi} \int_0^{2\pi} \int_0^{\infty} e^{-r^2/2} r\, dr\, d\theta \\ &= \frac{1}{2\pi} \int_0^{2\pi} \left[-e^{-r^2/2}\right]_0^{\infty} d\theta \\ &= \frac{1}{2\pi} \int_0^{2\pi} 1\, d\theta \;=\; 1. \;\checkmark \end{aligned}$$

APPENDIX II

SOME REAL ANALYSIS

In this appendix we develop some of the facts, tools, and notation from real analysis that are used in this text.

§44. Completeness and Some Consequences

Completeness. The real numbers possess the important property of *completeness*, which can be formulated in several equivalent ways.

The Dedekind Property. The most intuitive formulation is what we'll call the 'Dedekind property' (Richard Dedekind, 1831 – 1916). Suppose S and T are any two non-empty subsets of the reals such that: (i) $\mathbb{R} = S \cup T$; and (ii) if $s \in S$ and $t \in T$ then $s < t$. Condition (ii) implies that S and T are disjoint while (i) says that every real number lies in either S or T. Note also that if (i) holds an equivalent way to state (ii) is: if $t \in T$ and $t < t'$ then $t' \in T$. The Dedekind property says that under these circumstances there is a real number L — which we'll call the *cut point* — such that $x < L \implies x \in S$ and $x > L \implies x \in T$. By (i) either $L \in S$ or $L \in T$; if $L \in S$ then L is the largest element of S while if $L \in T$ then L is T's smallest element.

The Dedekind property essentially says that the real numbers have no 'gaps'. (This statement is not true for the rational numbers $\mathbb{Q}$, which have the same algebraic properties as the reals but lack completeness.)

Figure 44.1. S and T have a cut point $L \in \mathbb{R}$.

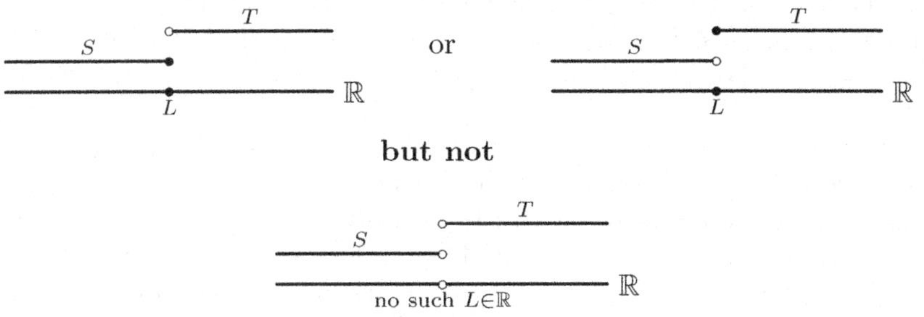

The Cauchy Property. Recall that a sequence of real numbers $a_1, a_2, \ldots$ *converges* to the *limit* L if, for any $\epsilon > 0$ (thought of as small), there is a number n_0

such that $|a_n - L| < \epsilon$ provided $n \geq n_0$. In this case we write $\lim_{n\to\infty} a_n = L$, or $a_n \to L$ as $n \to \infty$. Note that, for $n \geq n_0$ and $m \geq n_0$, we have via the triangle inequality that

$$|a_n - a_m| \leq |a_n - L| + |L - a_m| < 2\epsilon.$$

In other words, since the numbers in the sequence are getting close to a common limit, they are getting close to each other as well. This is the property of Cauchy: a sequence $a_1, a_2, \ldots$ is *Cauchy*, or *a Cauchy sequence*, if for any $\epsilon > 0$, there is a number n_0 such that $|a_n - a_m| < \epsilon$ provided $m \geq n \geq n_0$. Equivalently, $a_1, a_2, \ldots$ is Cauchy, if for any $\epsilon > 0$, there is a number n_0 such that $|a_{n_0} - a_n| < \epsilon$ provided $n \geq n_0$ (see exercises). Convergent sequences are therefore Cauchy. The Cauchy property states that the converse also holds: any Cauchy sequence of real numbers $(a_n : n \geq 1)$ has a limit $L \in \mathbb{R}$. The importance of the Cauchy condition is that it can be used to establish the existence of a limit without knowing what the limit is. The limit L does not appear in the Cauchy condition. This criterion for convergence is due to Augustin-Louis Cauchy, 1789 – 1857.

The Least Upper Bound Property. A third useful characterization will involve the notion of least upper bound. A set of numbers S is *bounded from above* if there is a number B such that $s \leq B$ for all $s \in S$. It is *bounded from below* if there is a number B such that $B \leq s$ for all $s \in S$. The set S is *bounded* if it is bounded both from above and below. Suppose that $S \subset \mathbb{R}$ is non-empty and bounded from above. The number B is the *least upper bound* for $S \subset \mathbb{R}$ if B is an upper bound for S and any number less than B is not an upper bound for S. This second condition means that, for any $\epsilon > 0$, there is an $s \in S$ with $s > B - \epsilon$. The *least upper bound property* states that any set S that is non-empty and bounded from above, in fact, has a least upper bound. This least upper bound is sometimes denoted l.u.b.(S); modern notation, which we shall use, is $\sup(S)$. ("Sup", pronounced like "soup", is short for "supremum", which is Latin for "the highest".) It is convenient to think of $\sup(S)$ as the largest element of S. This is not quite right, however, as $\sup(S)$ may not be an element of S. For example, take $S = \{x : x < 1\}$. Then $\sup(S) = 1 \notin S$. If, however, S does possess a largest element $s_{\max}$, then it is easy to see that $s_{\max} = \sup(S)$. For example, if $S = \{x : x \leq 1\}$, then 1 is the largest element of S and $\sup(S) = 1$.

If S is empty, then *any* real number B is vacuously an upper bound for S ($s \leq B$ for all $s \in S$ because there are no $s \in S$). In this case there is no smallest upper bound and we will take $\sup(\emptyset) = -\infty$. If a nonempty set S of real numbers has no upper bound, so S has arbitrarily large elements, we will take $\sup(S) = \infty$. With these latter conventions, the notion of least upper bound is defined for all subsets of real numbers.

If S is nonempty, the number B is the *greatest lower bound* for S if B is a lower bound for S and no larger number is a lower bound for S. This is denoted by $\inf(S)$ ("inf" is short for "infimum", which is Latin for "the lowest".) If S is empty, we take $\inf(S) = \infty$; if S has no lower bound, we take $\inf(S) = -\infty$. The least upper bound property is equivalent to the statement that all non-empty sets that are bounded from below have a greatest lower bound. This follows since B is a lower bound for S if and only if $-B$ is an upper bound for $-S = \{-s : s \in S\}$. (The set $-S$ is S reflected about 0.) Note that $\inf(S) = -\sup(-S)$ (see exercises).

The Increasing Sequence Property. Another useful characterization of completeness involves increasing sequences of numbers. The sequence $(a_n : n \geq 1)$ is *increasing* if $a_1 \leq a_2 \leq a_2 \leq \cdots$. Note that an increasing sequence is automatically bounded from below (by a_1). The *increasing sequence property* says that any bounded increasing sequence of real numbers has a limit $L \in \mathbb{R}$. Of course, if an increasing sequence is not bounded from above then it runs off to infinity. In either case it has a limit — in the first case as a real number, in the second case ∞.

The Sequential Compactness Property. If $x_1, x_2, \ldots$ is a sequence of numbers and $1 \leq n_1 < n_2 < n_3 < \cdots$ is an increasing sequence of integers then $x_{n_1}, x_{n_2}, \ldots$ is called a *subsequence* of $x_1, x_2, \ldots$. For example, $x_2, x_4, x_6, \ldots$ is a subsequence. Suppose $[a, b]$ is a closed interval and $x_1, x_2, \ldots$ is a sequence with each $x_n \in [a, b]$. The *sequential compactness property* says that there is a number $L \in [a, b]$ and a subsequence $x_{n_1}, x_{n_2}, \ldots$ such that $x_{n_k} \to L$ as $k \to \infty$. That is, there is a subsequence that converges to a number belonging to the interval.

The Equivalence of these Properties. Our first theorem states that these five properties are equivalent.

THEOREM 44.2 (EQUIVALENT CHARACTERIZATIONS OF COMPLETENESS FOR $\mathbb{R}$). *The real numbers $\mathbb{R}$ possess the following equivalent properties:* (i) *the Dedekind property;* (ii) *the Cauchy property;* (iii) *the least upper bound property;* (iv) *the increasing sequence property; and* (v) *the sequential compactness property.*

The fact that the reals possess any of these properties, which we do not prove here, is a fundamental property of their formal mathematical construction. Here we take (i) as our definition of completeness and prove that (i) $\Longrightarrow$ (ii) $\Longrightarrow$ (iii) $\Longrightarrow$ (iv) $\Longrightarrow$ (v). Equivalence will follow from the fact that (v) $\Longrightarrow$ (i), which we leave as an exercise.

PROOF OF THEOREM 44.2. (i) $\Longrightarrow$ (ii): Suppose the Dedekind property holds and that (a_n) is a Cauchy sequence. First note that all Cauchy sequences are bounded (see exercises) so (a_n) is bounded above and below. Now let

$$T = \{t \in \mathbb{R} : a_n > t \text{ for only finitely many } n\},$$

and let $S = T^c$, so $S \cup T = \mathbb{R}$. Furthermore S and T are both non-empty as any upper bound for (a_n) is in T and any lower bound minus one is not in T (if $a_n \geq B$ for every n then $a_n > B - 1$ for every n). Also, if $t \in T$ and $t < t'$ then $t' \in T$. Hence S and T satisfy the conditions for the Dedekind property — let L denote the cut point. For any $\epsilon > 0$ we claim that the interval $(L - \epsilon, L + \epsilon)$ contains a_n for infinitely many n. To see this, suppose the contrary: $(L - \epsilon, L + \epsilon)$ contains a_n for only finitely many n. Choose any x with $L < x < L + \epsilon$, say $x = L + \frac{\epsilon}{2}$. Then $x \in T$ and $a_n > x$ for only finitely many n. Since $(L - \epsilon, \infty) = (L - \epsilon, L + \epsilon) \cup (x, \infty)$, we have $a_n > L - \epsilon$ for only finitely many n. But then $L - \epsilon \in T$, a contradiction.

Next we construct a subsequence of (a_n) that converges to L. Take $n_1 = 1$ and suppose $n_1 < n_2 < \cdots < n_{k-1}$ have been selected. With $\epsilon = \frac{1}{k}$ pick $n_k > n_{k-1}$ such that $a_{n_k} \in (L - \epsilon, L + \epsilon)$. This is possible as this interval contains a_n for infinitely many n, and hence for some n exceeding n_{k-1}. Since $|a_{n_k} - L| < \frac{1}{k}$, we see that $a_{n_k} \to L$ as $k \to \infty$. But if a subsequence of a Cauchy sequence converges to a number L then the entire sequence converges to L (see exercises).

(ii) $\Longrightarrow$ (iii): Suppose all Cauchy sequences converge. Let S be a non-empty set with $s_0 \in S$ that is bounded from above by B_0. We must produce a least upper bound for S. First we construct an increasing sequence $s_0 \leq s_1 \leq s_2 \leq \cdots$ of elements of S and a decreasing sequence $B_0 \geq B_1 \geq B_2 \geq \cdots$ of upper bounds for S with the further property that $B_n - s_n \to 0$ as $n \to \infty$. We already have s_0 and B_0; suppose now that $s_0, \ldots s_n$ and $B_0, \ldots, B_n$ have already been constructed. Note that $s_n \leq B_n$ as B_n is an upper bound for S. Let $a = (s_n + B_n)/2$, so a is situated half way between s_n and B_n. If a is an upper bound for S, take $s_{n+1} = s_n$ and $B_{n+1} = a$. Otherwise, choose any $s \in S$ with $s > a$ and put $s_{n+1} = s$ and $B_{n+1} = B_n$. In either case, $s_n \leq s_{n+1}$ and $B_n \geq B_{n+1}$. Also, $B_{n+1} - s_{n+1} \leq (B_n - s_n)/2$, so, in general, $B_n - s_n \leq (B_0 - s_0)/2^n \to 0$ as $n \to \infty$.

Next we show that the sequence s_n is Cauchy. Fix any $\epsilon > 0$ and pick n_0 so that $(B_0 - s_0)/2^{n_0} < \epsilon$. If $m \geq n \geq n_0$, we have $s_m \geq s_n$, $s_m \leq B_{n_0}$, and $s_{n_0} \leq s_n$, so

$$|s_m - s_n| \;=\; s_m - s_n \;\leq\; B_{n_0} - s_{n_0} \;\leq\; (B_0 - s_0)/2^{n_0} \;<\; \epsilon,$$

as required. Since all Cauchy sequences converge, $s_n \to L \in \mathbb{R}$ as $n \to \infty$. Note also that $B_n \to L$ as $n \to \infty$, since $|B_n - s_n| \to 0$ (see exercises).

We claim that L is the least upper bound for S. First, if $\epsilon > 0$ then there is some s_n with $L - \epsilon < s_n \leq L$. (This holds since the sequence s_n increases up to L.) So, if L is an upper bound, it is the least upper bound. If some $s \in S$ satisfied $s > L$, we would have all the $B_n \geq s$ (since they are all upper bounds for S). But the B_n's decrease down to L — so this is not possible. Hence L is an upper bound for S, and therefore is the least upper bound.

(iii) $\Longrightarrow$ (iv): Now suppose that all non-empty sets that are bounded from above have a least upper bound. Let $a_1 \leq a_2 \leq \cdots$ be any increasing sequence that is bounded above. We must show that $a_n \to L$ for some $L \in \mathbb{R}$. Let $S = \{a_1, a_2, \dots\}$ and put $L = \sup(S)$. Note that $\sup S$ is guaranteed to exist as S is non-empty and bounded from above. Pick any $\epsilon > 0$. Then $L - \epsilon$ is not an upper bound for S and for some n_0 we have $a_{n_0} > L - \epsilon$. Then for any $n \geq n_0$, we have $L - \epsilon < a_{n_0} \leq a_n \leq L$, so $|a_n - L| < \epsilon$. It follows that $a_n \to L$.

(iv) $\Longrightarrow$ (v): Assume the increasing sequence property and suppose (x_n) is a sequence of numbers (not necessarily increasing) in the closed interval $[a, b]$. We must construct a convergent subsequence whose limit, L, is in $[a, b]$.

We construct a nested sequence of closed intervals

$$[a, b] = I_1 \supset I_2 \supset I_3 \supset \cdots$$

such that: (i) each I_k contains x_n for infinitely many n; and (ii) the length of I_k is $(b - a)/2^{k-1}$. Set $I_1 = [a_1, b_1]$ where $a_1 = a$ and $b_1 = b$. Then, by assumption, $x_n \in I_1$ for infinitely many n. Suppose that nested closed intervals $I_1, I_2, \dots, I_k$ have been constructed so that each of the I_j, $1 \leq j \leq k$, contains x_n for infinitely many n. Let a_k and b_k denote the lower and upper endpoints of I_k and let m_k denote its midpoint, i.e., $I_k = [a_k, b_k]$ and $m_k = (a_k + b_k)/2$. Then either the interval $[a_k, m_k]$ or the interval $[m_k, b_k]$ contains x_n for infinitely many n (possibly both do). In the first case, take $I_{k+1} = [a_k, m_k]$; otherwise take $I_{k+1} = [m_k, b_k]$. Then, by construction, I_{k+1} contains x_n for infinitely many n and the length of I_{k+1} is half the length of I_k. Thus (i) and (ii) are satisfied by this process.

By construction, we have $a \leq a_k \leq b$ and each $a_k \leq a_{k+1}$ (either $a_{k+1} = a_k$ or $a_{k+1} = m_k > a_k$). By the presumption of (iv), for some number L, $a_k \to L$ as $k \to \infty$. Since $a \leq a_k \leq b$ we also have $a \leq L \leq b$, i.e., $L \in [a, b]$.

We now construct positive integers $n_1 < n_2 < \cdots$, where $x_{n_k} \in I_k$, as follows. Take $n_1 = 1$, so $x_{n_1} \in I_1$ holds because $I_1 = [a, b]$. Suppose $n_1 < n_2 < \cdots < n_k$ have been constructed. Choose n_{k+1} so that $n_k < n_{k+1}$ and $x_{n_{k+1}} \in I_{k+1}$. We may do this since I_{k+1} contains x_n for infinitely many n, and hence for some n exceeding n_k. Since $x_{n_k} \in I_k$ we have $|x_{n_k} - a_k| \leq (b - a)/2^{k-1} \to 0$ as $k \to \infty$. Combining this with the fact that $a_k \to L$ we see that $x_{n_k} \to L$.

(v) $\Longrightarrow$ (i): See exercises. □

Some Consequences. We recall the definition of continuity. If $f(x)$ is defined on an interval I, it is *continuous at* $x \in I$ if for any $\delta > 0$ there is an $\epsilon > 0$ (whose value depends on ϵ and x) such that $|f(\tilde{x}) - f(x)| < \epsilon$ whenever $\tilde{x} \in I$ with $|\tilde{x} - x| < \delta$. It is *continuous on* I if it is continuous at each $x \in I$. It is *uniformly continuous on* I if δ may be chosen in a way that does not depend on x. That is, for any $\epsilon > 0$ there is a $\delta > 0$ such that $|f(\tilde{x}) - f(x)| < \epsilon$ whenever $|\tilde{x} - x| < \delta$ and $x, \tilde{x} \in I$. The following familiar facts from first semester calculus are consequences of the completeness of the reals.

THEOREM 44.3 (INTERMEDIATE VALUE THEOREM). *Suppose $f(x)$ is continuous on the interval $[a, b]$ with $f(a) < 0$ and $f(b) > 0$. Then for some number c, with $a < c < b$, $f(c) = 0$.*

THEOREM 44.4 (EXTREME VALUE THEOREM). *Suppose $f(x)$ is continuous on the interval $[a,b]$. Then $f(x)$ assumes both a maximum and a minimum value on $[a,b]$. Specifically: for some $x_{\max} \in [a,b]$, $f(x_{\max}) \geq f(x)$ for all $x \in [a,b]$; for some $x_{\min} \in [a,b]$, $f(x_{\min}) \leq f(x)$ for all $x \in [a,b]$.*

PROOF OF THEOREM 44.3. Let $S = \{x \in [a,b] : f(x) < 0\}$ so, in particular, $a \in S$ and S is bounded above by b. Let $c = \sup(S)$. We cannot have $c = a$, for otherwise $f(x) \geq 0$ for all $x > a$ and $f(a) = \lim_{x\to a^+} f(x) \geq 0$, contradicting the fact that $f(a) < 0$. Similarly we cannot have $c = b$, for otherwise $f(x) < 0$ for values of x arbitrarily close to b (since c is the *least* upper bound for S) and $f(b) = \lim_{x\to b^-}(x) \leq 0$, contradicting $f(b) > 0$. Hence $a < c < b$. Now fix any $\delta > 0$. The interval $I = (c-\delta, c+\delta)$ contains values of x to the left of c for which $f(x) < 0$ (otherwise $c - \delta$ would be an upper bound for S), while $f(x) \geq 0$ for all $x \in I$ to the right of c. Since this holds for any $\delta > 0$ and $f(x)$ is continuous at $x = c$, we must have $f(c) = 0$. □

PROOF OF THEOREM 44.4. Let $T = \sup\{f(x) : a \leq x \leq b\}$. Then $f(x) \leq T$ for any $x \in [a,b]$, and for any $t < T$, there is a number $x \in [a,b]$ with $f(x) > t$. (Otherwise t would be an upper bound for $\{f(x) : a \leq x \leq b\}$.) Therefore there is a sequence of points $x_1, x_2, \ldots$ in $[a,b]$ with $f(x_n) \to T$ as $n \to \infty$. By sequential compactness, we may extract a subsequence x_{n_k} that converges to some limit $x_{\max} \in [a,b]$. By construction, $f(x_{n_k}) \to T$ as $k \to \infty$. By continuity of $f(x)$, $f(x_{n_k}) \to f(x_{\max})$. It follows that $f(x_{\max}) = T$. A similar construction produces $x_{\min}$. □

Another important consequence of completeness (that is not usually treated in calculus) is:

THEOREM 44.5 (UNIFORM CONTINUITY THEOREM). *Suppose $f(x)$ is continuous on the interval $[a,b]$. Then $f(x)$ is uniformly continuous on $[a,b]$.*

PROOF. We must show that for any $\epsilon > 0$ there is a $\delta > 0$ such that $|f(\tilde{x}) - f(x)| < \epsilon$ whenever $|\tilde{x} - x| < \delta$ and $x, \tilde{x} \in [a,b]$. Suppose not. Then there would exist some number $\epsilon > 0$ for which there is sequence of pairs of numbers

$$(x_1, \tilde{x}_1), (x_2, \tilde{x}_2), \ldots, (x_n, \tilde{x}_n), \ldots$$

in $[a,b]$ with each $|x_n - \tilde{x}_n| \to 0$ as $n \to \infty$ but with $|f(x_n) - f(\tilde{x}_n)| \geq \epsilon$ for all n. By sequential compactness, we can extract a subsequence $n_1 < n_2 < \cdots$ with $x_{n_k} \to L \in [a,b]$ as $k \to \infty$. Since $|x_n - \tilde{x}_n| \to 0$, it follows also that $\tilde{x}_{n_k} \to L$ as $k \to \infty$. By the continuity of $f(x)$ on $[a,b]$ we obtain that both $f(x_{n_k}) \to f(L)$ and $f(\tilde{x}_{n_k}) \to f(L)$ as $k \to \infty$. But this contradicts the fact that each $|f(x_{n_k}) - f(\tilde{x}_{n_k})| \geq \epsilon$. This contradiction establishes the theorem. □

EXERCISES

1. Show that the sequence $(a_n; a \geq 1)$ is Cauchy if, for any $\epsilon > 0$, there is a number n_0 such that $|a_{n_0} - a_n| < \epsilon$ provided $n \geq n_0$.

2. Show that any Cauchy sequence is bounded.

3. Show that if any subsequence of a Cauchy sequence converges to L, then the entire sequence converges to L.

4. Show that if $a_n \to L$ and $|b_n - a_n| \to 0$ as $n \to \infty$, then $b_n \to L$ as $n \to \infty$.

5. Show that if $a_n \to L$ and each $a_n \leq a$, then $L \leq a$. Give an example where each $a_n < a$ but $L = a$.

6. If $a_n \to L_1$, $b_n \to L_2$, and $a_n \leq b_n$, show that $L_1 \leq L_2$.

7. If $A \subset B \subset \mathbb{R}$, show that $\inf(A) \geq \inf(B)$ and $\sup(A) \leq \sup(B)$.

8. If $S \subset \mathbb{R}$, show that $\inf(S) = -\sup(-S)$ where $-S = \{-s : s \in S\}$.

9. If A and B are subsets of $\mathbb{R}$ with $a < b$ whenever $a \in A$ and $b \in B$ show that $\sup A \leq \inf B$. Give an example where equality holds.

10. Assume that $a_n \to L$ as $n \to \infty$. Show that any subsequence a_{n_i} converges to the same limit.

11. Suppose $a_n \leq L \leq b_n$ and that $a_n \leq c_n \leq b_n$ for all $n \geq 1$. If $b_n - a_n \to 0$, show that $c_n \to L$.

12. Prove that (v) $\implies$ (i) in Theorem 44.2. Hint: construct a nested sequence of intervals $[s_k, t_k]$ with $s_k \in S$, $t_k \in T$, and $t_k - s_k \to 0$; then use sequential compactness to get a limit point L; show L is the cut point.

§45. Basics of Estimation

"Big Oh" and "Little Oh" Notation. Sometimes we wish to approximate an unwieldy function, call it $f(x)$, by a more analytically tractable function, say $g(x)$, near $x = 0$. If $\epsilon(x)$ — ϵ stands for "error" — is defined as $\epsilon(x) = f(x) - g(x)$, then we have $f(x) = g(x) + \epsilon(x)$. We seek a means of quantifying the order of magnitude of the error term $\epsilon(x)$.

Big Oh. We say that $\epsilon(x) \in O(x^n)$ (read $\epsilon(x)$ is big oh of x^n) if for some numbers $C > 0$ (C may be large) and $\delta > 0$ (δ may be small), we have $|\epsilon(x)| \leq C|x|^n$ whenever $|x| \leq \delta$. If this is true, we write "$f(x) = g(x) + O(x^n)$ as $x \to 0$". The larger n is, the better the approximation since $|x|^{n+1} \leq |x|^n$ if $|x| \leq 1$. In fact, $O(x^{n+1}) \subset O(x^n)$ for if $\epsilon(x) \in O(x^{n+1})$ then for positive numbers δ and C we have $|\epsilon(x)| \leq C|x|^{n+1}$ if $|x| \leq \delta$. With $\delta' = \min(\delta, 1) > 0$ we therefore also have $|\epsilon(x)| \leq C|x|^n$ whenever $|x| \leq \delta'$. If we write $f(x) = O(x^n)$, then implicitly $g(x) = 0$ and $f(x)$ itself is of order $|x|^n$ for small x.

EXAMPLE 45.1. Here we show that $\frac{1}{1+x} = 1 - x + O(x^2)$ as $x \to 0$. For $|x| < 1$, we have

$$\frac{1}{1+x} = 1 - x + x^2 - x^3 \pm \cdots,$$

as the right hand side is an infinite geometric series with ratio $-x$. Then $\epsilon(x) = x^2 - x^3 \pm \cdots$ and satisfies

$$\begin{aligned} |\epsilon(x)| &\leq |x|^2 + |x|^3 + \cdots \\ &= |x|^2(1 + |x| + |x|^2 + \cdots) = |x|^2 \frac{1}{1-|x|}. \end{aligned}$$

Now $1 \leq \frac{1}{1-|x|} \leq 2$, provided we further restrict x by requiring $|x| \leq \frac{1}{2}$. So $|\epsilon(x)| \leq 2|x|^2$ provided $|x| \leq \frac{1}{2}$. Here $C = 2$ and $\delta = \frac{1}{2}$ *realize* the big oh estimate.

EXAMPLE 45.2. Here we show that $\ln(1+x) = x - \frac{1}{2}x^2 + O(x^3)$. Let $h(x) = \frac{1}{1+x} - (1-x)$; from the previous examples we know that $|h(x)| \leq 2x^2$ provided $|x| \leq \frac{1}{2}$. Then

$$\begin{aligned} \ln(1+x) &= \int_1^{1+x} \frac{1}{t}\, dt \\ &= \int_0^x \frac{1}{1+s}\, ds \quad (\text{substitute } s = t-1) \\ &= \int_0^x (1 - s + h(s))\, ds \\ &= x - \frac{1}{2}x^2 + \int_0^x h(s)\, ds = x - \frac{1}{2}x^2 + \epsilon(x), \end{aligned}$$

where $\epsilon(x) = \int_0^x h(s)\, ds$. We must be a little careful in estimating $|\epsilon(x)|$, because x could be negative. Let's require that $|x| \leq \frac{1}{2}$. Then $0 \leq |s| \leq |x| \leq \frac{1}{2}$ on the range of integration, so $|h(s)| \leq 2|s|^2 \leq 2|x|^2$. Since the length of the interval of integration is $|x|$, we see that $|\epsilon(x)| \leq 2|x|^2 \cdot |x| = 2|x|^3$. Here the big oh estimate is realized by $\delta = \frac{1}{2}$ and $C = 2$.

EXAMPLE 45.3. In example 45.2 we saw that $\ln(1+x) = x - \frac{1}{2}x^2 + \epsilon(x)$, where $\epsilon(x) \in O(x^3) \subset O(x^2)$. Since $-\frac{1}{2}x^2 \in O(x^2)$, we get the cruder estimate that $\ln(1+x) = x + O(x^2)$. Here we use that if $\epsilon(x) \in O(x^n)$ and $\tilde{\epsilon}(x) \in O(x^n)$ then $\epsilon(x) + \tilde{\epsilon}(x) \in O(x^n)$ (see exercises).

This cruder estimate is sufficient to prove a fact that we use in several places:

THEOREM 45.4. *If $a_n \to a$ as $n \to \infty$, then $\left(1 + \frac{a_n}{n}\right)^n \to e^a$ as $n \to \infty$.*

PROOF. Let the positive numbers C and δ realize the $O(x^2)$ estimate in $\ln(1+x) = x + O(x^2)$ as $x \to 0$. Since $a_n \to a$, the sequence is bounded in absolute value by some number B and $\left|\frac{a_n}{n}\right| \le \frac{B}{n} \to 0$ as $n \to \infty$. If $\left|\frac{a_n}{n}\right| \le \delta$,

$$\ln\left[\left(1+\frac{a_n}{n}\right)^n\right] = n\ln\left(1+\frac{a_n}{n}\right) = n\left[\frac{a_n}{n} + \epsilon_n\right] = a_n + n\epsilon_n,$$

where $|\epsilon_n| \le C\left(\frac{a_n}{n}\right)^2 \le \frac{CB^2}{n^2}$. Since $\left|\frac{a_n}{n}\right| \le \delta$ holds for sufficiently large n we see that $n\epsilon_n \to 0$ as $n \to \infty$. Since $a_n \to a$, we get that

$$\ln\left[\left(1+\frac{a_n}{n}\right)^n\right] \to a,$$

as $n \to \infty$. Since the exponential function is continuous, the result follows by exponentiating both sides. □

Little Oh. A related means of quantifying the "goodness" of an approximation is the little oh notation. Again $f(x)$ is some unwieldy function defined near 0 and $g(x)$ is typically an analytically tractable approximation. With $\epsilon(x) = f(x) - g(x)$, we have $\epsilon(x) \in o(x^n)$ (read $\epsilon(x)$ is little oh of x^n) if $\left|\frac{\epsilon(x)}{x^n}\right| \to 0$ as $x \to 0$. This is a stronger condition than $O(x^n)$ as $O(x^n)$ simply means that $\left|\frac{\epsilon(x)}{x^n}\right|$ is *bounded* near $x = 0$. In fact, we have

$$o(x^n) \subset O(x^n) \subset o(x^{n-1}).$$

To see the second '$\subset$' note that if $\epsilon(x) \in O(x^n)$ then, on some interval $[-\delta, \delta]$ we have

$$\left|\frac{\epsilon(x)}{x^{n-1}}\right| \le \left|\frac{Cx^n}{x^{n-1}}\right| = C|x| \to 0,$$

as $x \to 0$. It is common to use the notation $\epsilon(x) \in o(1)$, or $\epsilon(x) = o(1)$, to mean that $\epsilon(x) \to 0$ as $x \to 0$. This is consistent with the definition of little oh as $x^0 = 1$.

EXAMPLE 45.5. From examples 45.1–3 we see that: $\frac{1}{1+x} = 1 - x + o(x)$; $\ln(1+x) = x - \frac{1}{2}x^2 + o(x^2)$; and $\ln(1+x) = x + o(x)$. Note also that $\ln(1+x) = o(1)$. This is all as $x \to 0$.

First and Second Order Approximations. Suppose $f(x)$ is defined on some open interval I containing the point x_0. In the discussion that follows it is always understood that the variable x is restricted to belong to I. We start with

LEMMA 45.6. *Suppose $f(x_0) = 0$ and $f'(x)$ is continuous on I with $|f'(x)| \le C$ for some bound $C > 0$. Then $|f(x)| \le C|x - x_0|$.*

PROOF. By the fundamental theorem of calculus,

$$f(x) \;=\; f(x) - f(x_0) \;=\; \int_{x_0}^{x} f'(s)\,ds. \tag{45.7}$$

The integrand is bounded by C and the length of the interval of integration is $|x - x_0|$ (x could be less than x_0 here). The conclusion follows. □

LEMMA 45.8. *Suppose $f(x_0) = f'(x_0) = 0$ and $f''(x)$ is continuous on I with $|f''(x)| \le C$. Then $|f(x)| \le C|x - x_0|^2$.*

PROOF. Lemma 45.6 as applied to f', rather than f, gives $|f'(x)| \le C|x - x_0|$. We refer to (45.7) again. This time the integrand is bounded by $C|s - x_0|$ which in turn is bounded by $C|x - x_0|$, since s is between x_0 and x. The length of the interval of integration in (45.7) is again $|x - x_0|$, so we get that that $|f(x)| \le |x - x_0| \cdot C|x - x_0| = C|x - x_0|^2$. □

LEMMA 45.9. *Suppose $f(x_0) = f'(x_0) = f''(x_0) = 0$ and $f'''(x)$ is continuous on I with $|f'''(x)| \le C$. Then $|f(x)| \le C|x - x_0|^3$.*

PROOF. Lemma 45.8 as applied to f' gives $|f'(x)| \le C|x - x_0|^2$. We refer to (45.7) yet again. This time the integrand is bounded by $C|s - x_0|^2$ which in turn is bounded by $C|x - x_0|^2$. Here we get that that $|f(x)| \le |x - x_0| \cdot C|x - x_0|^2 = C|x - x_0|^3$. □

These lemmas can be "upgraded" to the following. Here there are no restrictions on the values of $f(x_0)$, $f'(x_0)$, or $f''(x_0)$.

THEOREM 45.10. *Suppose $f'(x)$ is continuous on I with $|f'(x)| \le C$. Then*

$$|f(x) - f(x_0)| \;\le\; C|x - x_0|.$$

THEOREM 45.11. *Suppose $f''(x)$ is continuous on I with $|f''(x)| \le C$. Then*

$$|f(x) - [f(x_0) + f'(x_0)(x - x_0)]| \;\le\; C|x - x_0|^2.$$

THEOREM 45.12. *Suppose $f'''(x)$ is continuous on I with $|f'''(x)| \le C$. Then*

$$\left|f(x) - \left[f(x_0) + f'(x_0)(x - x_0) + \frac{1}{2}f''(x_0)(x - x_0)^2\right]\right| \;\le\; C|x - x_0|^3.$$

PROOF OF THEOREMS 45.10, 11, AND 12. For Theorem 45.10, apply Lemma 45.6 to $h(x) = f(x) - f(x_0)$; for Theorem 45.11, apply Lemma 45.8 to $h(x) = f(x) - f(x_0) - f'(x_0)(x - x_0)$; and for Theorem 45.12, apply Lemma 45.9 to $h(x) = f(x) - f(x_0) - f'(x_0)(x - x_0) - \frac{1}{2}f''(x_0)(x - x_0)^2$. Regarding Theorem 45.12, for example, we see that $h(x_0) = 0$. Also $h'(x) = f'(x) - f'(x_0) - f''(x_0)(x - x_0)$

so $h'(x_0) = 0$. Similarly, $h''(x) = f''(x) - f''(x_0)$ so $h''(x_0) = 0$. Finally, $h'''(x) = f'''(x)$, so $h'''(x)$ is bounded by C by hypothesis. The function h therefore meets the criteria of Lemma 45.9 and the conclusion follows. The details of Theorems 45.10 and 11 are easier and left to the reader. □

Let's re-name a few things and use the big oh notation. Let $x = x_0 + \Delta x$ so $x - x_0 = \Delta x$. Also, let $\Delta f = f(x) - f(x_0) = f(x_0 + \Delta x) - f(x_0)$. Then Theorem 45.10 implies that

$$f'(x) \text{ bounded near } x_0 \quad \Longrightarrow \quad \Delta f \; = \; O(\Delta x),$$

while Theorem 45.11 implies that

$$f''(x) \text{ bounded near } x_0 \quad \Longrightarrow \quad \Delta f \; = \; f'(x_0)\Delta x + O(\Delta x^2),$$

and Theorem 45.12 implies that

$$f'''(x) \text{ bounded near } x_0 \quad \Longrightarrow \quad \Delta f \; = \; f'(x_0)\Delta x + \frac{1}{2} f''(x_0)\Delta x^2 + O(\Delta x^3),$$

all as $\Delta x \to 0$. To say that the derivative $f^{(k)}$ is 'bounded near x_0' means $|f^{(k)}(x)| \le C$ for $|\Delta x| \le \delta$, for some positive C and δ. The same C and δ realize the big oh estimates on the right-hand sides.

Example 45.13. With $f(x) = e^{-x}$ we have $f'(x) = -e^{x}$ (so $f'(0) = -1$) and $f''(x) = e^{x}$. Since $f''(x)$ is bounded by e on the interval $[-1, 1]$ we obtain $e^{-x} = 1 - x + O(x^2)$ near $x = 0$. Here $\delta = 1$ and $C = e$ realize the big oh estimate. For $x \ge 0$ we can be more precise: specifically, $1 - x \le e^{-x} \le 1 - x + \frac{1}{2}x^2$. This follows from the identity

$$e^{-x} \; = \; 1 - \int_0^x e^{-t}\, dt.$$

Since $e^{-t} \le 1$ we get that $e^{-x} \ge 1 - \int_0^x 1\, dt = 1 - x$. Setting this inequality back in to our identity gives $e^{-x} \le 1 - \int_0^x (1 - t)\, dt = 1 - x + \frac{1}{2}x^2$. In §40 we use the following immediate consequences for $x \ge 0$: (i) $|e^{-x} - 1| \le x$; (ii) $|e^{-x} - (1 - x)| \le \frac{1}{2}x^2$; and (iii) $|(1 - e^{-x}) - x| \le \frac{1}{2}x^2$.

Functions of Two Variables. In developing Itô's formula, we will need to estimate quantities of the form $F(t, b) - F(t_0, b_0)$. We do this in two steps: first estimate the increment from (t_0, b_0) to (t_0, b) and then the increment from (t_0, b) to (t, b) (refer to Figure 45.14). Each such increment involves changing only one variable at a time and the results from the previous section may be used. We use the common notation $F_t(t, b) = \frac{\partial}{\partial t} F(t, b)$, $F_{tt} = \frac{\partial^2 F}{\partial t^2}$, $F_{tb} = \frac{\partial^2 F}{\partial b \partial t}$, $F_b = \frac{\partial F}{\partial b}$, $F_{bb} = \frac{\partial^2 F}{\partial b^2}$, and $F_{bbb} = \frac{\partial^3 F}{\partial b^3}$ to denote these partial derivatives. In the development that follows, we assume all derivatives are bounded as necessary for application of Theorems 45.10–12.

Figure 45.14. $F(t,b) - F(t_0,b_0)$ in two steps.

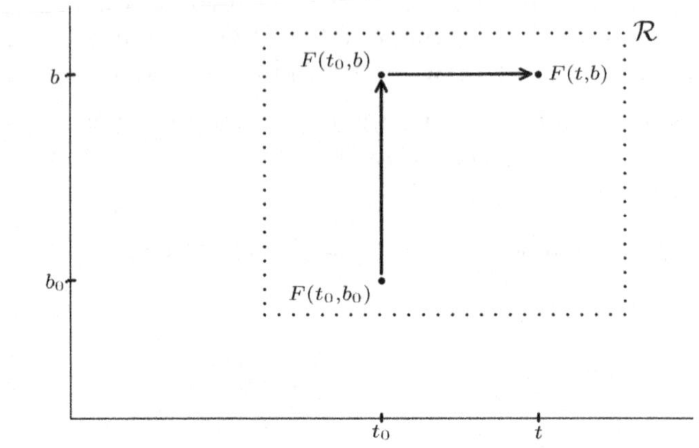

Then we have:

$$
\begin{aligned}
\Delta F &= F(t,b) - F(t_0,b_0) \\
&= F(t_0,b) - F(t_0,b_0) \quad + \quad F(t,b) - F(t_0,b) \\
&= F_b(t_0,b_0)\Delta b + \frac{1}{2}F_{bb}(t_0,b_0)\Delta b^2 + O(\Delta b^3) && (F_{bbb} \text{ is bounded}) \\
&\quad + F_t(t_0,b)\Delta t + O(\Delta t^2) && (F_{tt} \text{ is bounded}) \\
&= F_b(t_0,b_0)\Delta b + \frac{1}{2}F_{bb}(t_0,b_0)\Delta b^2 + O(\Delta b^3) \\
&\quad + \big(F_t(t_0,b_0) + O(\Delta b)\big)\Delta t + O(\Delta t^2) && (F_{tb} \text{ is bounded}) \\
&= F_t(t_0,b_0)\Delta t + F_b(t_0,b_0)\Delta b + \frac{1}{2}F_{bb}(t_0,b_0)\Delta b^2 \\
&\quad + O(\Delta b^3) + O(\Delta b)\Delta t + O(\Delta t^2).
\end{aligned}
$$

If F_{tt}, F_{tb}, and F_{bbb} are bounded in absolute value by C on some region $\mathcal{R}$ containing the action then we have

$$\Delta F = F_t(t_0,b_0)\Delta t + F_b(t_0,b_0)\Delta b + \frac{1}{2}F_{bb}(t_0,b_0)\Delta b^2 + \text{error},$$

where $|error| \le C \cdot (|\Delta b|^3 + |\Delta t \Delta b| + |\Delta t|^2)$.

EXERCISES

1. Show that if $\epsilon(x)$ and $\tilde{\epsilon}(x)$ are both $O(x^n)$, then so is $s(x) = a\epsilon(x) + b\tilde{\epsilon}(x)$ for any numbers a and b. You must compute the values of C and δ for $s(x)$ in terms of the C's and δ's for $\epsilon(x)$ and $\tilde{\epsilon}(x)$.

2. Show that if $f(x) \in O(x^n)$ and $g(x) \in O(x^m)$ then $p(x) = f(x)g(x) \in O(x^{n+m})$.

3. Prove examples 45.1 and 45.2 using Theorems 45.10 – 45.12.

BIBLIOGRAPHY

Undergraduate Probability and Stochastic Processes

G. Grimmett and D. Welsh. Probability: An Introduction. Oxford University Press, Oxford, 1986.

I. Miller and M. Miller. John E. Freund's Mathematical Statistics. Sixth edition. Prentice Hall, New Jersey, 1999.

S. Ross. A First Course in Probability. Fourth edition. Macmillan Publishing Co., New York, 1994.

S. Ross. Stochastic Processes. Second edition. John Wiley & Sons, New Jersey, 1996.

S. Ross. Introduction to Probability Models. Seventh edition. Academic Press, London, 2000.

Graduate Probability and Stochastic Processes

P. Billingsley. Probability and Measure. Third edition. John Wiley & Sons, New York, 1995.

R. Durrett. Probability: Theory and Examples. Second edition. Duxbury Press, Belmont, CA, 1995.

R. Durrett. Stochastic Calculus A Practical Approach. CRC Press, New York, 1996.

J. S. Rosenthal. A First Look at Rigorous Probability Theory. World Scientific, Singapore, 2000.

Real Analysis and Measure Theory

R. C. Buck. Advanced Calculus. Third edition. McGraw-Hill, New York, 1978.

W. Rudin. Principles of Mathematical Analysis. Third edition. McGraw Hill, New York, 1976.

W. Rudin. Real and Complex Analysis. Third edition. McGraw Hill, New York, 1987.

H. Royden. Real Analysis. Third edition. MacMillan, New York, 1988.

Financial Mathematics

F. Black and M. Scholes. "The Pricing of Options and Corporate Liabilities", J. Polit. Econom., **81** (1973), 637-659.

J. Hull. Options, Futures, and Other Derivatives. Sixth edition. Prentice Hall, NJ, 2006.

S. Ross. An Elementary Introduction to Mathematical Finance. Third edition. Cambridge University Press, New York, 2011.

D. Stefanica. A Primer for the Mathematics of Financial Engineering. FE Press, New York, 2008.

Other

Encyclopedic Dictionary of Mathematics. Second edition. Mathematical Society of Japan, Itô Kiyosi, editor. The MIT Press, Cambridge MA, 1993.

NDEX

Made in United States
Orlando, FL
04 November 2026

90200637R00138